PROGRESS IN CHEMISTRY AND BIOCHEMISTRY: KINETICS, THERMODYNAMICS, SYNTHESIS, PROPERTIES AND APPLICATIONS, VOLUME 1

(A FESTSCHRIFT IN HONOR OF THE 75TH BIRTHDAY OF PROFESSOR GENNADY E. ZAIKOV)

Progress in Chemistry and Biochemistry: Kinetics, Thermodynamics, Synthesis, Properties and Applications, Volume 1

(A Festschrift in Honor of the 75th Birthday of Professor Gennady E. Zaikov)

Eli M. Pearce
and
G. E. Zaikov
Editors

Nova Science Publishers, Inc.
New York

LIBRARY OF CONGRESS CATALOGING-IN-PUBLICATION DATA

Progress in chemistry and biochemistry : kinetics, thermodynamics, synthesis, properties, and applications / editors, Eli M. Pearce, G.E. Zaikov.
p. cm.
Includes bibliographical references and index.
ISBN 978-1-60692-344-3 (hardcover : alk. paper)
1. Polymers. 2. Composite materials. 3. Nanostructured materials. 4. Polymeric composites. 5. Nanostructured materials--Synthesis. 6. Nanostructured materials--Thermal properties. 7. Adhesion. I. Pearce, Eli M. II. Zaikov, Gennadii Efremovich.
QD381.P695 2009
540--dc22 2008046450

Published by Nova Science Publishers, Inc. ✦ New York

CONTENTS

PREFACE

"All is afraid of time and only time is afraid of pyramids"
Proverb in Ancient Egypt

"Nothing is for free in this life"
Napoleon Bonaparte, Moscow 1812

"Greatness of the country is defined by
quantity of trans-continental rockets,
quantity and quality of Olympic medals
and a limousine of the President of the country"
Robert Kennedy
Minister of Justice of the USA
Washington D.C., 1962

One may agree with that definition which was given by Robert Kennedy concerning the importance of the country in the world. However I have included Science in this list, as a must in the country's development on the nearest and the long-term future. Speaking about a science, it's logical to recollect the scientists who brought and keep carrying on the worthy contribution to development of a science. It's pertinent to do on the occasion of any anniversary and, in particular, in connection with the 75th anniversary.

The given compilation is devoted to such an anniversary of Prof. Gennady E. Zaikov. His scientific biography (with social impregnations) you can read in the first chapter of the given compilation. The compilation itself has been prepared by those scientists with whom G. E. Zaikov worked during more than half a century of his scientific career.

The editor of the compilation did not intend to include in the book only articles (reviews and original publications) by G. E. Zaikov (although he has written something for this compilation). The basic idea of the compilation is to show readers a circle of his interests and a circle of his friends, whom he has walked with (and keeps walking) throughout his life.

The proverb of Ancient Greece says: "If God will conceive to punish someone, first he will take away his/her mind". We see that God does not punish G. E. Zaikov. He is full of energy and creative plans.

I think that this collection will be interesting to the broad audience of scientists and students working in the area of organic chemistry, and to engineers and technicians employed in manufacturing.

We expect to publish three volumes with the same title. This is the first volume.

Prof. Eli M. Pearce
Brooklyn Polytechnic University
Six Metrotech Center
Brooklyn, New York, USA

In: Progress in Chemistry and Biochemistry
Editors: E. M. Pearce and G. E. Zaikov

ISBN: 978-1-60692-344-3

Chapter 1

PROFESSOR GENNADY EFREMOVICH ZAIKOV: MORE THAN HALF A CENTURY IN SCIENCE

Eli M. Pearce
Herman F. Mark Polymer Research Institute;
Polytechnic University; 333 Jay St., Six Metrotech Center
Brooklyn, NYC, USA

January 7, 2010 will be the 75th birthday of Prof. G. E. Zaikov and more than 50 years of his scientific activity. Zaikov was born in Omsk, Siberia (USSR) where he graduated from their primary, middle, and high school. He also graduated from a musical professional school where he studied violin and pianoforte. However, his parents Efrem and Matrena decided that it might be better for their son to continue his education by following in the footsteps of his mother - a chemistry teacher in High School and Omsk's Medical Institute (his father was a mathematician and land-surveyor). Therefore, in 1952 Gennady moved to Moscow where he entered the Moscow State University (MSU), and he graduated with a chemistry degree in December 1957. His bachelor's degree dealt with the problem of separating Li6 and Li7 isotopes. After this he joined the Institute of Chemical Physics (ICP) in Moscow in February 1958. In 1996 this Institute was split into two parts: N. N. Semenov Institute of Chemical Physics (ICP) and N. M. Emanuel Institute of Biochemical Physics. At the present time Prof. G. E. Zaikov is working at the N. M. Emanuel Institute of Biochemical Physics (IBP). So, G.E. Zaikov never changed the place of his job.

Gennady was originally invited to ICP by Professor Nikolai Markovich Emanuel. Under his guidance, G. E. Zaikov defended in 1963 his Ph.D. thesis entitled "Comparison of the Kinetics and Mechanism of Oxidation of the Organic Compounds in Gaseous and Liquid Phases" in 1963. These results were the foundation for industrial application. A plant floor was built at Moscow petrochemical plant (Kapotnya district) for production of 10000 tons/year acetic acid and 5000 tons/year methylethylketone by oxidation of n-butane in liquid phase in critical conditions (50 atm, 150 °C). The main contributors to this plant floor were N.M. Emanuel, E.A. Blumberg, Z.K. Maizus, M.G. Bulygin, E.B. Chizhov and G.E. Zaikov. In 1968, he defended a Doctor of Science thesis entitled "The Role of Media in Radical-Chain Oxidation Reactions". In 1970 he became a full professor.

In 1966, Gennady began to become involved with polymer science. N. M. Emanuel charged Zaikov with the organization of work on problems associated with aging and stabilization of polymers, and, later, with the combustion of polymeric materials. In the 1970s there were about 1000 scientists (about 50 research centers) in the U.S.S.R. working on these problems including 200 scientists from ICP under Zaikov's leadership. The research was conducted on all aspects of these polymer problems thermal degradation, oxidation, ozonolysis, photodegradation and radiation degradation, hydrolysis, biodegradation, mechanical degradation, pyrolysis, and flammability. Scientists from synthetic laboratories of this division (Prof. V.V. Ershov, E.G. Rozantsev, K.M. Dyumaev) prepared several very important and original stabilizers for polymers and organized production of these stabilizers.

After "perestroika and degradation" of the U.S.S.R. in 1991, the new Russian government decreased the financial support of science significantly. So, G. E. Zaikov has now with him in the N. M. Emanuel Institute only 15 coworkers (instead of 200 in 1970–1980s).

He compensated for the decrease of scientists in his Institute by increasing cooperation with another research centers in Russia and abroad.

Now G.E.Zaikov has scientific cooperation with:

- Prof. Victor Manuel de Matos Lobo and dr. Artur Valente (Coimbra University, Coimbra, Portugal),
- Prof. Alfonso Jimenez (Alicante University, Alicante, Spain),
- Dr. Nekane Guarrotxena Arlunduaga (Institute of Polymer Science and Technology, Madrid, Spain),
- Prof. Alberto D'Amore (Second Naples University, Naples, Italy),
- Dr. Antonio Ballada (Former Vice-President of Himont Co., Milan, Italy),
- Prof. Goerg Michler (Martin Luther University, Halle-Saale, Germany),
- Dr. Frank Pudel (OHMI Consulting Co., Magdeburg, Germany),
- Prof. Ryszard Kozlowski (Institute of Natural Fibers, Poznan, Poland),
- Prof. Jan Pielichowski (Cracow University of Technology, Cracow, Poland),
- Dr. Daniel Horak (Institute of Macromolecular Science, Prague, Czeck Republic),
- Prof. Slavi Kirillov Rakovsky (Institute of Catalysis, Sofia, Bulgaria),
- Prof. Cornelia Vasile (Polymer Research Institute, Iassi, Romania),
- Prof. Richard A. Pethrick (University of Strathclyde, Glasgow, Scotland, UK),
- Dr. Gerald Kirshenbaum (Brooklyn Polytechnic University, Brooklyn, New York, USA),
- Prof. David Schiraldi (Case Western Reserve University, Clevelend, Ohio, USA),
- Prof. Bob Howell (Central Michigan University, Mount Pleasant, Michigan, USA),
- Dr. James Summers (Former Head of Division of PolyOne Co., Cleveland, Ohio, USA),
- Dr. LinShu Liu (US Department of Agriculture, Windmoor, Pennsylvania, USA),
- Dr.Raijesh Ananjiwala, Research Textile Institute, Port Elisabeth, South Africa).
- He has also cooperation with CIS countries (former republics of the USSR):
- Prof. Anatolii A. Turovskii, Prof. Roman G. Makitra and Prof Yurii G. Medvedevskikh (Pisarzhevskii Institute of Physical Chemistry, L'viv Division and Institute of Coal, L'viv, Ukraine),
- Prof. Nikolai A. Turovskii (Donetsk State University, Donetsk, Ukraine),
- Prof. Alexandr I. Burya (Dnepropetrovsk State Agriculture University, Dnepropetrovsk, Ukraine),

- Prof. Nodare G.Lekishvili and Prof. Omari Mukbaniani (I. Javakhishvili Tbilisi State University, Georgia),
- Prof. Jimsher N. Aneli (Institute of Kibernetic, Tbilisi, Georgia),
- Prof. Jenis A. Djamanbaev (Institute of Organic Chemistry, Bishkek, Kirgisia),
- Prof. Nikolai R. Prokopchuk (Belorussian State Technical University, Minsk, Belorussia).
- Prof. Norair M. Beylerian (Institute of Chemical Physics, Erevan, Armenia)

G.E. Zaikov has also cooperation with scientists from many Research Centers of Russia. I would like to give the names only some of them (the most important):

- Prof. A.A. Berlin, Prof. A.L. Iordanskii and Dr. K.Z. Gumargalieva (N.N. Semenov Institute of Chemical Physics, Moscow);
- Dr. N.A. Sivov (D.I. Topchiev Institute of Pethrochemical Synthesis, Moscow);
- Dr. N.N. Komova, Dr. A.A. Ol'khov, Prof. B. Tsoi (M.V. Lomonosov Moscow State Academy Fine Chemical Technology, Moscow);
- Prof. V.S. Osipchik (D.I. Mendeleev Russian Chemical-Technical University, Moscow);
- Prof. Yu.A. Ershov (The Second Moscow State Medical University, Moscow);
- Prof. N.Ya. Yaroshenko (Institute of Pure Chemical Compounds, Moscow);
- Prof. Yu.G. Yanovsky (Institute of Applied Mathematic, Moscow);
- Dr. O.A. Legon'kova (Moscow State University of Applied Biotechnology, Moscow);
- Prof. A.K. Mikitaev (L.Ya. Karpov Physico-Chemical Instotute, Moscow);
- Prof. A.M. Egorov (Oncology Center, Moscow);
- Dr. E.V. Kalugina (Plastic Company, Moscow);
- Dr. G.V. Kozlov, Prof. M.Kh. Ligidov, Prof. N.I. Mashukov (K.Kh. Berbekov Kabardino-Balkarian State University, Nal'chik, Kabardino-Balkaria);
- Prof. Yu.B. Monakov (Institute of Organic Chemistry, Ufa, Bashkortostan);
- Prof. M.I. Abdullin, Prof. V.P. Zakharov, Prof. S.V. Kolesov, Prof. R.Z. Biglova (Bashkirian State University, Ufa, Bashkortostan);
- Prof. S.S. Zlotsky (Ufa State Technological Oil University, Ufa, Bashkortostan);
- Prof. F.F. Niyazi (Kursk State University, Kursk);
- Prof. V.A. Babkin (Volgograd State Technical University, Volgograd);
- Prof. A.I. Rakhimov (Institute of Ecology, Volgograd);
- Prof. V.F. Kablov (Branch of Volgograd State Technical University, Volzhsk, Volgograd district);
- Prof. T.N. Lomova (Research Institute of Solutions, Ivanovo).

So, he has scientific cooperation (publication of original papers, reviews, books and volumes) with 20 Research abroad centers, 8 centers in CIS – cooperation and 20 – inside of Russia.

Zaikov left his position as a head of laboratory on September, 2007 but he became a head of Polymer Division (PD) in IBP. PD included 3 laboratories (about 50 scientists).

Figure 1 shows changing of amount of staff members in his laboratory on time and figure 2 shows amount of books which he published (mostly in English).

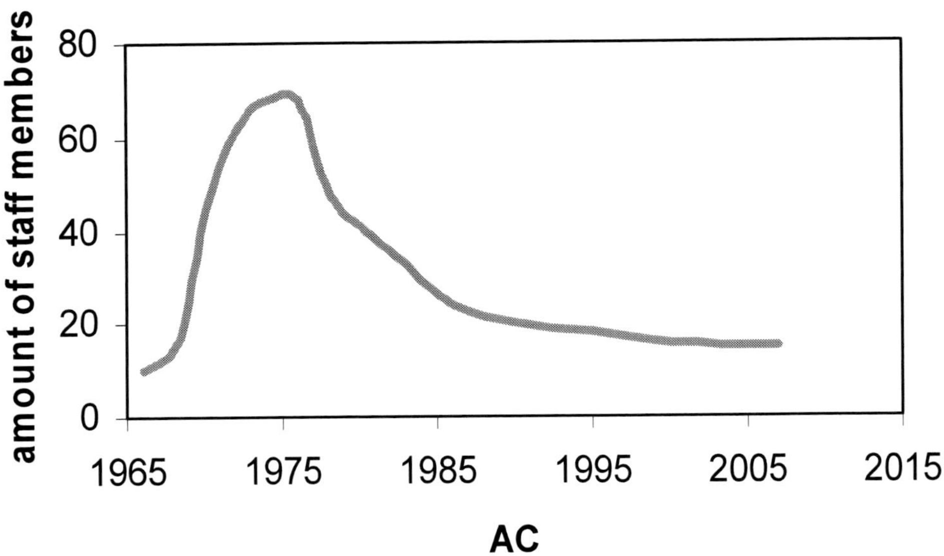

Figure 1. Amount of staff members in laboratory of chemical resistance of polymers in different time.

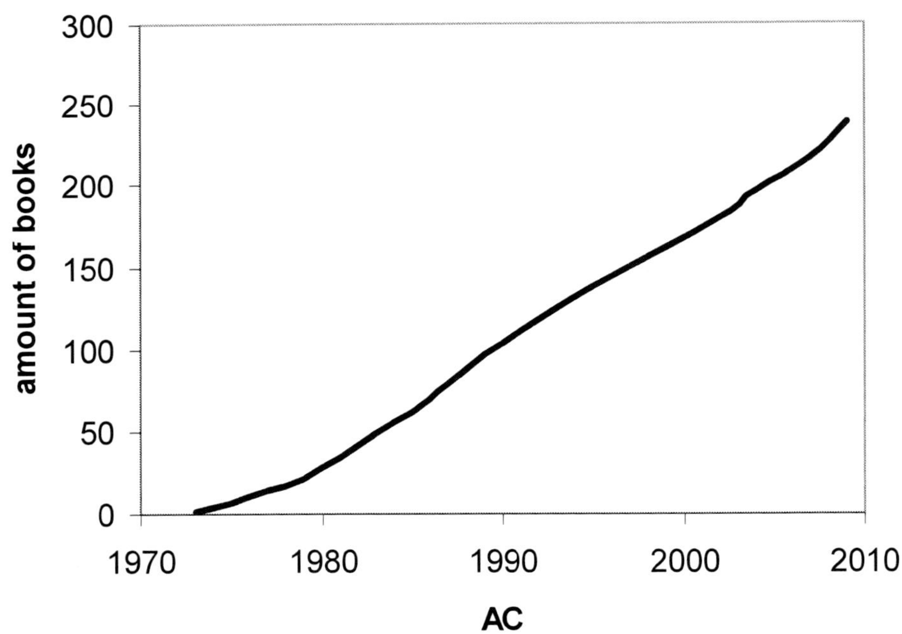

Figure 2. Integral amount of books published by G.E. Zaikov.

G. E. Zaikov is an outstanding scientist with expertise in wide areas of chemistry: chemical and biological kinetics, chemistry and physics of polymers, history of chemistry, biochemistry. In addition to his position at the N. M. Emanuel Institute, he is a lecturer at the Moscow State Academy of Fine Chemical Technology and he is researcher at Volzhsk Branch of Volgograd State Technological University. He taught his students from his own books: Degradation and Stabilization of Polymers, Physical Methods in Chemistry, and Acid Rains and Environmental Problems. G. E. Zaikov has written about 2500 original articles, 230 monographs (30 in Russian and 200 in English), and 270 chapters in 60 volumes. It is apparent from this work that he has made valuable contributions to the theory and practice of polymers—aging and development of new stabilizers for polymers, organization of their industrial production, life-time predictions for use and storage, and the mechanisms of oxidation, ozonolysis, hydrolysis, biodegradation, and decreasing of polymer flammability. New methods of polymer modification using the processes of degradation were introduced into practice by Zaikov. These methods allow the production of new polymeric materials with improved properties. Most recently, he is also very active in the field of semiconductors and electroconductive polymers, polymer blends, and polymer composites including nanocomposites.

G. E. Zaikov is a member of many editorial boards of journals published in Russia, Poland, Bulgaria, the U.S.A. and England. Below I would like to give a list of his activity in this field:

- Chemistry International, UK, 1987 – 1991;
- Russian Journal of Highmolecular Compounds, 1970 - 1984;
- Polymer Degradation and Stability, UK, 1982 - 2004;
- Polymer News, USA, 1988 - 2002;
- International Journal of Polymeric Materials, USA, associate editor, 1989–2000; Member of Editorial Board: 2001–2002;
- Polymers in Medicine, Poland, 1982–1998;
- Polymer Yearbook, associate editor, Gordon and Breach, UK, 1985–2000;
- Polymer Yearbook, Co-editor, Rapra Technology, UK, 2000 - 2003;
- Polymer Yearbook, Co-editor, Nova Science Publishers, USA, 2005 – to data;
- Polymer and Polymer Composites, UK, 1994–2000;
- Journal of Chemical and Biochemical Kinetics, USA, editor-in-chief, 1992–2000;
- Russian Journal of Textile Chemistry, 1992–to date;
- Oxidation Communications, Sofia, Bulgaria, 1994–to date;
- Associate editor of series "Polymer Science and Engineering," Gordon and Breach Publ., USA, 1990–2000;
- Editor of series "Polymer Books for the 21st Century," Nova Science Publ., New York, USA, 1990–2006;
- Editor of series "Chemistry and Biochemistry," Nova Science Publishers, New York, USA, 2002–to date;
- Editor-in-chief of series "New Achievement in Polymer Science", VSP International Sci. Publ., Leiden and Brill Academic Publishers, Amsterdam, the Netherlands, 1990 – 2004;
- Editor of series "Chemical and Biochemical Physics on the Edge of XXI Century," Nova Science Publ., USA 2000 - 2006;
- Russian Polymer News Journal, Associate Editor, New Jersey, USA, 1996–2003;
- Journal of Balkan Tribological Association, Sofia, Bulgaria, 2001–to date;
- Polymer Plastic Technology and Engineering, USA, 1997–2001;
- Member of Research Board and Advisers, American Biographical Institute, Inc., NC, USA, 2001–to date;
- Member of Editorial Board of Polymer International, 2004 – to data;

- Member of Editorial Board of Journal of Chemical Physics and Mesoscopy, Russian Academy of Sciences, Izhevsk, Russia, 2004–to date;
- Member of Editorial Board of Journal of Natural Fibers, Poznan, Poland, 2005 – to data;
- Member of Editorial Board of D.I. Mendeleev Journal of Russian Chemical Society, Moscow, 2006 – to data;
- Member of Editorial Board of Encyclopedia of Engineer-Chemist, Moscow, 2006 – to data;
- Member of Editorial Board of Journal of Coatings, Moscow, 1990 – 2000.

He is member of Academy of Creation (San Diego, USA – Moscow, Russia), International Academy of Sciences (Munich, Germany), American Chemical Society, Plastic Engineering Society (USA) and Royal Chemical Society (UK).

In the former Soviet Union (after academician N. M. Emanuel's death), he headed the team dealing with the problem of polymer aging in the U.S.S.R. and the Eastern European countries in cooperation with the Soviet Academy of Sciences. His present position is head of Division, member of directorium, and deputy of department of the N. M. Emanuel Institute of Biochemical Physics, Russian Academy of Sciences, Professor of Polymer Chemistry in the Moscow State Academy of Fine Chemical Technology, Professor of Polymer Chemistry in Volzhsk Branch of Volgograd State Technical University and field of interest Chemical Physics, Chemical Kinetics, Flammability, Degradation and Stabilization of Polymers, Diffusion, Polymer Materials, Kinetics in Biology, History of Chemistry, Jokes.

A few words about the personal life of G. E. Zaikov. In addition to his parents (discussed previously), he has a sister, Zinaida E. Zaikova, who was also a teacher of mathematics in high school, and she passed away on February 2008 in Omsk city. Two of his sisters (Klara and Inna) died during the Stalin collectivization period at the end of the 1920th - 1930th from starvation. Zaikov's wife, Marina Izrailevna Artsis is a member of the N. M. Emanuel Institute of Biochemical Physics and is a Ph.D.in Chemistry. His son, Vadim G. Zaikov is a Sr. Research Chemist of Avery Dennison Co (Ohio, USA). He is a Ph.D. in Chemistry twice. The first one he received in USSR (Moscow Textile Institute, Prof. Leonid S. Gal'braikh), the second one he received in the College of William and Mary (Williamsburg, VA) in the laboratory of Prof. William H. Starnes. His daughter, Julia, is an anesthesiologist (in Russia). His granddaughter Alexandra (19 years old) is a student (USA), and his grandson Denis (10 years old) is a schoolboy (USA).

On his 75th birthday anniversary, G. E. Zaikov is in the prime of his life. Although support for scientists and research is now at a low point for many in Russia, he is hopeful that for the sake of his country and its future that this will improve (probably far future).

The practice of good science still exists in Russia and G. E. Zaikov has been and is a significant contributor. We wish him a most happy birthday.

Figure 3. Professor G.E. Zaikov (November, 9 2007) with his new book "Chemical and Biochemical Physics" (Nova Science Publishers, New York). Above him (right side) – the picture of his teacher Professor N.M. Emanuel (1915 – 1984) and his grandson's Denis drawing (left side).

Figure 4. Staff of laboratory of chemical resistance of polymers: seating – Alexei A. Borodin (graduated student); first row (from the left to the right) – Lidia A. Zimina, Dr. Olga V. Alexeeva, Marina L. Konstantinova, Prof. Gennady E. Zaikov (Head of Division), Larisa L. Madyuskina; second row (from the left to the right) – Dr. Sergei M. Lomakin (Head of Laboratory), Prof. Stanislav D. Razumovskii, Prof.. Vladimir M. Gol'dberg, Prof.. Alexander A. Volod'kin, Dr. Nikolai N. Madyuskin, Dr. Marina I. Artsis, Vyacheslav V. Podmaster'ev. November, 9 2007. N.M. Emanuel Institute of Biochemical Physics Russian Academy of Sciences.

In: Progress in Chemistry and Biochemistry
Editors: E. M. Pearce and G. E. Zaikov
ISBN: 978-1-60692-344-3

Chapter 2

THE NANOADHESION INFLUENCE ON STRUCTURE AND PROPERTIES OF PARTICULATE-FILLED POLYMER NANOCOMPOSITES

G.V. Kozlov[1], A.I. Burya[1], Yu.G. Yanovskii[2] and G.E. Zaikov[3]
[1] State Agrarian University,
49027, Ukraine, Dnepropetrovsk, Voroshilov st., 25
I_dolbin@mail.ru
[2] Institute of Applied Mechanics of Russian Academy of Sciences,
119991, Russia, Moscow, Leninskii pr., 32 a
[3] N.M. Emanuel Institute of Biochemical Physics,
Russian Academy of Sciences,
119334, Russia, Moscow, Kosygin st., 4
Chembio@sky.chph.ras.ru

ABSTRACT

The results of the present paper have shown the existence in polymer particulate-filled nanocomposites nanoadhesion effect, which consists of an enhanced level of polymeric matrix-nanofiller interaction. This effect has a purely dimensional origin and was realized for nanofiller particles with a diameter less than 80 nm. The nanoadhesion effect influences strongly on macroscopic properties of those nanocomposites, in which it was realized. So, essential thermal stability enhancement and considerable thermal expansion coefficient reduction of nanocomposites on the base of phenylone at introduction 1 mass. % aerosil only are due to the nanoadhesion effect. Appreciation of nanoadhesion effect allows qualitatively and quantitatively to describe exactly nanocomposites reinforcement degree.

Keywords: Polymer nanocomposite, interfacial adhesion, thermal expansion, nanoadhesion, structure, thermal stability.

INTRODUCTION

At present polymer nanocomposites are some of the most perspective engineering materials [1]. Moreover the main attention is paid to nanocomposites filled by layered silicates (organoclays) [2] or nanotubes [3], whereas particulate-filled nanocomposites were studied to a far smaller extent. Nevertheless, possibilities of the last class of polymer nanocomposites are far from being exhausted [4].

As it is well known [5], for multiphase materials the interaction level between phases plays a defining role in properties formation of these materials. What was said above relates completely to polymer composites, too (nanocomposites). So, the authors [6] have demonstrated that the filler contents increase in the polymer composites results in good interfacial adhesion case of polymeric matrix-filler to modulus elasticity enhancement, but interfacial adhesion level determination by experimental methods in polymer composites case comes up on definite difficulties. So, the adhesional strength value determined by standard methods can differ sufficiently from the same for composites in virtue of the existence of various kinds of nonaccounted effects in real composites: filler particles aggregation [7], polymeric matrix structure change due to filler introduction [8] and a number of other factors. The indicated above importance of the interfacial phenomena consists in their influence practically on all polymer composites properties. Therefore the most suitable, from a practical point of view, method of the interfacial adhesion level determination in bulk composites is the estimation of this factor according to composite macroscopic properties, for which a number of techniques (for example, [9]) was elaborated. The purpose of the present paper is to determine the interfacial adhesion level and the factors, influencing on it, and the influence of the indicated level on thermal and mechanical properties of particulate-filled nanocomposites on the base of phenylone as well.

EXPERIMENTAL

As polymer binding of the studied nanocomposites the linear heterochain copolymer phenylone S-2 of the following chemical constitution was used:

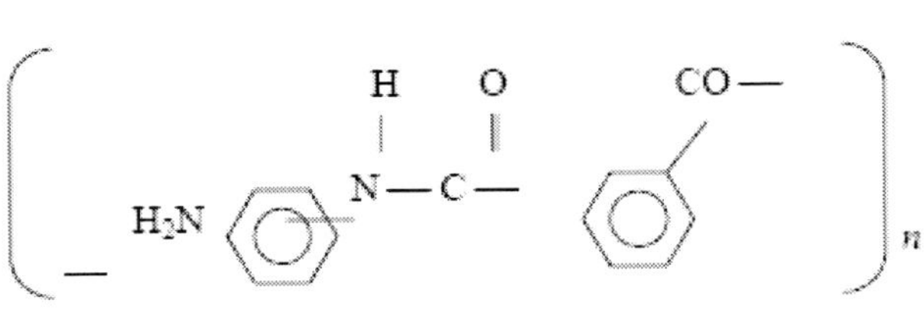

As particulate nonafiller the aerosil (A) with particles average diameter 15 nm and specific surface 170 m^2/g and oxinitride silicium-yttrium (OSY) Si_3N_4-Y_2O_3 with average diameter of particles 64 nm an specific surface 43 m^2/g were used. The contents of the first from the indicated nanofillers makes up to 0.3, 0.5 and 1.0 mass. %, the second one is varied within the range 0.2-10 mass. %.

The nanofiller introduction in polymeric matrix was realized in rotating electromagnetic field with the aid of nonequiaxed ferromagnetic particles with the ratio of length to diameter 4-5; the ferromagnetic particles volume, loading in an apparatus reactor, was within the range 0.04-0.05 from electromagnetic field action volume. The electromagnetic induction value of this field was within the range 0.08-0.12 Tl. At the indicated parameters it was found out experimentally that the optimal processing duration of nanocomposites in electromagnetic field was 270-300 s. Nanocomposites samples preparation was realized by compressional pressing method at temperature 537-616 K, pressure 40-100 MPa and exposure at the indicated parameters at a rate of 1 min/mm per product thickness.

The processing of powder-like compositions phenylone/A and phenylone/OSY in rotating electromagnetic field allows to inhibite (or decrease) nanofiller particles aggregation. For checking of the achievement of this effect the phenylone/A nanocomposite samples were processed by simple mechanical blending. Further the samples of nanocomposites, filled by aerosil and obtained with rotating electromagnetic field application, will be designated as phenylone/A-MF and obtained by mechanical blending – as phenylone/A. Nanocomposites filled by oxinitride silicium-yttrium were designated as phenylone/OSY.

The linear coefficient of thermal expansion of nanocomposites was determined on dilatometer DKV-SAM, intended for automatic registration of dilatometric curves of different solids, within the temperature range 298-498 K. For the tests the prismatic (6×4 mm) samples with the length 50±3 mm were used. The nonparallelism of sample grinded butt-ends consists of not more than ±0.02 mm. The number of samples for tests for each nanocomposite was no less than three.

For the surface structure study the three-dimensionally reflecting surface structure analyzer Zygo NewView 5022 was used.

The nanocomposites study by method of thermogravimetric analysis (TGA) was made on derivatograph Q-1500 of Paulik-Erdei system of the firm MOM (Hungary). The studies were carried out within the temperature range 298-873 K with the temperature rise rate 10 K/min, mass of a sample was equal to 100 mg. The curves recording was made with maximal sensitivity.

The mechanical tests on compression were made on testing machine FP-100 at temperature 293 K and strain rate 10^{-3} s^{-1}. The testing samples have diameter 10±0.5 mm and height 15±0.5 mm. For each nanocomposite five samples were tested.

Results and Discussion

In figure 1 the curves 1-3 represent three main types of the dependences of linear coefficient of thermal expansion of the studied nanocomposites α_n on nanofiller volume contents φ_n. The straight line 1 illustrates the case, when between two phases of nanocomposites adhesion is absent and at $\alpha_m > \alpha_n$ (α_m, α_n are thermal expansion coefficients of polymeric matrix and nanofiller, respectively) and the absence in interfacial layers residual compression strain polymeric matrix at heating will be expanded irrespective of filler particles and in this case $\alpha_n = \alpha_m$ [10]. The straight line 2 corresponds to a simple rule of mixtures [10]:

$$\alpha_n^m = \alpha_m(1 - \varphi_n) + \alpha_n \varphi_n, \quad (1)$$

where α_n^m is the thermal expansion coefficient of mixture, φ_n is the volume fraction of filler.

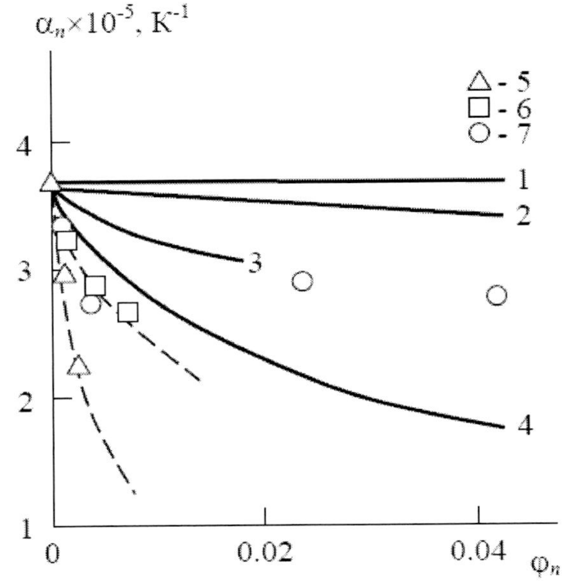

Figure 1. The dependences of linear coefficient of thermal expansion α_n on nanofiller volume contents φ_n. 1 – adhesion absence on interfacial boundary; 2 – mixtures rule; 3 – the Terner equation; 5-7 – experimental data for nanocomposites phenylone/A-MF (1, 2, 3, 5), phenylone/A (1, 2, 3, 6) and phenylone/OSY (1, 2, 4, 7).

The rule (1) is correct for the ideal case only, when each phase expands independently one from another. And lastly, the straight line 3 corresponds to Terner equation [10]:

$$\alpha_n^T = \frac{\alpha_m(1 - \varphi_n)K_m + \alpha_n \varphi_n K_n}{(1 - \varphi_n)K_m + \varphi_n K_n}, \quad (2)$$

where K_m and K_n are a bulk elasticity modulus of polymeric matrix and nonofiller, respectively.

At the equation (2) derivation the strains equality method was used for calculation of mixtures thermal expansion coefficient depending on density, elasticity modulus, thermal expansion coefficient and mass relationship of composing components. If the made assumptions are correct, then the formula (2) will also be applicable for polymer nanocomposites.

Besides, in figure 1 values α_n determined experimentally are shown by points for two series of the studied nanocomposites including phenylone, for which $\alpha_n = \alpha_m$. Let's note, that in Terner equation Young modulus E magnitudes were used instead of the bulk elasticity modulus K. As it was known [10], such replacement allows to obtain a lower boundary of α_n

value for polymer composites. From the data of figure 1 it follows that α_n values for both nanocomposites series, filled by aerosil, are disposed substantially lower than the curve calculated according to Terner equation, i.e. they are substantially smaller than the lower limiting values α_n for polymer composites. Besides α_n values for nanocomposites phenylone/A-MF are significantly smaller than the corresponding values for nanocomposites phenylone/A. In other words, inhibition (or weakening) of aerosil particles aggregation process substantially decreases α_n value. In case of nanocomposites phenylone/OSY the values α_n at small φ_n are somewhat smaller than the ones calculated according to Terner equation, although they are close to them, and at φ_n>0.05 the values α_n for them are disposed between the curves, calculated according to the mixtures rule and Terner equation, i.e. they have α_n values intermediate on absolute value and typical for polymer composites [10].

Let's consider physical significance of the described above α_n changes for nanocomposites, studied in this paper. It is possible to estimate polymeric matrix-filler interaction level (interfacial adhesion level) with the aid of the parameter *b*, determined according to the equation [10]:

$$\alpha_n = \alpha_n^m - b\left(\alpha_n^m - \alpha_n^T\right). \tag{3}$$

The larger the value *b* the higher interfacial adhesion level. For a large enough number of polymer composites with different matrixes and fillers *b* value varies from -0.19 to 1.39 [10]. Calculation according to the equation (3) for the studied nanocomposites shows that for phenylone/A-MF $b\approx12$, phenylone/A $b\approx6$ and for phenylone/OSY value *b* varies within the range ~ 5.40-0.35, decreasing at φ_n increase (figure 2). Therefore, polymeric matrix-filler interaction level in the studied nanocomposites can exceed substantially (on one order of magnitude) this value for microcomposites. Following to the authors [11] we called this effect nanoadhesion. Let's remind that in paper [11] the adhesion level for two-layered films of polystyrene on micro- and nanolevel was measured. It turns out, that in the last case adhesional strength is substantially higher in comparison with the first one. The authors explained [11] the indicated effect of polymers nanoadhesion by different degree of interfacial layer formation completion between two films of polystyrene.

In the nanocomposites case the nanoadhesion cause is a small contact area of polymeric mztrix-nanofiller. From the comparison of aerosil and OSY particles size and figure 2 data one can make a conclusion that two factors as minimum influence on value b: nanofiller particles size and their aggregation level, i.e. in the long run nanofiller aggregates size as well. So, value *b* for aerosil is higher than for oxinitride silicium-yttrium (particles size are 15 and 64 nm, respectively) and for nanocomposites phenylone/A-MF value *b* is substantially larger (twice) than for phenylone/A as that was mentioned above, processing of phenylone and aerosil powder-like blend in rotating electromagnetic field reduces aerosil particles aggregation level. Besides, the effect of *b* decrease at φ_n increase for nanocomposites phenylone/OSY (figure 2) is also due to enhancement of oxinitride silicium-yttrium particles aggregation level at its contents rise.

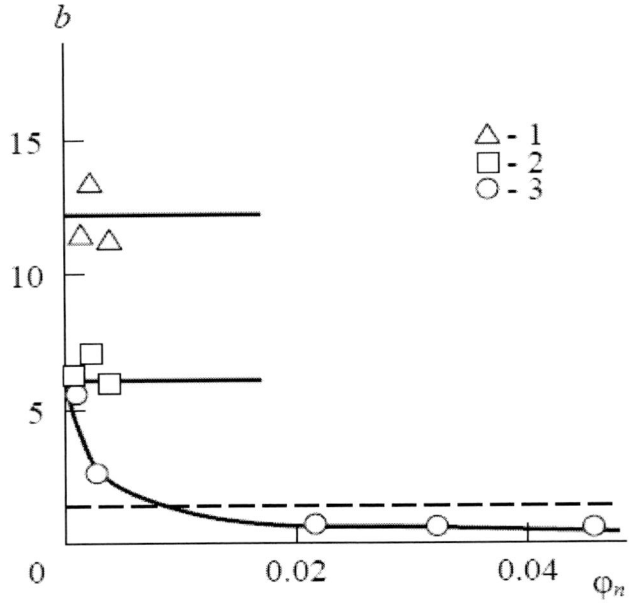

Figure 2. The dependences of parameter b on nanofiller volume contents φ_n for nanocomposites phenylone/A-MF (1), phenylone/A (2) and phenylone/OSY (3). The horizontal shaded line indicates upper limiting value b for microcomposites.

In figure 3 the micrographs of the surface of phenylone/A (figure 3a, b, c) and phenylone/A-MF (figure 3d, e, f) with aerosil content 1 mass. % samples are shown, from which one can see the aggregation process inhibition for the last nanocomposites series. Figure 2d is particularly outstanding in this sense, where one can see "cloud" of aerosil particles, which could not unite in an aggregate owing to the application of processing in electromagnetic field. And on the contrary, figure 2a, b, c clearly demonstrate substantial aggregation of aerosil particles in large enough aggregates. The obtained data assume that the studied nanoadhesion effect is purely dimensional effect or true nanoeffect [12]. It is well known that in adhesional strength measurements of different material pairs it is impossible to achieve the full contact of their surfaces. That is why the indicated contact is realized according to some hillocks only, that decreases contact area and, correspondingly, adhesional strength [13]. At very small (nanometer scale) contact area sizes the existence of large hillocks and depressions on the contact surface is scarcely probable that results to sharp adhesion level enhancement. This condition can be expressed more specifically: the nanoadhesion is achieved at the same scale of its participant sizes – nanoparticle and macromolecular coil.

In figure 4 the dependence of parameter b on nanofiller particles surface average area S_p for nanocomposites phenylone/aerosil and phenylone/OSY is shown. As one can see, the decrease b or adhesion level reduction at the increase S_p (nanofiller particles size) is observed and at $S_p \approx 2\times10^4$ nm^2, that corresponds to particles diameter ~ 80 nm, this dependence tries to attain $b\approx1$, i.e. to perfect microadhesion level ($\alpha_n = \alpha_n^T$, that follows from the equation (3)).

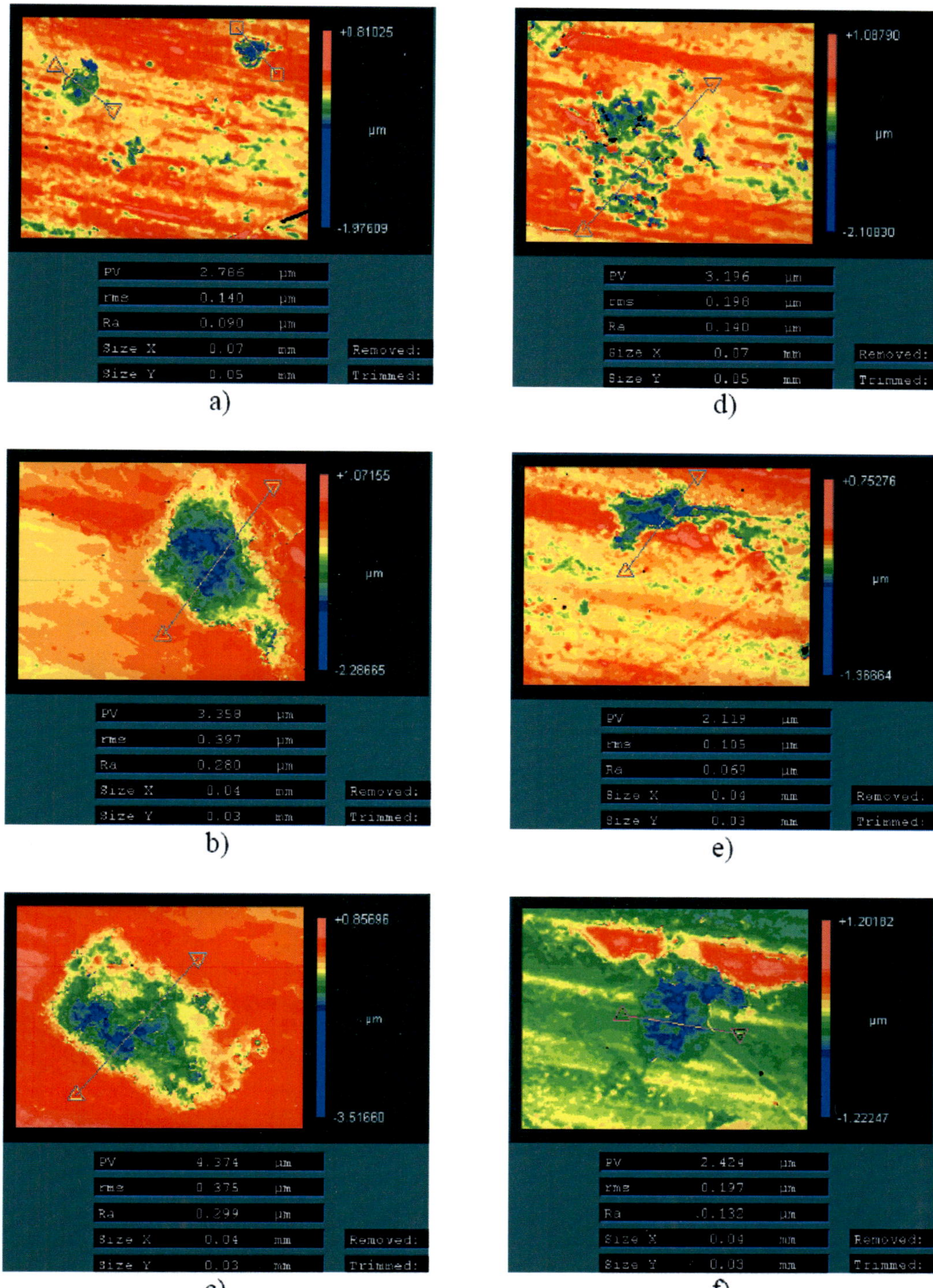

Figure 3. Electron micrographs of surfaces of nanocomposites phenylone/A (a, b, c) and phenylone/A-MF (d, e, f).

The nanoadhesion effect influences strongly on polymer nanocomposites macroscopic properties. As follows from the data of figure 1, the introduction 1 mass. % of aerosil in phenylone results to α_n decrease in about 3 times (for nanocomposites phenylone/A-MF). In case of microcomposites such α_n decrease requires the introduction of filler with volume contents φ_f=0.3-0.7 [10], i.e. practically on two orders more. Further as an example let's consider nanoadhesion influence on thermal and mechanical properties of the studied nanocomposites.

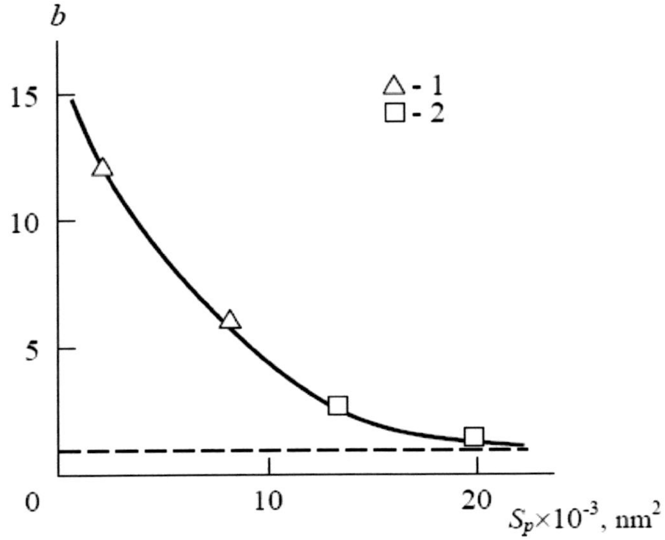

Figure 4. The dependence of parameter b on average area of nanofiller particles surface S_p for nanocomposites filled by aerosil (1) an oxinitride silicium-yttrium (2). The horizontal shaded line indicates perfect microadhesion level (b=1).

At present a large number of kinetic equations for the description of thermogravimetric analysis (TGA) data is known [14, 15]. However, all the indicated models do not take into consideration the polymer melt structure and the character of diffusive processes proceeding in thermooxidative degradation course. The last are important in the respect of controlling of oxidizer (for example, oxygen) access to reactive centers of polymeric macromolecules [16]. As a rule, the temperature of thermal degradation beginning in TGA tests (for example, temperature of 20 %-th sample mass loss $T_{20\%}$) is higher than temperature of the so-called transition "liquid 1 – liquid 2" T_{ll}, which can be estimated as follows [17]:

$$T_{ll} \approx (1.20 \pm 0.05) T_g, \tag{4}$$

where T_g is glass transition temperature of polymer.

At T_{ll} there occurs a transition of the polymer melt from "liquid with fixed structure" (where residual structural ordering is observed [17]) to true liquid state or "structureless liquid". Nevertheless, melt "structureless" at $T > T_{ll}$ relates to the absence of the supermolecular structure, but the structure of a macromolecular coil in melt remains an important structural factor (in essence, unique at $T > T_{ll}$) [16]. Therefore structural analysis of thermal stability of polymer nanocomposites on the base of phenylone, filled by aerosil, will be given fur ther.

In figures 5 and 6 the TGA curves for nanocomposites phenylone/A and phenylone/A-MF are shown, respectively. As the plots of these figures show, on the first stage for the studied materials within the temperatures range 293-383 K sample mass gradual decrease on 1.5-2.0 is observed, connected with moisture loss. Then, up to T=573 K sample mass remains practically nonchanged. The intensive thermooxidative degradation of phenylone and nanocomposites on its base, accompanied by a considerable mass loss, begins in the air above 663 K.

The experimentally determined temperatures of sample mass 20 %-th loss $T_{20\%}$ for all studied materials are listed in table 1. As one can see, the introduction of only one percent of aerosil into phenylone rises its thermal stability on 76-80 K. Let's consider the physical fundamentals of this strong enough effect.

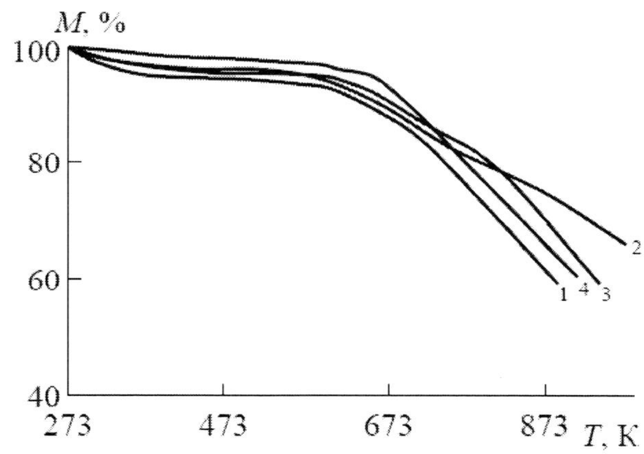

Figure 5. The TGA curves of phenylone (1) and nanocomposites phenylone/A with aerosil contents 0.3 (2), 0.5 (3) and 1.0 (4) mass. %.

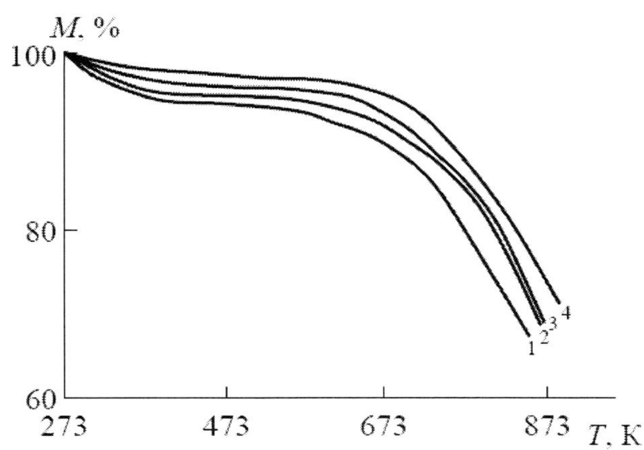

Figure 6. The TGA curves of phenylone (1) and nanocomposites phenylone/A-MF with aerosil contents 0.3 (2), 0.5 (3) and 1.0 (4) mass. %.

Table 1. The comparison of experimental $T_{20\%}$ and calculated according to the equation (8) $T^T_{20\%}$ temperatures of sample 20 %-th mass loss for nanocomposites phenylone/aerosil

Material	The aerosil contents, mass. %	$T_{20\%}$, K	$T^T_{20\%}$, K, the equation (16)	$T^T_{20\%}$, K, the equation (14)
Phenylone S-2	-	773	799	799
Phenylone/A	0.3	775	803	805
	0.5	798	810	805
	1.0	849	825	804
Phenylone/A-MF	0.3	798	809	805
	0.5	818	823	805
	1.0	853	837	804

As it was known [18], within the framework of a strange (anomalous) diffusion conception on fractal object it's possible to select two its main types: slow and fast diffusion. The dependence of mobile displacement s on time t is assumed as a basis of such division [18]:

$$s \sim t^{\beta}, \tag{5}$$

where for classical case β=1/2, for slow diffusion β<1/2 and for the fast one - β>1/2.

The most precisely the macromolecular coil structure, which is fractal object, can be characterized with the aid of its fractal (Hausdorff) dimension Δ_f, describing the coil elements distribution in space [19]. Earlier within the framework of fractional derivatives theory the interrelation Δ_f and β was shown, which is expressed analytically as follows [18]:

$$\beta = \frac{\Delta_f - 1}{4} \tag{6}$$

for slow diffusion and

$$\beta = \frac{\Delta_f - 1}{\Delta_f} \tag{7}$$

for the fast one. The structural boundary between the mentioned kinds of diffusion one can consider the value Δ_f=2.5 at its general variation $2.0 \le \Delta_f < 3.0$: at Δ_f<2.5 (less compact macromolecular coils) the fast diffusion of oxidizer (oxygen) is realized, at Δ_f>2.5 – a slow one.

For theoretical estimation of value $T_{20\%}$ ($T^T_{20\%}$) the following equation was used [20]:

$$\Delta_f = c\left(T_{20\%}^T - T_g\right)^\beta, \tag{8}$$

where c is constant, accepted equal to 0.093 for fast diffusion and 0.305 – for slow one [20].

The equation (8) defines three factors, influencing on polymer materials thermal stability: chemical constitution of a polymer, characterized by T_g value, polymer melt structure, characterized by dimension Δ_f, and a type (intensity) of oxidizer diffusion, connected with the structure and characterized by exponent β [20].

To carry out the estimation of dimension Δ_f is possible with the aid of a number of methods. However, in the present case the most obvious one will be the following. As it is known [21, 22], the introduction of filler in polymer can change its effective fractal dimension Δ_f (let's remind, that for polymer materials melt the condition Δ_f=d_f was accepted), which can be calculated according to the equation [23]:

$$d_f = 3 - 6\left(\frac{\varphi_{den}}{SC_\infty}\right)^{1/2}, \tag{9}$$

where φ_{den} is relative fraction of nanocomposites densely-packed regions, in which relaxation processes are inhibited, S is a cross-sectional area of a macromolecule (for phenylone S=17.9 Å^2 [24]), C_∞ is characteristic ratio, which is equal to 3 for phenylone [24].

To densely-packed regions of nanocomposites clusters in polymeric matrix, nanofiller and interfacial regions with relative fractions φ_{cl}, φ_n and φ_{if}, respectively can be attribute. The value φ_{cl} for phenylone can be obtained according to the equation (9) at d_f=2.447 [24]. The relation between φ_{if} and φ_n can be obtained as follows. As it is known [7] both polymeric matrix and aerosil particles surface, interacting at interfacial regions formation, are fractal objects. At interaction of such objects there is the only linear scale l, defining their interpenetration distance [25]. As in polymer nanocomposites nanofiller elasticity modulus, as a rule, is larger than the corresponding parameter for polymeric matrix, then it is assumed [7], that in this case aerosil penetration in polymeric matrix occurs and then l is equal to interfacial layer thickness l_{if}. Further it can be written [25]:

$$l_{if} \sim a\left(\frac{r_n}{a}\right)^{2(d-d_{surf})/d}, \tag{10}$$

where a is a lower linear scale of polymeric matrix fractal behavior, r_n is a nanofiller particles radius, d is dimension of Euclidean space, in which fractal is considered (obviously, in our case d=3), d_{surf} is fractal dimension of nanofiller particles surface. The proportionality coefficient in the relationship (10) is accepted equal to 1.0 [22].

For polymers the value a is accepted equal to statistical segment length l_{st} [7], determined according to the formula [26]:

$$l_{st} = l_0 C_\infty, \tag{11}$$

where l_0 is the length of skeletal bond of the main chain, which is equal to 1.40 Å for phenylone [24].

The value d_{surf} can be determined according to the following equation [27]:

$$S_u = 400 r_n^{d_{surf}-d}, \qquad (12)$$

where specific surface of aerosil particles S_u is given in m^2/g and r_n – in nm.

The equations (11) and (12) give l_{st}=0.42 nm, d_{surf}=2.95 and then l_{if}=2.74. Further the following equation can be used [22]:

$$\varphi_{if} = \varphi_n \left[\left(\frac{r_n + l_{if}}{r_n} \right)^3 - 1 \right], \qquad (13)$$

which at r_n=7.5 nm and l_{if}=2.74 nm gives the following relationship between φ_{if} and φ_n for aerosil:

$$\varphi_{if} = 1.55\varphi_n. \qquad (14)$$

Let's note, that for the layered silicates (organoclays) the equation (14) has a form [21, 22]:

$$\varphi_{if} = 1.91\varphi_n. \qquad (15)$$

As nanocomposites reinforcement degree is defined by sum (φ_n+φ_{if}) (see the equation (18)), then the comparison of the equations (14) and (15) shows, that by its reinforcing action nonaggregated aerosil is close to exfoliated organoclays. The equations (14) and (15) were obtained for perfect adhesion polymeric matrix-nanofiller, i.e. for the condition b=1. At nanoadhesion effect availability (b>1) the value l_{if} is proportional to b [11] and at the first approximation the equation (14) is necessary to write as follows:

$$\varphi_{if} = 1.55\varphi_n b. \qquad (16)$$

Then for the value φ_{den} in the equation (9) in case of nanocomposites phenylone/aerosil let's obtain finally:

$$\varphi_{den} = \varphi_{cl} + \varphi_n + 1.55\varphi_n b. \qquad (17)$$

Further one can calculate theoretical values $T_{20\%}^T$, using the equations (7), (8), (9) and (17). The comparison of experimental $T_{20\%}$ and calculated by the indicated method $T_{20\%}^T$

values of sample 20 %-th mass loss temperatures in TGA tests is shown in table 1. As one can see, the good enough correspondence of theory and experiment is obtained (the average discrepancy values $T_{20\%}$ and $T^T_{20\%}$ makes up 2.2 %). In this table the similar calculation of $T^T_{20\%}$ without accounting of nanoadhesion effect is given, i.e. at the calculation the equation (15) was used instead of the equation (17). In this case theory and experiment correspondence grows substantially worse. This means, that the indicated above $T_{20\%}$ enhancement at the introduction of particulate nanofiller (aerosil) small amounts by 76-80 K is due to nanoadhesion effect. At the indicated effect absence such enhancement would not exceed 32 K.

The equation (8) gives a strict and physically substantiated definition of temperature $T_{20\%}$. However, the thermal stability structural analysis can be made by empirical more obvious method. In figure 7 the dependence $T_{20\%}$ on value φ_{den} is shown, which demonstrates linear increase $T_{20\%}$ at φ_{den} enhancement. The extrapolation of linear plot $T_{20\%}(\varphi_{den})$ to limiting values φ_{den}=0 and φ_{den}=1.0 gives the values $T_{20\%}$=616 and 1040 K, respectively. Theoretically these values $T_{20\%}$ can be calculated according to the equation (8) at the condition Δ_f=3 (slow diffusion) and Δ_f=2 (fast diffusion), respectively. Such calculation gives the following values $T^T_{20\%}$: 645 and 1040 K, respectively, i.e. close enough to the one obtained by extrapolation. Hence, as follows from the equations (16) and (17), such empirical treatment assumes also nanoadhesion effect influence on the value $T_{20\%}$.

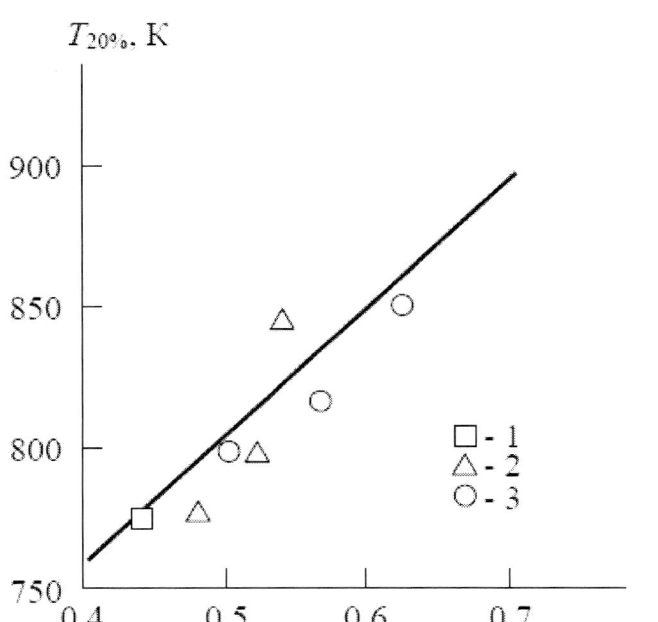

Figure 7. The dependence of sample 20 %-th mass loss temperature $T_{20\%}$ on relative fraction of densely-packed regions φ_{den} for phenylone (1) and nanocomposites phenylone/A (2), phenylone/A-MF (3).

Let's consider briefly nanoadhesion influence on nanocomposites phenylone/aerosil mechanical properties, namely, on their reinforcement degree. Compression tests have shown, that the elasticity modulus E_m of initial matrix phenylone is equal to 3.05 GPa and nanocomposite phenylone/A-MF with aerosil content 0.3 mass. % - 3.29 GPa. As it is known [22], nanocomposites reinforcement degree is defined by the equation:

$$\frac{E_n}{E_m} = 1 + 11(\varphi_n + \varphi_{if})^{1.7}, \qquad (18)$$

where E_n is nanocomposite elasticity modulus.

By using in the equation (18) the estimation φ_{if} according to the formula (14) let's obtain theoretical value E_n (E_n^T) equal to 3.06 GPa and by using formula (16) E_n^T =3.33 GPa. It is not difficult to see, that the second value E_n^T, obtained with nanoadhesion effect accounting, corresponds to the experiment much better (the discrepancy of E_n and E_n^T makes 1 %), than obtained without taking into consideration the indicated effect (the discrepancy is about 10 %). Therefore, nanoadhesion effect results to nanocomposites elasticity modulus increase by 0.263 GPa even at insignificant aerosil content of 0.3 mass. %.

Figure 8 gives the obvious notion of nanoadhesion effectivity action, where the curves of stress-strain (σ-ε) for initial phenylone and nanocomposite phenylone/A-MF with aerosil content 0.3 mass. % are shown. As one can see, the introduction of aerosil indicated amount changes not only mechanical characteristics of matrix polymer, but its fracture type.

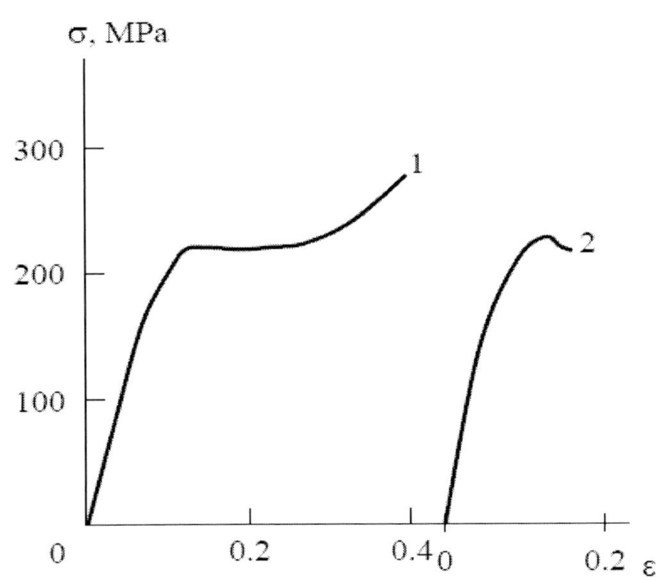

Figure 8. The diagrams stress-strain (σ-ε) for phenylone (1) and nanocomposite phenylone/A-MF with aerosil content 0.3 mass. % (2) in compression tests.

Let's consider in more detail nanoadhesion effect influence on nanocomposites reinforcement degree on the example of phenylone/OSY systems. For the indicated nanocomposites the equations (10)-(14) give the following relationship between φ_{if} and φ_n:

$$\varphi_{if} = 0.102\varphi_n . \tag{19}$$

In figure 9 the comparison of dependences $E_n(\varphi_n)$ calculated according to the equations (18) and (19) is adduced and is also obtained experimentally. As one can see, the indicated curves $E_n(\varphi_n)$ correspondence leaves much to be desired. Therefore let's use the equation (19) modification similar to the formula (16):

$$\varphi_{if} = 0.102\varphi_n b . \tag{20}$$

In figure 9 the comparison of the dependences $E_n(\varphi_n)$, obtained experimentally and calculated according to the equations (18), (20) is also shown. As follows from figure 9 data, in this case much better correspondence of theory and experiment (their average discrepancy makes 1.9 %) is obtained. But the main merit of the equation (20) does not even consist in this (using of the equations (18) and (19) gives also acceptable average discrepancy of theory and experiment, namely, 4.2 %). Firstly, the calculation with due regard to the nanoadhesion effect (the equation (20)) allows to reproduce the main qualitative features of the experimental curve $E_n(\varphi_n)$, namely, to increase ("hump") of E_n at small φ_n ($b>1$, nanoadhesion effect) and slower enhancement of E_n at $\varphi_n \geq 0.023$ ($b<1$ – adhesion level reduction due to OSY particles aggregation). Secondly, the equations (16) and (20) allow quantitative accounting interfacial adhesion level on nanocomposites reinforcement degree.

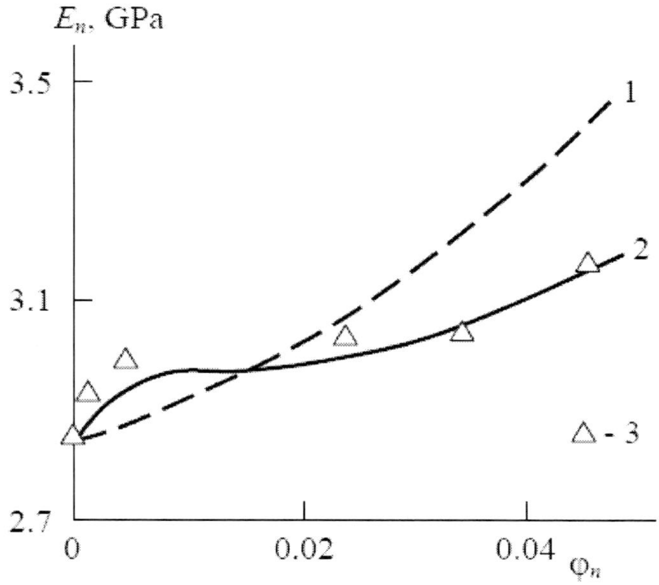

Figure 9. The dependences of elasticity modulus E_n on nanofiller volume contents φ_n for nanocomposites phenylone/OSY. 1 – the calculation according to the equations (18) and (19); 2 – according to the equations (18) and (20); 3 – experimental data.

Conclusions

Therefore, the present paper results have shown the existence in polymer particulate-filled nanocomposites nanoadhesion effect, which consists in the enhanced level of polymeric matrix-nanofiller interaction. This effect has purely dimensional origin and is realized for nanofiller particles with diameter less than 80 nm. The nanoadhesion effect influences strongly on macroscopic properties of those nanocomposites, in which it is realized. So, the essential thermal stability enhancement and the considerable thermal expansion coefficient reduction of nanocomposites on the base of phenylone at the introduction only 1 mass. % aerosil are due to the nanoadhesion effect. Appreciation of the nanoadhesion effect allows qualitatively and quantitatively an exact description of nanocomposites reinforcement degree.

References

[1] Pul C., Owens F. The Nanotechnologies. Moscow, *Technosfera,* 2005, 336 p.

[2] Sheng N., Boyce M.C., Parks D.M., Rutledge G.C., Abes J.I., Cohen R.E. *Polymer,* 2004, v. 45, № 2, p. 487-506.

[3] Buchachenko A.L. *Uspekhi Khimii,* 2003, v. 72, № 5, p. 419-437.

[4] Okhlopkova A.A., Vinogradov A.V., Pinchyk L.S. Plastics Filled by Ultradisperse Inorganic Particles. Gomel, *IMMS NASB,* 1999, 164 p.

[5] Lipatov Yu.S. The Interfacial Phenomena in Polymers. Kiev, *Naukova Dumka,* 1980, 260 p.

[6] Khunyants N.N., Lyapunova M.A., Manevich L.I., Oshmyan V.G., Shaulov A.Yu. *Mekhanika Kompozithykh Materialov,* 1986, v. 22, № 2, p. 231-234.

[7] Kozlov G.V., Yanovskii Yu.G., Lipatov Yu.S. *Mekhanika Kompozitsionnykh Materialov I Konstruktsii,* 2002, v. 8, № 1, p. 111-149.

[8] Kozlov G.V., Lipatov Yu.S. *Composite Interfaces,* 2002, v. 9, № 6, p. 509-527.

[9] Kubat J., Rigdahl M., Welander M. *J. Appl. Polymer Sci.,* 1990, v. 39, № 5, p. 1527-1539.

[10] Holliday L., Robinson J. In book: Polymer Engineering Composites. Ed. Richardson M.O.W. London, Applied Science Publishers LTD, 1978, p. 241-283.

[11] Tanaka K., Nagamura T. Mater. Intern. Conf. on Polymer and Advanced Materials. POLYMEX-2006, Huatulko, Mexico, November 5-9, 2006, Session 1, p. 51.

[12] Andrievskii R.A. *Rossiiskii Khimichesk. Zhurnal,* 2002, v. 46, № 5, p. 50-56.

[13] Wu S. Polymer Interface and Adhesion. New York, *Marcel Dekker,* 1982, 492 p.

[14] Wendlandt U. Thermal Methods of Analysis. Moscow, *Mir,* 1978, 526 p.

[15] Zuru A.A., Whitehead R., Griffiths D.L. *Thermochim. Acta,* 1990, v. 164, № 3, p. 285-305.

[16] Kozlov G.V., Zaikov G.E. The Structural Stabilization of Polymers: Fractal Models. Leiden-Boston, Brill Academic Publishers, 2006, 345 p.

[17] Bershtein V.A., Egorov V.M. Differential Scanning Calorymetry in Physics-Chemistry of Polymers. Leningrad, *Khimiya,* 1990, 256 p.

[18] Shogenov V.Kh., Akhkubekov A.A., Akhkubekov R.A. Izvestiya VUZov, *Severo-Kavkazsk. region, estestv. nauki,* 2004, № 1, p. 46-50.

[19] Vilgis T.A. *Physica A,* 1988, v. 153, № 2, p. 341-354.
[20] Dolbin I.V., Burya A.I., Kozlov G.V. *Teplofizika Vysokikh Temperatur,* 2007, v. 45, № 3, p. 355-358.
[21] Kozlov G.V., Malamatov A.Kh., Antipov E.M., Karnet Yu.N., Yanovskii Yu.G. *Mekhanika Kompozitsionnykh Materialov I Konstruktsii,* 2006, v. 12, № 1, p. 99-140.
[22] Malamatov A.Kh., Kozlov G.V., Mikitaev M.A. The reinforcement Mechanisms of Polymer Nanocomposites. Moscow, Publishers RKhTU, 2006, 240 p.
[23] Kozlov G.V., Novikov V.U. Uspekhi Fizichesk. *Nauk,* 2001, v. 171, № 7, p. 717-764.
[24] Kozlov G.V., Burya A.I., Dolbin I.V. *Prikladnaya Fizika,* 2006, № 1, p. 14-18.
[25] Hentschel H.G.E., *Deutch J.M. Phys. Rev. A,* 1986, v. 29, № 12, p. 1609-1611.
[26] Wu S. J. *Polymer Sci.: Part B: Polymer Phys.,* 1989, v. 27, № 4, p. 723-741.
[27] Kozlov G.V., Shustov G.B. *Khimicheskaya Technologiya,* 2006, № 1, p. 24-26.

In: Progress in Chemistry and Biochemistry
Editors: E. M. Pearce and G. E. Zaikov

ISBN: 978-1-60692-344-3

Chapter 3

POLYMERIC NANOCOMPOSITES BASED ON THE ORGANOMODIFIED LAYERED SILICATES: STRUCTURE, MANUFACTURE AND PROPERTIES

A.K. Mikitaev[1*], A.Yu. Bedanokov[2] and M.A. Mikitaev[3]
[1] Kabardino – Balkarian State University, Nalchik, Russia
[2] D.I.Mendeleev Russian University for Chemical Technology, Moscow, Russia
[3] L.Ya.Karpov Research Institute, Moscow, Russia

ABSTRACT

The analytic review of recent advances in the field of synthesis and study of the structures and properties of the polymeric nanocomposites on the basis of the organomodified layered silicates is given.

Keywords: montmorillonite, organoclays, nanocomposites.

1. INTRODUCTION

The polymeric nanocomposites are the polymers filled with nanoparticles which interact with the polymeric matrix on the molecular level in contrary to the macrointeraction in composite materials. Mentioned nanointeraction results in high adhesion hardness of the polymeric matrix to the nanoparticles [1,52].

Usual nanoparticle is less than 100 nanometers in any dimension, 1 nanometer being the billionth part of a meter [1,2].

The analysis of the reported studies tells that the investigations in the field of the polymeric nanocomposite materials are very promising.

The first notion of the polymeric nanocomposites was given in patent in 1950 [3]. Blumstain pointed out in 1961 [4] that polymeric clay–based nanocomposites had increased

* Author to whom all the correspondence should be sent: 360004, Nalchik, Chernyshevskogo street, 173

thermal stability. It was demonstrated using the data of the thermogravimetric analysis that the polymethylmetacrylate intercalated into the Na^+ - methylmetacrylate possessed the temperature of destruction 40-50°C higher than the initial sample.

This branch of the polymeric chemistry did not attract much attention until 1990 when the group of scientists from the Toyota Concern working on the polyamide–based nanocomposites [5-9] found a two–times increase in the elasticity modulus using only 4.7weight% of the inorganic compound and 100°C increase in the temperature of destruction, both discoveries widely extending the area of application of the polyamide. The polymeric nanocomposites based on the layered silicates began being intensively studied in state, academic and industrial laboratories all over the world only after that.

2. Structure of the Layered Silicates

The study of the polymeric nanocomposites on the basis of the modified layered silicates (broadly distributed and well–known as various clays) is of much interest. The natural layered inorganic structures used in producing the polymeric nanocomposites are the montmorillonite [10,11,12], hectorite [13], vermiculite [14], kaolin, saponine [15] and others. The sizes of inorganic layers are about 220 and 1 nanometers in length and width respectively [16,17].

The perspective ones are the bentonite breeds of clays which include at least 70% of the minerals from the montmorillonite group.

Montmorillonite $(Na,K,Ca)(Al,Fe,Mg)[(Si,Al)_4O_{10}](OH)_2 \times nH_2O$, named after the province Montmorillion in France, is the high–dispersed layered aluminous silicate of white or gray color in which appears the excess negative charge due to the non-stoichiometric replacements of the cations of the crystal lattice, charge being balanced by the exchange cations from the interlayer space. The main feature of the montmorillonite is its ability to adsorb ions, generally cations, and to exchange them. It produces plastic masses with water and may enlarge itself 10 times. Montmorillonite enters the bentonite clays (the term "bentonite" is given after the place Benton in the USA).

The inorganic layers of clays arrange the complexes with the gaps called layers or galleries. The isomorphic replacement within the layers (such as Mg^{2+} replacing Al^{3+} in octahedral structure or Al^{3+} replacing Si^{4+} in tetrahedral one) generates the negative charges which electrostatically are compensated by the cations of the alkali or alkali-earth metals located in the galleries (figure 1) [18]. The bentonite is very hydrophilic because of this. The water penetrates the interlayer space of the montmorillonite, hydrates its surface and exchanges cations what results in the swelling of the bentonite. The further dilution of the bentonite in water results in the viscous suspension with bold tixotropic properties.

The more pronounced cation–exchange and adsorption properties are observed in the bentonites montmorillonite of which contains predominantly exchange cations of sodium.

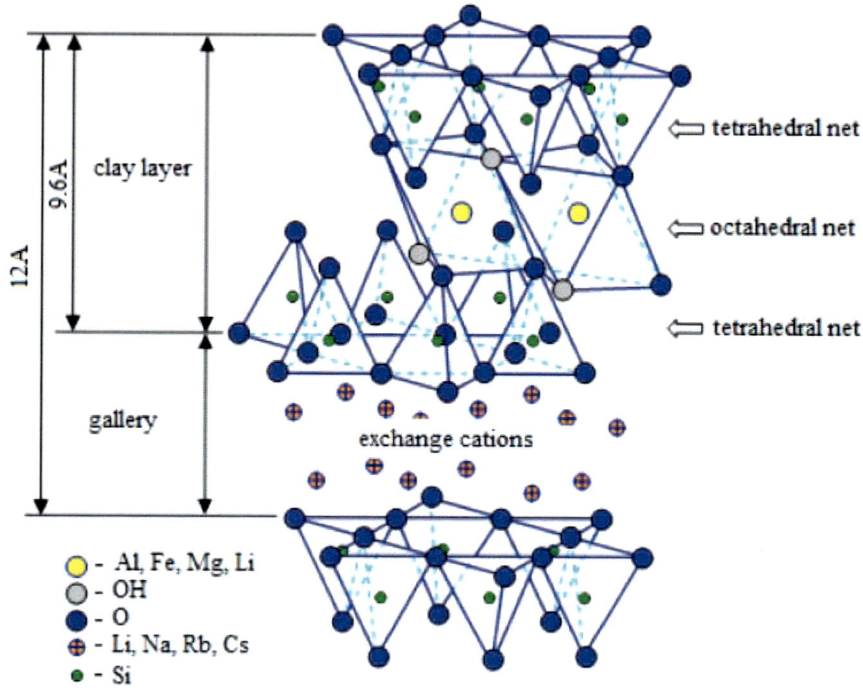

Figure 1. In the paper "Polymeric nanocomposites based on the organomodified layered silicates: structure, manufacture and properties"; by Mikitaev A.K., Bedanokov A.Yu. and Mikitaev M.A.

3. Modification of the Layered Silicates

Layered silicates possess quite interesting properties – sharp drop of hardness at wetting, swelling at watering, dilution at dynamical influences and shrinking at drying.

The hydrophility of aluminous silicates is the reason of their incompatibility with the organic polymeric matrix and is the first hurdle need to be overridden at producing the polymeric nanocomposites.

One way to solve this problem is to modify the clay by the organic substance. The modified clay (organoclay) has at least two advantages: (1) it can be well dispersed in polymeric matrix [19] and (2) it interacts with the polymeric chain [13].

The modification of the aluminous silicates can be done with the replacement of the inorganic cations inside the galleries by the organic ones. The replacing by the cationic surface – active agents like bulk ammonium and phosphonium ions increases the room between the layers, decreases the surface energy of clay and makes the surface of the clay hydrophobic. The clays modified such a way are more compatible with the polymers and form the layered polymeric nanocomposites [52]. One can use the non-ionic modifiers besides the organic ones which link themselves to the clay surface through the hydrogen bond. Organoclays produced with help of non – ionic modifiers in some cases become more chemically stable than the organoclays produced with help of cationic modifiers (figure 2.I) [20].

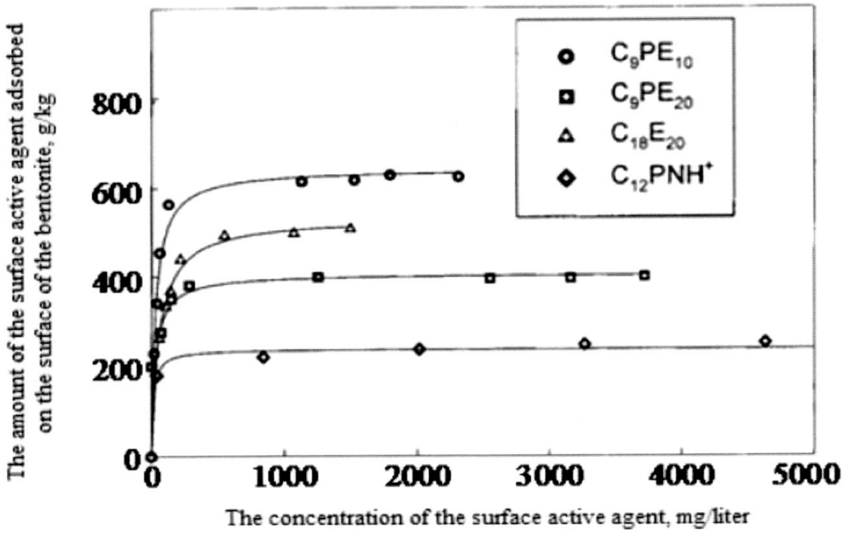

Figure 2.I. In the paper "Polymeric nanocomposites based on the organomodified layered silicates: structure, manufacture and properties"; by Mikitaev A.K., Bedanokov A.Yu. and Mikitaev M.A.

The least degree of desorption is observed for non – ionic interaction between the clay surface and organic modifier (figure 2.II). The hydrogen bonds between the ethylenoxide grouping and the surface of the clay apparently make these organoclays more chemically stable than organoclays produced with non-ionic mechanism.

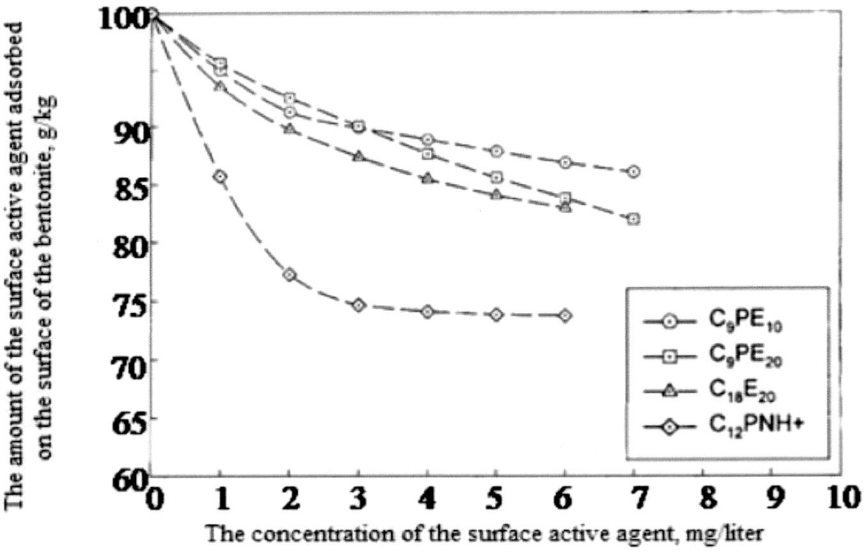

Figure 2.II. In the paper "Polymeric nanocomposites based on the organomodified layered silicates: structure, manufacture and properties"; by Mikitaev A.K., Bedanokov A.Yu. and Mikitaev M.A.

4. STRUCTURE OF THE POLYMERIC NANOCOMPOSITES ON THE BASIS OF THE MONTMORILLONITE

The study of the distribution of the organoclay in the polymeric matrix is of great importance because the properties of composites obtained are in the direct relation from the degree of the distribution.

According Giannelis [21], the process of the formation of the nanocomposite goes in several intermediate stages (figure 3). The formation of the tactoid happens on the first stage – the polymer surrounds the agglomerations of the organoclay. The polymer penetrates the interlayer space of the organoclay on the second stage. Here the gap between the layers may reach 2-3 nanometers [22]. The further separation of the layers, third stage, results in partial dissolution and disorientation of the layers. Exfoliation is observed when polymer shifts the clay layers on more than 8-10 nanometers.

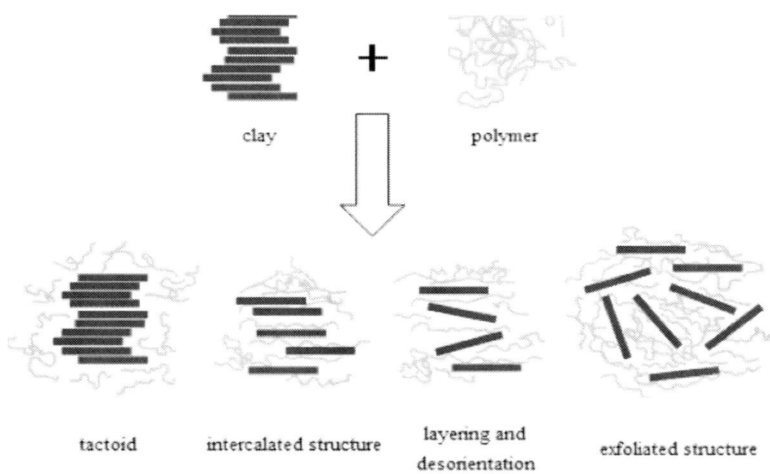

Figure 3. In the paper "Polymeric nanocomposites based on the organomodified layered silicates: structure, manufacture and properties"; by Mikitaev A.K., Bedanokov A.Yu. and Mikitaev M.A.

All mentioned structures may be present in real polymeric nanocomposites in dependence from the degree of distribution of the organoclay in the polymeric matrix. Exfoliated structure is the result of the extreme distribution of the organoclay. The excess of the organoclay or bad dispersing may born the agglomerates of the organoclays in the polymeric matrix what finds experimental confirmation in the X-ray analysis [11,12,21,23].

In the following subsections we describe a number of specific methods used at studying the structure of the polymeric nanocomposites.

4.1. Determination of the Interlayer Space

The X-ray determination of the interlayer distance in the initial and modified layered silicates as well as in final polymeric nanocomposite is one of the main methods of studying the structure of the nanocomposite on the basis of the layered silicate. The peak in the small – angle diapason (2θ=6-8°C) is characteristic for pure clays and responds to the order of the structure of the silicate. This peak drifts to the smaller values of the angle 2θ in organomodified clays. If clay particles are uniformly distributed in the bulk of the polymeric matrix then this peak disappears, what witnesses on the disordering in the structure of the layered silicate. If the amount of the clay exceeds the certain limit of its distribution in the polymeric matrix, then the peak reappears again. This regularity was demonstrated on the instance of the polybutylenterephtalate (figure 4) [11].

The knowledge of the angle 2θ helps to define the size of the pack of the aluminous silicate consisting of the clay layer and interlayer space. The size of such pack increases in a row from initial silicate to polymeric nanocomposite according to the increase in the interlayer space. The average size of that pack for montmorillonite is 1.2-1.5 nanometers but for organomodified one varies in the range of 1.8-3.5 nanometers.

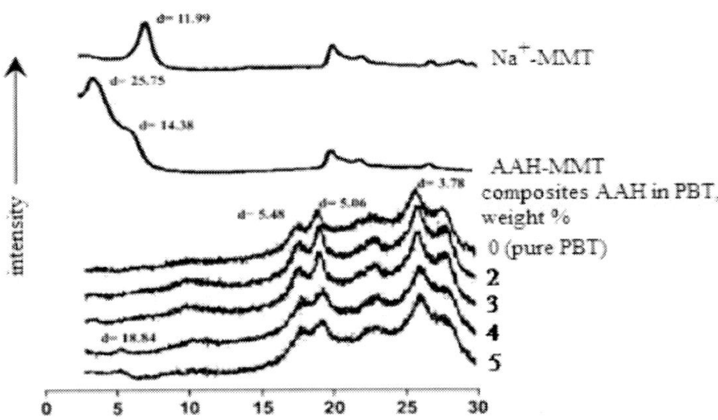

Figure 4. In the paper "Polymeric nanocomposites based on the organomodified layered silicates: structure, manufacture and properties"; by Mikitaev A.K., Bedanokov A.Yu. and Mikitaev M.A.

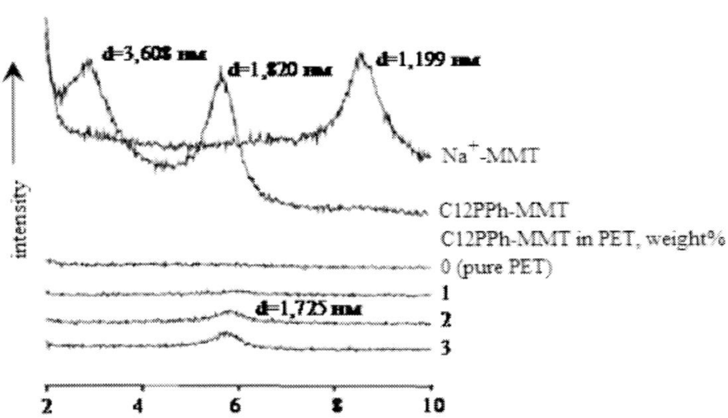

Figure 5. In the paper "Polymeric nanocomposites based on the organomodified layered silicates: structure, manufacture and properties"; by Mikitaev A.K., Bedanokov A.Yu. and Mikitaev M.A.

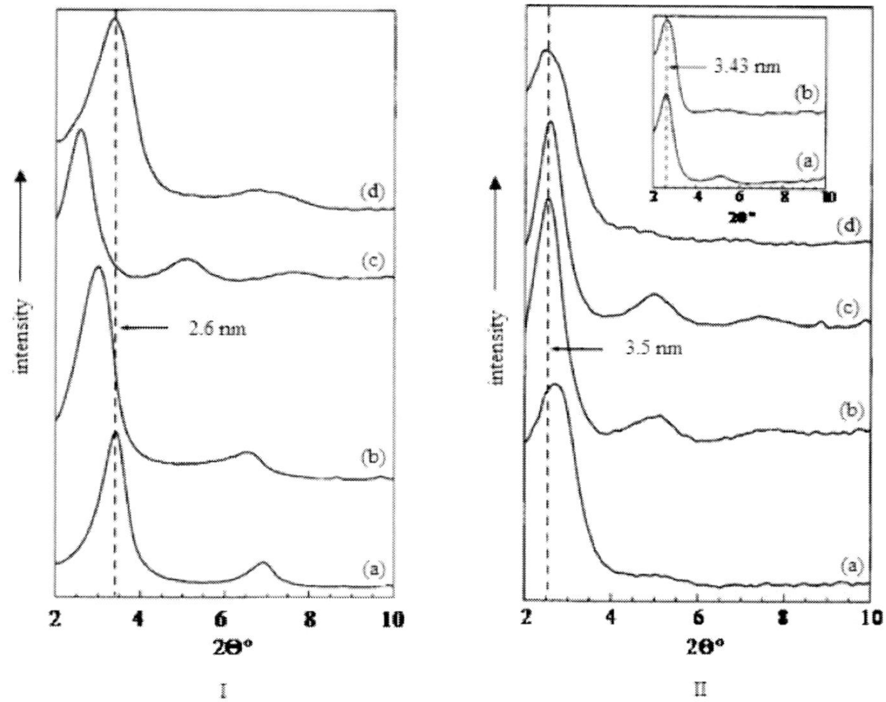

Figure 6. In the paper "Polymeric nanocomposites based on the organomodified layered silicates: structure, manufacture and properties"; by Mikitaev A.K., Bedanokov A.Yu. and Mikitaev M.A.

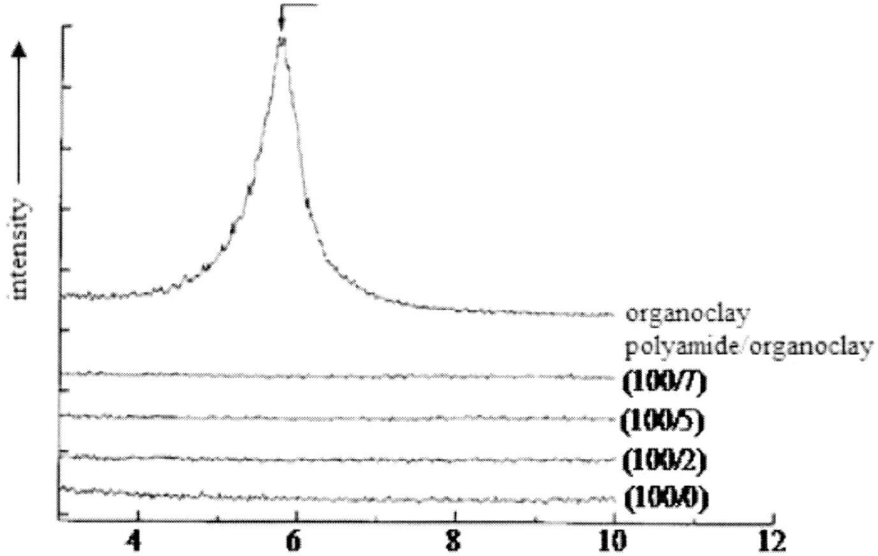

Figure 7. In the paper "Polymeric nanocomposites based on the organomodified layered silicates: structure, manufacture and properties"; by Mikitaev A.K., Bedanokov A.Yu. and Mikitaev M.A.

Summing up we conclude that comparing the data of the X-ray analysis for the organoclay and nanocomposite allows for the determination of the optimal clay amount need be added to the composite. The data from the scanning tunneling (STM) and transmission electron (TEM) microscopes [27, 28] can be used as well.

4.2. The Degree of the Distribution of the Clay Particles in the Polymeric Matrix

The two structures, namely the intercalated and exfoliated ones, could be distinguished with the respect to the degree of the distribution of the clay particles, figure 8. One should note that clay layers are quite flexible though they are shown straight in the figure. The formation of the intercalated or exfoliated structures depends on many factors, for example the method of the production of the nanocomposite or the nature of the clay etc [29].

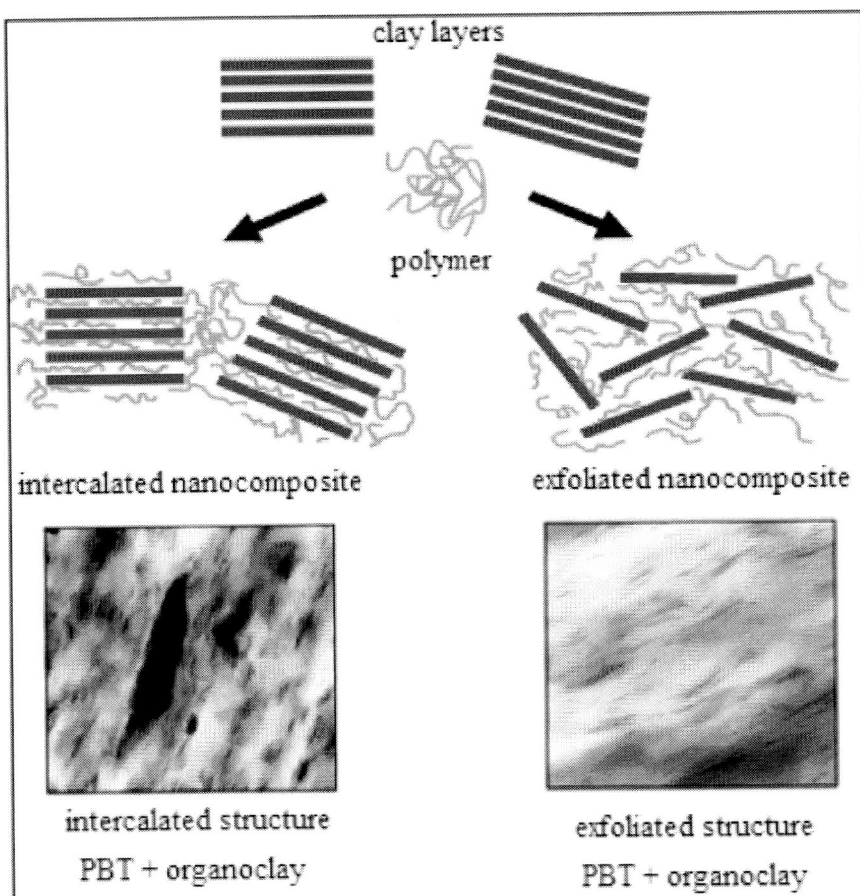

Figure 8. In the paper "Polymeric nanocomposites based on the organomodified layered silicates: structure, manufacture and properties"; by Mikitaev A.K., Bedanokov A.Yu. and Mikitaev M.A.

The TEM images of the surface of the nanocomposites can help to find out the degree of the distribution of the nanosized clay particles, see plots (a) to (d) in the figure 9.

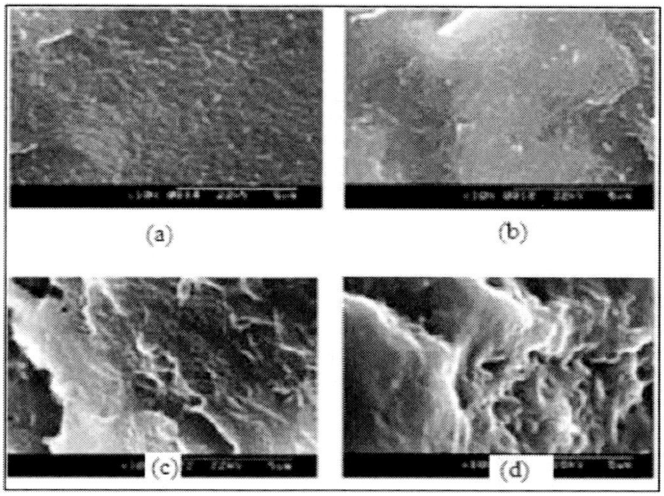

Figure 9. In the paper "Polymeric nanocomposites based on the organomodified layered silicates: structure, manufacture and properties"; by Mikitaev A.K., Bedanokov A.Yu. and Mikitaev M.A.

The smooth surface tells about the uniform distribution of the organoclay particles. The surface of the nanocomposite becomes deformed with the increasing amount of the organoclay, see plots (a) to (d) in the figure 10. Probably, this is due to the influence of the clay agglomerates [30,31].

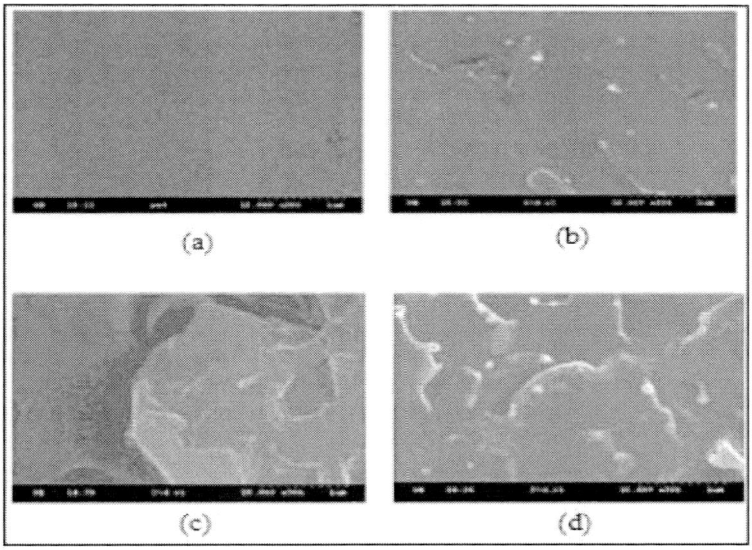

Figure 10. In the paper "Polymeric nanocomposites based on the organomodified layered silicates: structure, manufacture and properties"; by Mikitaev A.K., Bedanokov A.Yu. and Mikitaev M.A.

Also one can use the STM images to judge on the degree of the distribution of the organoclay in the nanocomposite, figures 11 and 12. If the content of the organoclay is 2-3weight% then the clay layers are separated by the polymeric layer of 4 to 10 nanometers width, figure 11. If the content of the organoclay reaches 4-5weight% then the majority of the clay becomes well distributed however the agglomerates of 4 to 8 nanometers may appear.

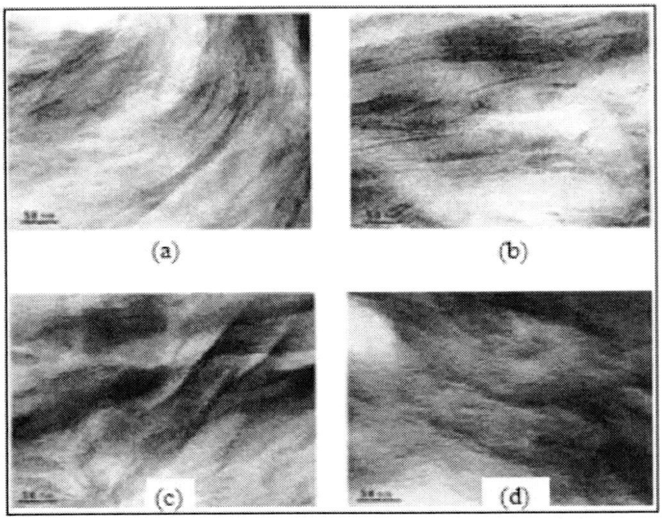

Figure 11. In the paper "Polymeric nanocomposites based on the organomodified layered silicates: structure, manufacture and properties"; by Mikitaev A.K., Bedanokov A.Yu. and Mikitaev M.A.

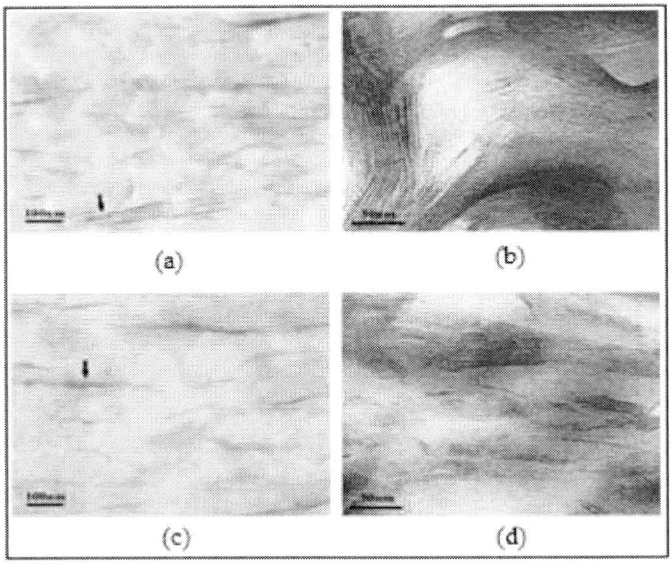

Figure 12. In the paper "Polymeric nanocomposites based on the organomodified layered silicates: structure, manufacture and properties"; by Mikitaev A.K., Bedanokov A.Yu. and Mikitaev M.A.

So the involvement of the X-ray analysis and the use of the microscopy data tell that the nanocomposite consists of the exfoliated clay at the low content (below 3weight%) of the organoclay.

5. Production of the Polymeric Nanocomposites on the Basis of the Aluminous Silicates

Different groups of authors [32-35] offer following methods for obtaining nanocomposites on the basis of the organoclays: (1) in the process of the synthesis of the polymer [33,36,37], (2) in the melt [38,39], (3) in the solution [40-46] and (4) in the sol-gel process [47-50].

The most popular ones are the methods of producing in melt and during the process of the synthesis of the polymer.

The producing of the polymeric nanocomposite in situ is the intercalation of the monomer into the clay layers. The monomer migrates through the organoclay galleries and the polymerization happens inside the layers [19,51], figure 13.

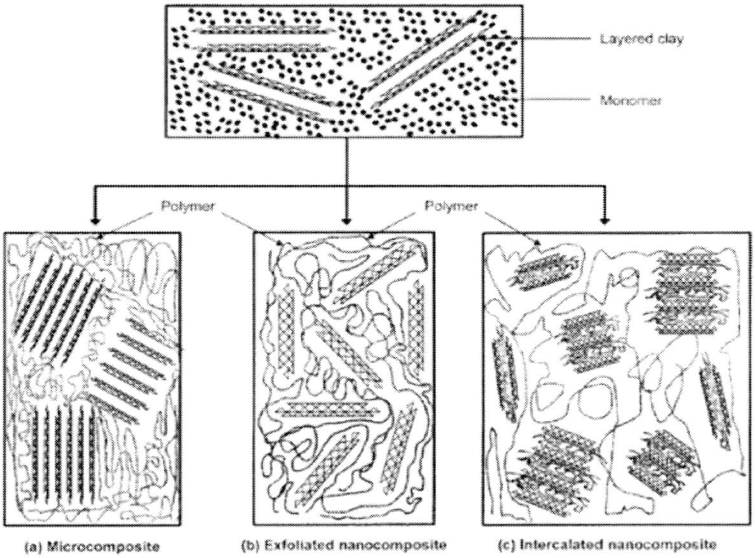

Figure 13. In the paper "Polymeric nanocomposites based on the organomodified layered silicates: structure, manufacture and properties"; by Mikitaev A.K., Bedanokov A.Yu. and Mikitaev M.A.

The polymerization may be initiated by the heat, irradiation or other source. Obviously, the best results on the degree of the distribution of the clay particles in the polymeric matrix must emerge if using given method. This is associated with the fact that the separation of the clay layers happens in the very process of the inclusion of the monomer in the interlayer space. In other words, the force responsible for the separation of the clay layers is the growth of the polymeric chain whereas the main factor for reaching the necessary degree of the clay distribution in solution or melt is just satisfactory mixing. The most favorable condition for synthesizing the nanocomposites is the vacuuming or the flow of the inert gas. Besides, one has to use the fast speeds of mixing for satisfactory dispersing of the organoclay in the polymeric matrix.

The method of obtaining the polymeric nanocomposites in melt (or the method of extrusion) is the mixing of the polymer melted with the organoclay. The polymeric chains lose the considerable amount of the conformational entropy during the intercalation. The probable motive force for this process is the important contribution of the enthalpy of the interaction between the polymer and organoclay at mixing. One should add that polymeric nanocomposite on the basis of the organoclays could be successively produced by the extrusion [22]. The advantage of the extrusion method is the absence of any solvents what excludes the unnecessary leaks. Moreover the speed of the process is several times more and the technical side is simpler. The extrusion method is the best one in the industrial scales of production of the polymeric nanocomposites what acquires the lesser source expenses and easier technological scheme.

If one produces the polymer – silicate nanocomposite in solution then the organosilicate swells in the polar solvent such as toluene or N,N-dimethylformamide. Then the added is the solution of the polymer which enters the interlayer space of the silicate. The removing of the solvent by means of evaporation in vacuum happens after that. The main advantage of the given method is that "polymer – layered silicate" might be produced from the polymer of low polarity or even non-polar one. However this method is not widely used in industry because of much solvent consumption [52].

The sol-gel technologies find application at producing nanocomposites on the basis of the various ceramics and polymers. The initial compounds in these technologies are the alcoholates of specific elements and organic oligomers.

The alcoholates are firstly hydrolyzed and obtained hydroxides being polycondensated then. The ceramics from the inorganic 3D net is formed as a result. Also the method of synthesis exists in which the polymerization and the formation of the inorganic glass happen simultaneously. The application of the nanocomposites on the basis of ceramics and polymers as special hard defensive coverages and like optic fibers [53] is possible .

6. Properties of the Polymeric Nanocomposites

Many investigations in physics, chemistry and biology have shown that the jump from macroobjects to the particles of 1-10 nanometers results in the qualitative transformations in both separate phases and systems from them [54].

One can improve the thermal stability and mechanical properties of the polymers by inserting the organoclay particles into the polymeric matrix. It can be done by means of joining the complexes of properties of both the organic and inorganic substances, that is combining the light weight, flexibility and plasticity of former and durability, heat stability and chemical resistance of latter.

Nanocomposites demonstrate essential change in properties if compared to the non-filled polymers. So, if one introduces modified layered silicates in the range of 2 to 10weight% into the polymeric matrix then he observes the change in mechanical (tensile, compression, bending and overall strength), barrier (penetrability and stability to the solvent impact), optical and other properties. The increased heat and flame resistance even at low filler content is among the interesting properties too. The formation of the thermal isolation and negligible penetrability of the charred polymer to the flame provide for the advantages of using these materials.

The organoclay as a nanoaddition to the polymers may change the temperature of the destruction, refractoriness, rigidity and rupture strength. The nanocomposites also possess the increased rigidity modulus, decreased coefficient of the heat expansion, low gas-penetrability, increased stability to the solvent impact and offer broad range of the barrier properties [54]. In table 1 we gather the characteristics of the nylon-6 and its derivative containing 4.7weight% of the organomodified montmorillonite.

Table 1. in the paper "Polymeric nanocomposites based on the organomodified layered silicates: structure, manufacture and properties"; by Mikitaev A.K., Bedanokov A.Yu. and Mikitaev M.A. The properties of the nylon-6 and composite based on it [54]

	Rigidity modulus, GPa	Tensile strength, MPa	Temperature of the deformation, °C	Impact viscosity, kJ/m^2	Water consumption, weight %	Coefficient of the thermal expansion (x,y)
Nylon-6	1.11	68.6	65	6.21	0.87	13×10^{-5}
Nanocomposite	1.87	97.2	152	6.06	0.51	6.3×10^{-5}

It is important that the temperature of the deformation of the nanocomposite increases on 87°C.

The thermal properties of the polymeric nanocomposites with the varying organoclay content are collected in table 2.

Table 2. in the paper "Polymeric nanocomposites based on the organomodified layered silicates: structure, manufacture and properties"; by Mikitaev A.K., Bedanokov A.Yu. and Mikitaev M.A. The main properties of the polymeric nanocomposites

Property	Composition								
	Polybutyleneterephtalate + AAX-montmorillonite					Polyethyleneterephtalate + C_{12}PPh-montmorillonite			
	Organoclay content, %								
	0	2	3	4	5	0	1	2	3
Viscosity, dliter/g	0.84	1.16	0.77	0.88	0.86	1.02	1.26	0.98	1.23
T_g, °C	27	33	34	33	33	---	---	---	---
T_m, °C	222	230	230	229	231	245	247	245	246
T_d, °C	371	390	388	390	389	370	375	384	386
W_{tR}^{600C}, %	1	6	7	7	9	1	8	15	21
Strength limit, MPa	41	50	60	53	49	46	58	68	71
Rigidity modulus, GPa	1.37	1.66	1.76	1.80	1.86	2.21	2.88	3.31	4.10
Relative enlargement, %	5	7	6	7	7	3	3	3	3

The inclusion of the organoclay into the polybutyleneterephtalate leads to the increase in the glass transition temperature (T_g) from 27 to 33 degrees centigrade if the amount of the clay raises from zero to 2weight%. That temperature does not change with the further increase of the organoclay content. The increase in T_g may be the result of two reasons [56-59]. The first is the dispersion of the small amount of the organoclay in the free volume of the polymer, and the second is the limiting of the mobility of the segments of the polymeric chain due to its interlocking between the layers of the organoclay.

The same as the T_g, the melting temperature T_m increases from the 222 to 230°C if the organoclay content raises from 0 to 2weight% and stays constant up to 5weight%, see table 2. This increase might be the consequence of both complex multilayer structure of the nanocomposite and interaction between the organoclay and polymeric chain [60,61]. Similar regularities have been observed in other polymeric nanocomposites also.

The thermal stability of the nanocomposites polybutyleneterephtalate, briefly PBT, (or polyethyleleterephtalate) / organoclay determined by the thermogravimetric analysis is presented in table 3 and in figures 14 and 15 [11,12].

Table 3. in the paper "Polymeric nanocomposites based on the organomodified layered silicates: structure, manufacture and properties"; by Mikitaev A.K., Bedanokov A.Yu. and Mikitaev M.A. The thermal properties of the fibers from PET with varying organoclay content

Organoclay content, weight %	$\eta_{inh}{}^{a}$	T_m (°C)	$\Delta H_m{}^{b}$ (J/g)	$T^i_D{}^{c}$ (°C)	$Wt_R{}^{600d}$ (%)
0 (pure PET)	1.02	245	32	370	1
1	1.26	247	32	375	8
2	0.98	245	33	384	15
3	1.23	246	32	386	21

a – viscosities were measured at 30°C using 0.1 gram of polymer on 100 milliliters of solution in mix phenol / tetrachlorineethane (50/50);
b – change in enthalpy of melting;
c – initial temperature of decomposition;
d – weight percentage of the coke remnant at 600°C.

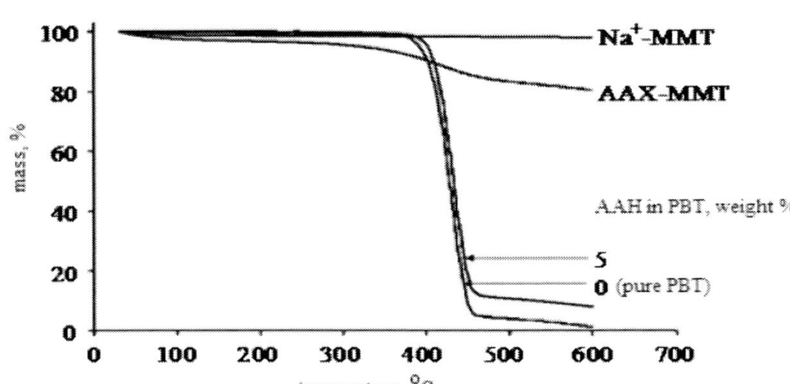

Figure 14. In the paper "Polymeric nanocomposites based on the organomodified layered silicates: structure, manufacture and properties"; by Mikitaev A.K., Bedanokov A.Yu. and Mikitaev M.A.

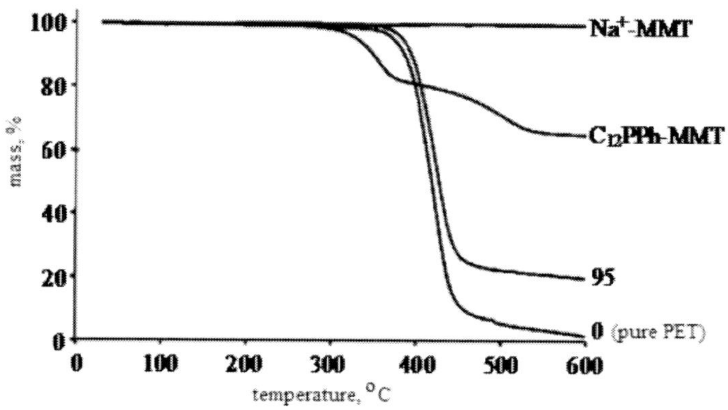

Figure 15. In the paper "Polymeric nanocomposites based on the organomodified layered silicates: structure, manufacture and properties"; by Mikitaev A.K., Bedanokov A.Yu. and Mikitaev M.A.

The temperature of the destruction, T_D, increases with the organoclay content up to 350°C in case of the composite PBT / organoclay. The thermogravimetrical curves for pure and composite PBTs have similar shapes below 350°C. The values of temperature T_D depends on the amount of organoclay above 350°C. The organoclay added becomes a barrier for volatile products being formed during the destruction [61,62]. Such example of the improvement of the thermal stability was studied in papers [63,64]. The mass of the remnant at 600°C increases with organoclay content.

Following obtained data authors draw the conclusion that the optimal results for thermal properties are being obtained if 2weight% of the organoclay is added [11,12,19].

The great number of studies on the polymeric composite organoclay – based materials show [11,12,13,19] that the inclusion of the inorganic component into the organic polymer improves the thermal stability of the latter, see tables 3 and 4.

Table 4. in the paper "Polymeric nanocomposites based on the organomodified layered silicates: structure, manufacture and properties"; by Mikitaev A.K., Bedanokov A.Yu. and Mikitaev M.A. The basic properties of the nanocomposite based on PBT with varying organoclay content

Organoclay content, weight %	I.V.[a]	T_g	T_m (°C)	T^i_D [b] (°C)	Wt_R^{600c} (%)
0 (pure PBT)	0,84	27	222	371	1
2	1,16	33	230	390	6
3	0,77	34	230	388	7
4	0,88	33	229	390	7
5	0,86	33	231	389	9

a – viscosities were measured at 30°C using 0.1 gram of polymer on 100 milliliters of solution in mix phenol / tetrachlorineethane (50/50);

b – initial temperature of the weight loss;

c – weight percentage of the coke remnant at 600°C.

The values of the melting temperature increase from 222 to 230°C if the amount of the organoclay added reaches 2weight% and then stay constant. This effect can be explained by both thermal isolation of the clay and interaction between the polymeric chain and organoclay [43,64]. The increase in the glass transition temperature also occurs what can be a consequence of several reasons [55,56,58]. One of the main among them is the limited motion of the segments of the polymeric chain in the galleries within the organoclay.

If the organoclay content in the polymeric matrix of the PBT reaches 2weight% then both the temperature of the destruction increases and the amount of the coke remnant increase at 660°C and then both stay practically unchanged with the further increase of the organoclay content up to 5weight%. The loss of the weight due to the destruction of the polymer in pure PBT and its composites looks familiar in all cases below 350°C. The amount of the organoclay added becomes important above that temperature because the very clays possess good thermal stability and make thermal protection by their layers and form a barrier preventing the volatile products of the decomposition to fly off [43,60]. Such instance of the improvement in thermal properties was observed in many polymeric composites [64-68]. The weight of the coke remnant increases with the rising organoclay content up to 2weight% and stays constant after that. The increase of the remnant may be linked with the high thermal stability of the organoclay itself. Also it is worth noticing that the polymeric chain closed in interlayer space of the organoclay has fewer degrees of oscillatory motion at heating due to the limited interlayer space and the formation of the abundant intermolecular bonds between the polymeric chain and the clay surface. And the best result is obtained at 2-3weight% content of the organoclay added to the polymer.

If one considers the influence of the organoclay added to the polyetheleneterephtalate, briefly PET, [69] then the temperature of the destruction increases on 16°C at optimal amount of organoclay of 3weight%. The coke remnant at 600°C again increases with the rising organoclay content, see table 3.

Regarding the change in the temperature of the destruction in the cases of PET and PBT versus the organoclay content one can note that both trends look similar. However the coke remnant considerably increases would the tripheyldodecylphosphonium cation be present within the clay. The melting temperature does not increase in case of the organoclay added into the PET in contrary to the case of PBT. Apparently this may be explained by the more crystallinity of the PBT and the growth of the degree of crystallinity with the organoclay content.

It becomes obvious after analyzing the above results that the introduction of the organoclay into the polymer increases the thermal stability of the latter according the (1) thermal isolating effect from the clay layers and (2) barrier effect in relation to the volatile products of destruction.

The studying of the mechanical properties of the nanocomposites, see table 2, have shown that the limit of the tensile strength increases with the organoclay added up to 3weight% for the majority of the composites. Further addition of the organoclay, up to 5weight%, results in the decreasing limit of the tensile strength. We explain this by the fact that agglomerates appear in the nanocomposite when the organoclay content exceeds the 3weight% value [61,70,71]. The proof for the formation of the agglomerates have been obtained from the X-ray study and using the data from electron microscopes.

Nevertheless the rigidity modulus increases with the amount of the organoclay added into the polymeric matrix, the resistance of the clay itself being the explanation for that. The oriented polymeric chains in the clay layers also participate in the increase of the rigidity modulus [72]. The percentage of enlargement at breaking became 6-7weight% for all mixes.

Using data of the table 2 we explain the improvement in the mechanical properties of the nanocomposites with added organoclay up to 3weight% by the good degree of distribution of the organoclay within the polymeric matrix. The degree of the improvement also depends on the interaction between the polymeric chain and clay layers.

The study of the influence of the degree of the extract of fibers on the mechanical properties has shown that the limit of strength and the rigidity modulus both increase in PBT whereas they decrease in nanocomposites, table 5. This can be explained by the breaking of the bonds between the organoclay and PBT at greater degree of extract. Such phenomena have been observed in numerous polymeric composites [73-75].

Table 5. in the paper "Polymeric nanocomposites based on the organomodified layered silicates: structure, manufacture and properties"; by Mikitaev A.K., Bedanokov A.Yu. and Mikitaev M.A. The ability to stretch of the nanocomposites PBT/organoclay at varying degrees of extract

Organoclay content, weight %	Limit of strength, MPa			Rigidity modulus, GPa		
	DR=1	DR=3	DR=6	DR=1	DR=3	DR=6
0 (pure PBT)	41	50	52	1.37	1.49	1.52
3	60	35	29	1.76	1.46	1.39

The first notions on the lowered flammability of the polymeric nanocomposites on the organoclay basis appeared in 1976 in the patent on the composite based on the nylon-6 [5]. The serious papers in the field were absent till the 1995 [76].

The use of the calorimeter is very effective for studying the refractoriness of the polymers. It can help at measuring the heat release, the carbon monoxide depletion and others. The speed of the heat release is one of the most important parameters defining the refractoriness [77]. The data on the flame resistance in various polymer / organoclay systems such as layered nanocomposite nylon-6 / organoclay, intercalated nanocomposites polystyrene (or polypropylene) / organoclay were given in paper [78] in where the lowered flammability was reported, see table 6. And the lowered flammability have been observed in systems with low organoclay content, namely in range from 2 to 5weight%.

Table 6. in the paper "Polymeric nanocomposites based on the organomodified layered silicates: structure, manufacture and properties"; by Mikitaev A.K., Bedanokov A.Yu. and Mikitaev M.A.
Calorimetric data

Sample	remnant (%)±0.5	Peak of the HRR (Δ%) (kW/m^2)	Middle of the HRR (Δ%) (kW/m^2)	Average value H_c (MJ/kg)	Average value SEA (m^2/kg)	Average CO left (kg/kg)
Nylon-6	1	1010	603	27	197	0.01
Nylon-6 / organoclay, 2%, delaminated	3	686 (32%)	390 (35%)	27	271	0.01
Nylon-6 / organoclay, 5%, delaminated	6	378 (63%)	304 (50%)	27	296	0.02
Polystyrene	0	1120	703	29	1460	0.09
PS / organoclay, 3%, bad mixing	3	1080	715	29	1840	0.09
PS / organoclay, 3%, intercalated / delaminated	4	567 (48%)	444 (38%)	27	1730	0.08
PSw/DBDPO/Sb_2O_3, 30%	3	491 (56%)	318 (54%)	11	2580	0.14
Polypropylene	0	1525	536	39	704	0.02
PP / organoclay, 2%, intercalated	5	450 (70%)	322 (40%)	44	1028	0.02

H_c – heat of combustion;
SEA – specific extinguishing area;
DBDPO – dekabrominediphenyloxide;
HRR – speed of the heat release.

The curve of the heat release for the polypropylene and the nanocomposite on its basis (organoclay content varying from 2 to 4weight%) is given in the figure 16 from which one can see that the speed of the heat release for the nanocomposite enriched with the 4weight% organoclay (the interlayer distance 3.5 nanometers) is 75% less than for pure polypropylene.

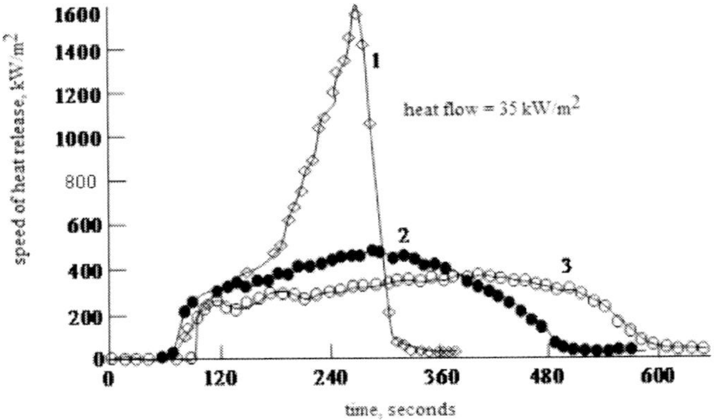

Figure 16. In the paper "Polymeric nanocomposites based on the organomodified layered silicates: structure, manufacture and properties"; by Mikitaev A.K., Bedanokov A.Yu. and Mikitaev M.A.

The comparison of the experimental data for the nanocomposites on the basis of the nylon-6, polypropylene and polystyrene gathered in table 7 show that the heat of combustion, the smoke release and the amount of the carbon monoxide are almost constant at varying organoclay content. So we conclude that the source for the increased refractoriness of these materials is the stability of the solid phase and not the influence of the vapor phase. The data for the polystyrene with the 30% of the dekabrominediphenyloxide and Sb_2O_3 are given in table 6 as the proof of the influence of the vapor phase of bromine. The incomplete combustion of the polymeric material in the latter case results in low value of the heat of the combustion and high quantity of the carbon monoxide released [79].

One should note that the mechanism for the increased fire resistance of the polymeric nanocomposites on the basis of the organoclays is not, in fact, clear at all. The formation of the barrier from the clay layers during the combustion at their collapse is supposed to be the main mechanism. That barrier slows down the combustion [80]. In our paper we study the influence of the nanocomposite structure on the refractoriness. The layered structure of the nanocomposite expresses higher refractoriness comparing to that in intercalated nanocomposite, see figure 17.

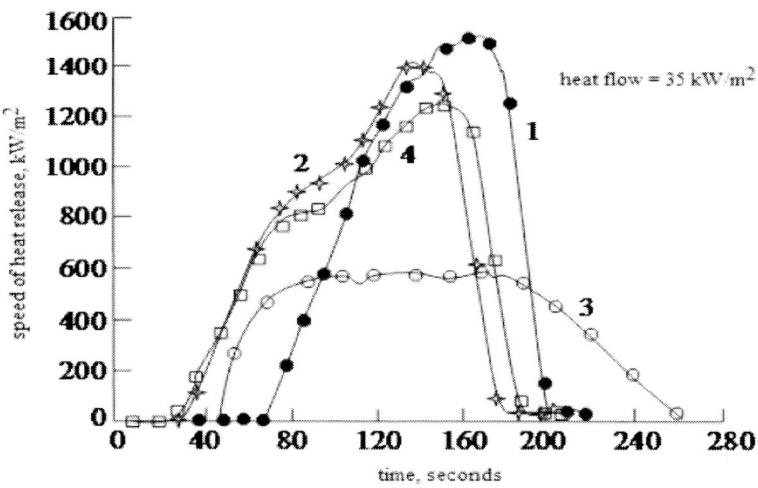

Figure 17. In the paper "Polymeric nanocomposites based on the organomodified layered silicates: structure, manufacture and properties"; by Mikitaev A.K., Bedanokov A.Yu. and Mikitaev M.A.

The data on the polymeric polystyrene – based nanocomposites presented in figure 17 are for (1) initial ammoniumfluorine hectorite and (2) quaternary ammonium montmorillonite. The intercalated nanocomposite was produced in first case whereas the layered-intercalated nanocomposite was produced in the second one. But because the chemical nature and the morphology of the organoclay used was quite different it is very difficult to draw a unique conclusion about the flame resistance in polymeric nanocomposites produced. Nonetheless, one should point out that good results of the same quality were obtained for both layered and intercalated structures when studying the aliphatic groupings of the polyimide nanocomposites based on these clays. The better refractoriness is observed in case of polystyrene embedded in layered nanocomposite while intercalated polystyrene – based nanocomposite (with MMT) also exhibits increased refractoriness.

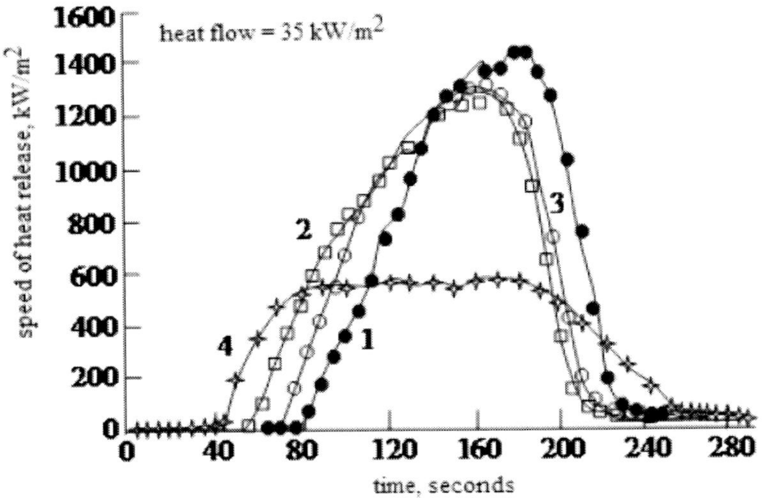

Figure 18. In the paper "Polymeric nanocomposites based on the organomodified layered silicates: structure, manufacture and properties"; by Mikitaev A.K., Bedanokov A.Yu. and Mikitaev M.A.

As one can see the from the figure 18, the speed of the heat release for the nanocomposite produced in nitrogen atmosphere at 170°C is much lower than for other samples. Probably, the reason for the low refractoriness of the nanocomposite produced in extruder without the vacuuming at 180°C is the influence of the high temperature and of the oxygen from the air what can lead to the destruction of the polymer in such conditions of the synthesis.

It is impossible to give an exact answer on the question about how the refractoriness of organoclay – based nanocomposites increases basing on only the upper experimental data but the obvious fact is that the increased thermal stability and refractoriness are due to the presence of the clays existing in the polymeric matrix as nanoparticles and playing the role of the heat isolators and elements preventing the flammable products of the decomposition to fly off.

There are still many problems unresolved in the field but indisputably polymeric nanocomposites will take the leading position in the chemistry of the advanced materials with high heat and flame resistance. Such materials can be used either as itself or in combination with other agents reducing the flammability of the substances.

The processes of the combustion are studied for the number of a polymeric nanocomposites based on the layered silicates such as nylon-6.6 with 5weight% of Cloisite 15A – montmorillonite being modified with the dimethyldialkylammonium (alkyls studied C_{18}, C_{16}, C_{14}), maleinated polypropylene and polyethylene, both (1.5%) with 10weight% Cloisite 15A. The general trend is two times reduction of the speed of the heat release. The decrease in the period of the flame induction is reported for all nanocomposites in comparison with the initial polymers [54].

The influence of the nanocomposite structure on its flammability is reflected in the table 7. One can see that the least flammability is observed in delaminated nanocomposite based on the polystyrene whereas the flammability of the intercalated composite is much higher [54].

Table 7. in the paper “Polymeric nanocomposites based on the organomodified layered silicates: structure, manufacture and properties”; by Mikitaev A.K., Bedanokov A.Yu. and Mikitaev M.A.

Flammability of several polymers and composites

Sample	Coke remnant, weight%	Max speed of heat release, kW/m^2	Average value of heat release, kW/m^2	Average heat of combustion, MJ/kg	Specific smoke release, m^2/kg	CO release, kg/kg
Nylon-6	1	1010	603	27	197	0.01
Nylon-6 + 2% of silicate (delaminated)	3	686	390	27	271	0.01
Nylon-6 + 5% of silicate (delaminated)	6	378	304	27	296	0.02
Nylon-12	0	1710	846	40	387	0.02
Nylon-12 + 2% of silicate (delaminated)	2	1060	719	40	435	0.02
Polystyrene	0	1562	803	29	1460	0.09
PS + 3% of silicate Na-MMT	3	1404	765	29	1840	0.09
PS + 3% of silicate C14-FH (intercalated)	4	1186	705	28	1790	0.09
PS + 3% of silicate 2C18-MMT (delaminated)	4	567	444	28	1730	0.08
Polypropylene	0	1525	536	39	704	0.02
PP + 2% of silicate (intercalated)	3	450	322	40	1028	0.02

The optical properties of the nanocomposites are of much interest too. The same materials could be either transparent or opaque depending on certain conditions. For example in figure 19 we see transparency, plot (a), and turbidity, plot (c), of the material in dependence of the frequency of the current applied.

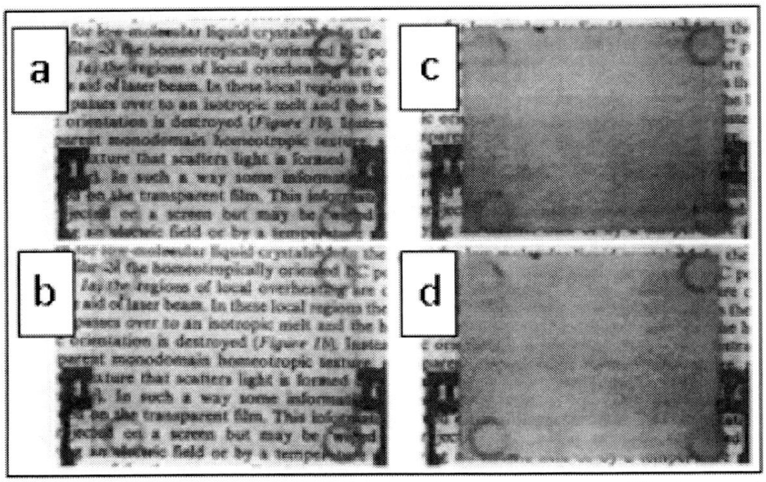

Figure 19. in the paper "Polymeric nanocomposites based on the organomodified layered silicates: structure, manufacture and properties"; by Mikitaev A.K., Bedanokov A.Yu. and Mikitaev M.A.

The effect in figure 19 is reversible and can be innumerately repeated. The transparent and opaque states exhibit the memory effect after the applied current switched off, plots (b) and (d) in figure 19. The study of the intercalated nanocomposites based on the smectite clays reveals that the optical and elecrooptical properties depend on the degree of intercalation [81].

CONCLUSIONS

The quantity of the papers in the field of the nanocomposite polymeric materials has grown multiple times in recent years. The possibility to use almost all polymeric and polycondensated materials as a matrix is shown. The nanocomposites from various organoclays and polymers have been synthesized. Here is just a small part of the compounds for being the matrix referenced in literature: polyacrylate [83], polyamides [82,84,85], polybenzoxasene [86], polybutyleneterephtalate [11,82,87], polyimides [88], polycarbonate [89], polymethylmetacrylate [90], polypropylene [91,92], polystyrene [90], polysulphones [93], polyurethane [94], polybuthyleneterephtalate and polyethyleneterephtalate [10,65,68,79,95,99-107], polyethylene [96], epoxies [97].

The organomodified montmorillonite is of the special interest because it can be an element of the nanotechnology and it can also be a carrier of the nanostructure and of asymmetry of length and width in layered structures. The organic modification is being usually performed using the ion-inducing surface-active agents. The non-ionic hydrophobisation of the surface of the layered structures have been reported either. The general knowledge about the methods of the study is being formed and the understanding of the structure of the nanocomposite polymeric materials is becoming clear. Also scientists come closer to the realizing of the relations between the deformational and strength properties and the specifics of the nanocomposite structure. The growth of researches and their direction into the nanoarea forecasts the fast broadening of the industrial involvement to the novel and attractive branch of the materials science.

REFERENCES

[1] Romanovsky B.V., Makshina E.V. // *Sorosovskii obrazovatelniu zhurnal,* 8(2) 2004: 50-55 [in Russian]
[2] Golovin Yu.I. // *Priroda,* 1 2004 [in Russian]
[3] Carter L.W., Hendrics J.G., Bolley D.S. United States Patent №2,531,396; 1950.
[4] Blumstain A. *Bull. Chem. Soc.* 1961:899-905.
[5] Fujiwara S., Sakamoto T. Japanese Application № 109,998; 1976.
[6] Usuki A., Kojima Y., Kawasumi M., Okada A., Fukushima Y., Kurauchi T., Kamigatio O. *Journal of Appl. Polym. science.* 1995;55:119.
[7] Usuki A., Koiwai A., Kojima Y., Kawasumi M., Okada A., Kurauchi T., Kamigaito O. *Journal of Appl. Polym. science.* 1995;55:119
[8] Okada A., Usuki A. *Mater Sci. Engng.* 1995;3:109.
[9] Okada A., Fukushima Y., Kawasumi M., Inagaki S., Usuki A., Sugiyama S., Kurauchi T., Kamigaito O. United States Patent №4,739,007; 1988.
[10] Mikitaev M.A., Lednev O.B., Kaladjian A.A., Beshtoev B.Z., Bedanokov A.Yu., Mikitaev A.K. // *Second International Conference* (Nalchik 2005)
[11] J.-H. Chang, Y.U. An, S.J. Kim, S. Im. *Polymer*, 2003;44:5655-5661.
[12] A.K.Mikitaev, A.Y.Bedanokov, O.B.Lednev, M.A.Mikitaev Polymer/silicate nanocomposites based on organomodified clays / Polymers, Polymer Blends, Polymer Composites and Filled Polymers. Synthesis, Properties, Application. Nova Science Publishers. New York 2006.
[13] D.M. Delozier, R.A. Orwoll, J.F. Cahoon, N.J. Johnston, J.G. Smith, J.W. Connell. *Polymer,* 2002;43:813-822.
[14] Kelly P., Akelah A., Moet A. *J. Mater. Sci.* 1994. V.29. P.2274-2280.
[15] J.-H. Chang, Y.U. An, D. Cho, E.P. Giannelis. *Polymer,* **2003**;44:3715-3720.
[16] Yano K, Usuki A, Okada A. J Polym Sci, Part A: *Polym. Chem.* 1997;35:2289.
[17] Garcia-Martinez JM, Laguna O, Areso S, Collar EP. *J. Polym. Sci.* Part B: Polym Phys 2000;38:1564.
[18] Giannelis E. P., Krishnamoorti R., Manias E. Advances in Polymer Science, Vol.138 Springer-Verlag Berlin Heidelberg 1999.
[19] D.M. Delozier, R.A. Orwoll, J.F. Cahoon, J.S. Ladislaw, J.G. Smith, J.W. Connell. *Polymer.* 2003;44:2231-2241.
[20] Y.-H. Shen. *Chemosphere.* 2001;44:989-995.
[21] E.P. Giannelis, *Adv. Mater.* 8 (1996) 29–35.
[22] H.R. Dennis, D.L. Hunter, D. Chang, S. Kim, J.L. White, J.W. Cho, D.R. Paul. *Polymer.* 2001;42:9513-9522.
[23] X. Kornmann, H. Lindberg, L.A. Berglund, *Polymer.* 42 (2001) 1303–1310.

[24] T.D. Fornes; D.R. Paul. Formation and properties of nylon 6 nanocomposites. Polímeros vol.13 no.4 São Carlos Oct./Dec. 2003.
[25] D. Voulgaris, D. Petridis. *Polymer*. 2002;43:2213-2218.
[26] H.-L. Tyan, Y.-C. Liu, K.-H. Wei. *Polymer*. 1999;40:4877-4886.
[27] Davis CH, Mathias LJ, Gilman JW, Schiraldi DA, Shields JR, Trulove P, Sutto TE, Delong HC. J Polym Sci, Part B: *Polym. Phys*. 2002;40:2661.
[28] Morgan AB, Gilman JW. *J. Appl. Polym. Sci.* 2003;87:1329.
[29] John N. Hay and Steve J. Shaw. Organic-inorganic hybridssthe best of both worlds? *Europhysics News.* (2003) Vol.34 No.3.
[30] Chang JH, An YU, Sur GS. *J. Polym. Sci. Part B: Polym. Phys.* 2003;41:94.
[31] Chang JH, Park DK, Ihn KJ. *J. Appl. Polym. Sci.* 2002;84:2294.
[32] Pinnavaia TJ. *Science.* 1983;220:365.
[33] Messersmith PB, Giannelis EP. *Chem. Mater.* 1993;5:1064.
[34] Vaia RA, Ishii H, Giannelis EP. *Adv. Mater.* 1996;8:29.
[35] Gilman JW. *Appl. Clay Sci.* 1999;15:31.
[36] Fukushima Y, Okada A, Kawasumi M, Kurauchi T, Kamigaito O. *Clay Miner.* 1988;23:27.
[37] Akelah A, Moet A. *J. Mater. Sci.* 1996;31:3589.
[38] Vaia RA, Ishii H, Giannelis EP. *Adv. Mater.* 1996;8:29.
[39] Vaia RA, Jandt KD, Kramer EJ, Giannelis EP. *Macromolecules.* 1995;28:8080.
[40] Greenland DG. *J. Colloid. Sci.* 1963;18:647.
[41] Chang JH, Park KM. *Polym. Engng. Sci.* 2001;41:2226.
[42] Greenland DG. *J. Colloid. Sci.* 1963;18:647.
[43] Chang JH, Seo BS, Hwang DH. *Polymer.* 2002;43:2969.
[44] Vaia RA, Jandt KD, Kramer EJ, Giannelis EP. *Macromolecules.* 1995;28:8080.
[45] Fukushima Y, Okada A, Kawasumi M, Kurauchi T, Kamigaito O. *Clay Miner.* 1988;23:27.
[46] Chvalun S.N. // *Priroda.* 2000; 7 [in Russian]
[47] Brinker C.J., Scherer G.W. *Sol-Gel Science.* Boston, 1990.
[48] Mascia L, Tang T. *Polymer.* 1998;39:3045.
[49] Tamaki R, Chujo Y. *Chem. Mater.* 1999;11:1719.
[50] Serge Bourbigot e.a. Investigation of Nanodispersion in Polystyrene–Montmorillonite Nanocomposites by Solid-State NMR. *Journal of Polymer Science: Part B: Polymer Physics.* Vol. 41, 3188–3213 (2003).
[51] O.B. Lednev, A.A. Kaladjian, M.A. Mikitaev, M.A. Tlenkopatchev. New polybutylene terephtalate and organoclay nanocomposite materials. // *Abstracts of the International Conference on Polymer materials.* (México, 2005)
[52] Tretiakov A.O. // Oborudovanie I instrument dlia professionalov №02(37) 2003 [in Russian]
[53] Sergeev G.B. // Ros. Chem. J. (The journal of the D.I.Mendeleev Russian chemical society), 2002, Vol. XLVI, №5 [in Russian]
[54] Lomakin S.M., Zaikov G.E. // *Visokomol. Soed.* B. 2005, 47(1) 104-120 [in Russian]
[55] Xu H, Kuo SW, Lee JS, Chang FC. *Macromolecules*. 2002;35:8788.
[56] Haddad TS, Lichtenhan JD. *Macromolecules.* 1996;29:7302.
[57] Mather PT, Jeon HG, Romo-Uribe A, Haddad TS, Lichtenhan JD. *Macromolecules.* 1996;29:7302.

[58] Hsu SLC, Chang KC. *Polymer.* 2002;43:4097.
[59] Chang JH, Seo BS, Hwang DH. *Polymer.* 2002;43:2969.
[60] Fornes TD, Yoon PJ, Hunter DL, Keskkula H, Paul DR. *Polymer.* 2002;43:5915.
[61] Chang JH, Seo BS, Hwang DH. *Polymer.* 2002;43:2969.
[62] Fornes TD, Yoon PJ, Hunter DL, Keskkula H, Paul DR. *Polymer.* 2002;43:5915.
[63] Wen J, Wikes GL. *Chem. Mater.* 1996;8:1667.
[64] Zhu ZK, Yang Y, Yin J, Wang X, Ke Y, Qi Z. *J. Appl. Polym. Sci.* 1999;3:2063.
[65] Mikitaev M.A., Lednev O.B., BeshtoevB.Z., Bedanokov A.Yu, Mikitaev A.K. // Second International conference "Polymeric composite materials and covers" (Yaroslavl 2005, may) [in Russian]
[66] Fischer HR, Gielgens LH, Koster TPM. *Acta Polym.* 1999;50:122.
[67] Petrovic XS, Javni L, Waddong A, Banhegyi GJ. *J. Appl. Polym. Sci.* 2000;76:133.
[68] Lednev O.B., Beshtoev B.Z., Bedanokov A.Yu, Alarhanova Z.Z., Mikitaev A.K. // *Second International Conference.* (Nalchik 2005) [in Russian]
[69] J.-H. Chang, S.J. Kim, Y.L. Joo, S. *Im. Polymer.* 2004;45:919-926.
[70] Lan T, Pinnavaia TJ. *Chem. Mater.* 1994;6:2216.
[71] Masenelli-Varlot K, Reynaud E, Vigier G, Varlet J. *J. Polym. Sci. Part B: Polym. Phys.* 2002;40:272.
[72] Yano K, Usuki A, Okada A. *J. Polym. Sci. Part A: Polym. Chem.* 1997;35:2289.
[73] Shia D, Hui Y, Burnside SD, Giannelis EP. *Polym. Engng. Sci.* 1987;27:887.
[74] Curtin WA. *J. Am. Ceram. Soc.* 1991;74:2837.
[75] Chawla KK. Composite materials science and engineering. New York: Springer; 1987.
[76] Burnside, S., D. and Giannelis, E. P., *Chem. Mater.* 7, 1995, 4597.
[77] Babrauskas, V., Peacock, R. D., *Fire Safety Journal,* 18, 1992, 225.
[78] Gilman, J., Kashiwagi, T., Lomakin, S., Giannelis, E., Manias, E., Lichtenhan, J., and Jones, P., in. Fire Retardancy of Polymers: the Use of Intumescence. *The Royal Society of Chemistry,* Cambridge, 1998, 203-221.
[79] Mikitaev A.K., Kaladjian A.A., Lednev O.B., Mikitaev M.A. // *Plastic masses.* 2004; 12:45-50 [in Russian]
[80] Gilman, J., Morgan A. 10th Annual BCC Conference, May 24-26, 1999.
[81] John N. Hay and Steve J. Shaw. Organic-inorganic hybrids: the best of both worlds? *Europhysics News.* (2003) Vol. 34 No. 3.
[82] D.M. Delozier, R.A. Orwoll, J.F. Cahoon, N.J. Johnston, J.G. Smith, J.W. Connell. *Polymer.* 2002;43:813-822.
[83] Chen Z, Huang C, Liu S, Zhang Y, Gong K. *J. Apply Polym. Sci.* 2000;75:796-801.
[84] Okado A, Kawasumi M, Kojima Y, Kurauchi T, Kamigato O. *Mater Res. Soc. Symp. Proc.* 1990;171:45.
[85] Leszek A. Utracki, Jorgen Lyngaae-Jorgensen. *Rheologica Acta.* 2002;41: 394-407.
[86] R. Wagener, T.J.G. Reisinger. *Polymer.* 2003;44:7513-7518.
[87] Li X, Kang T, Cho WJ, Lee JK, Ha CS. *Macromol Rapid Commun.*
[88] H.-L. Tyan, Y.-C. Liu, K.-H. Wei. *Polymer.* 1999;40:4877-4886.
[89] Vaia R, Huang X, Lewis S, Brittain W. *Macromolecules.* 2000;33:2000-4.
[90] Okamoto M, Morita S, Taguchi H, Kim Y, Kotaka T, Tateyama H. *Polymer.* 2000;41:3887-90.
[91] W.S. Chow, Z.A. Mohd Ishak, J. Karger-Kocsis, A.A. Apostolov, U.S. Ishiaku. *Polymer.* 2003;44:7427-7440.

[92] Antipov E.M., Guseva M.A., Gerasin V.A., Korolev Yu.M., Rebrov A.V., Fisher H.R., Razumovskaya I.V. // *Visokomol. soed. A.* 2003; 45(11): 1885-1899 [in Russian].
[93] Sur G, Sun H, Lyu S, Mark J. *Polymer.* 2001;42:9783-9.
[94] Wang Z, Pinnavaia T. *Chem. Mater.* 1998;10:3769-71.
[95] Bedanokov A.Yu., Beshtoev B.Z. / Malij polimernij congress (Moscow 2005) [in Russian]
[96] Antipov E.M., Guseva M.A., Gerasin V.A., Korolev Yu.M., Rebrov A.V., Fisher H.R., Razumovskaya I.V. // *Visokomol. Soed. A.* 2003; 45(11): 1874-1884 [in Russian].
[97] Lan T., Kaviartna P., Pinnavaia T. Proceedings of the ACS PMSE 1994;71:527-8.
[98] Kawasumi et al. Nematic liquid crystal/clay mineral composites. *Science & Engineering C,* Vol 6, pp 135-143, 1998.
[99] Lednev O.B., Kaladjian A.A., Mikitaev M.A. // *Second International Conference* (Nalchik 2005) [in Russian]
[100] Mikitaev A.K., Kaladjian A.A., Lednev O.B., Mikitaev M.A., Davidov E.M. // *Plastic masses.* 2005;4:26-31[in Russian]
[101] Eid A., Mikitaev M.A., A.Y.Bedanokov, Mikitaev A.K. Recycled Polyethylene Terephthalate/Organo-Montmorillanite Nanocomposites, Formation And Properties // *The first Afro-Asian Conference on Advanced Materials Science and Technology.* (AMSAT 06), Egypt, 2006.
[102] A.K. Mikitaev, A.Y.Bedanokov, O.B.Lednev, M.A.Mikitaev Polymer/silicate nanocomposites based on organomodified clays/ Polymers, Polymer Blends, Polymer Composites and Filled Polymers. Synthesis, Properties, Application. Nova Science Publishers. New York 2006.
[103] Malamatov A.H., Kozlov G.V., Mikitaev M.A. Mechanismi uprochnenenia polimernih nanokompositov, (Moscow, RUChT 2006) 240 p. [in Russian]
[104] Eid A. Doctor Thesis, (Moscow, RUChT 2006) 121 p. [in Russian]
[105] Lednev O.B. Doctor Thesis (Moscow, RUChT 2006) 128 p. [in Russian]
[106] Malamatov A.H. Professor Thesis (Nalchik, KBSU 2006) 296 p. [in Russian]
[107] Borisov V.A., Bedanokov A.Yu., Karmokov A.M., Mikitaev A.K., Mikitaev M.A., Turaev E.R. // *Plastic masses.* 2007; 5 [in Russian]

In: Progress in Chemistry and Biochemistry
Editors: E. M. Pearce and G. E. Zaikov

ISBN: 978-1-60692-344-3

Chapter 4

KINETIC MODEL OF COPOLYMERIZATION AT LINEAR CHAIN TERMINATION

Yu. G. Medvedevskikh, G. I. Khovanets', I. Yu. Yevchuk and G. E. Zaikov*

Department of Physical Chemistry of Combustible Minerals
L. M. Lytvynenko Institute of Physico-organic Chemistry and
Coal Chemistry NAS of Ukraine 79060 Lviv Naukova St., 3a,
e-mail: vfh@org.lviv.net
*N. M. Emanuel Institute of Biochemical Physics RAS
119334 Moscow, Kosygin St. 4, e-mail: chembio@sky.chph.ras.ru

ABSTRACT

Numerous experimental data on kinetics of photoinitiated stationary and nonstationary polymerization of mono- and di(meth)acrylates to high conversions show the possibility of the change of chain termination mechanism in reaction zones, where translation mobility of radicals (and, respectively, the rate of bimolecular chain termination) abruptly decreases. In this case linear from the formal stoichiometric point of view chain termination mechanism, which appears as an act of self-burial of active radicals at their propagation in polymer matrix, becomes determinative. Suggested earlier kinetic models with taking into consideration possible role of linear chain termination are in satisfactory agreement with all experimental data on homopolymerization, but need to be extended on copolymerization processes. In present article we propose kinetic model of copolymerization at linear chain termination, which appears as an act of self-burial of active radical in reaction of its propagation.

Keywords: kinetic model, copolymerization, linear chain termination.

INTRODUCTION

Process of polymerization to high conversions is accompanied by the change of phase state of polymerizing system. It affects all factors, which determine kinetics of polymerization. But initiation of reaction and chain propagation, changing kinetic parameters, retain the essence of their elementary acts. Conversely, chain termination can change its mechanism. It is related to the fact that at abrupt increasing of diffusion braking of bimolecular chain termination, linear (i. e. monomolecular on radicals) mechanism of chain termination goes into effect, which is the act of chain propagation leading to the capture of macroradical by polymer matrix [1] or to its freezing [2, 3].

Therefore, if the rate of bimolecular chain termination, which is accompanied by active radicals decay, becomes smaller than the rate of "self-burial" of active radicals at their propagation in polymeric matrix, what is possible at abrupt decreasing of macroradicals translation mobility, mechanism of chain termination from formal stoichiometric point of view becomes linear. It affects increasing of the order of reaction on initiator at high degrees of polymerization and achieving of high concentration of radicals till the end of the process [4, 5].

According to the concept of different reaction zones [6, 7], which is the development of the concept of microheterogeneity of polymerizing system [8], process of polymerization of bifunctional monomers is concentrated in two reaction zones already at early stages: in monomer phase and at interface liquid monomer – micrograins of polymer matrix at practically limited conversion on monomer. Kinetic model of photoinitiated polymerization of di(meth)acrylates is in good agreement with all experimental data at assumption, that at the stage of autoacceleration the main contribution into the summary kinetics of polymerization is made by the process at the interface layer, where because of the abrupt decreasing of segmental and, respectively, translation macroradicals mobility the chain termination is linear [6, 7, 9].

Investigations of kinetics of nonstationary process of postpolymerization of di(meth)acrylates, i. e. dark polymerization after *UV* irradiation switching off, confirmed this assumption and allowed to ascertain, that linear chain termination obeys the Kolraush stretched exponent law, and to obtain the numerical estimations of the chain termination rate constant k_t over the range $(1\div5)\cdot10^{-5}$ m^3/mol·s. Theoretical derivation [10] of the stretched exponent law was suggested on the base on the supposition that linear chain termination is the act of chain propagation leading to the trap.

At photoinitiated polymerization of monofunctional (meth)acrylates in bulk theoretical describing of the process to high conversions [9] satisfactorily agrees with experimental data at assumption that in monomer-polymer phase and at interface between it and polymer-monomer matrix the chain termination is bimolecular, but in polymer-monomer matrix it is linear [9].

Studing of dark postpolymerization kinetics of (meth)acrylates at conversions ≥ 0.7, which correspond to maximum of polymerization rate at the stage of autoacceleration, confirms this assumption [9, 11]. At that, similar to di(meth)acrylates, the linear chain termination obeys the stretched exponent law, and the rate constant is of the same order of 10^5 m^3/mol·s [9, 11].

Hence, kinetic models of homopolymerization to high conversions both bi- and mono(meth)acrylates are in satisfactory agreement with experimental data at assumption that at least at the stage of autoacceleration of these processes the linear chain termination as an act of self-burial of active radicals in reactions of chain propagation is of a great significance. Since considerable part of technologically valuable materials are copolymers, properties of which depend on their syntheses conditions in many respects, it was necessary to derive the kinetic model of copolymerization at linear chain termination.

KINETIC MODEL DERIVATION

Let us use the method of routes, developed both for catalytic and radical-chain processes [11, 12, 13]. According to it, the rate *r(v, w, u)* on *i*-route in the basis *w* of the group *w* of the graph $G = \{R, S\}$, what represents topology of kinetic scheme on multiplicity of radicals $R = \{R_i\}$ and elementary reactions $S = \{S_i\}$, which join radicals, is described by expression

$$r(v,w,u) = L(\sum_i M_i)^{-1} M(v)\Delta(v,w,u)^{\cdot} \qquad (1)$$

Here: L is total radical concentration; M_i is additional minor of a single group $v = \{R_i\}$: $M_i = (-1)^{n-1} \left| B_i \right|$, where $\left| B_i \right|$ is matrix determinant, which is obtained from initial matrix B of the n order by deleting of *i*-line and *i*-column; $M(v)$ is additional minor of the group v of power l, i. e. which consists of l tops of the graph $G = \{R, S\}$: $M(v) = (-1)^{n-l} \left| B(v) \right|$, where $\left| B(v) \right|$ is determinant of matrix $B(v)$, which is obtained from initial matrix B by deleting of those lines and columns, determinants of which coincide with determinants of the tops of the graph $G = \{R, S\}$, represented in given group; $\Delta(v, w, u)$ is Δ-factor of the route, which is determined by product of frequencies of elementary reactions, which form cyclic sequence of the route in given basis w of the group v.

According to the assumption, that linear chain termination is the act of self-burial of active radical R_i in reaction of its propagation, as a result it transforms into inactive radical R_z, kinetic scheme of propagation and linear chain termination at copolymerization of monomers A_1 and A_2 becomes as follows

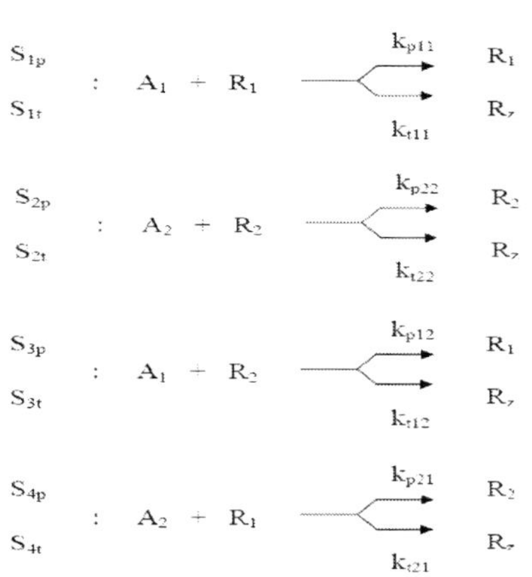

(2)

The graph $G = \{R, S\}$, which presents topology of kinetic scheme (2) with taking into account only elementary reactions of chain propagation, is shown on figure 1.

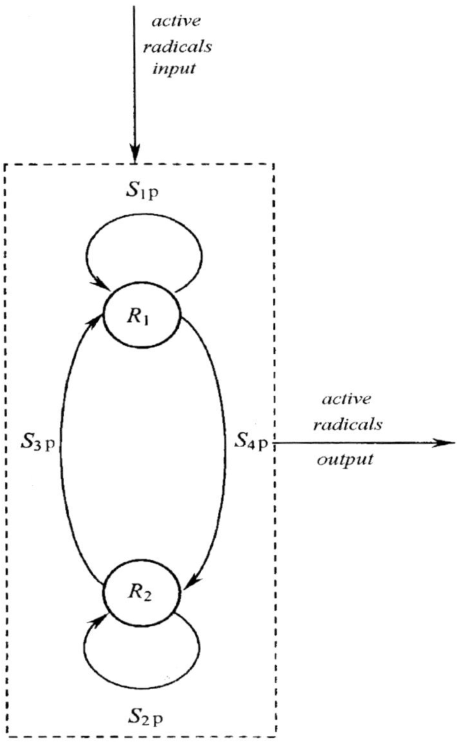

Figure 1. The graph $G = \{R, S\}$, which presents topology of kinetic scheme of chain propagation at copolymerization.

On kinetic matrix determination $B = (b_{ij})$ its diagonal elements obey the relation $b_{ij} = -\sum_{i \neq j} b_{ij}$, and nondiagonal ones b_{ij} are the intensities of transformation from the state R_j into the state R_i. At present case the matrix order is equal $n = 2$, and its elements are determined only by the intensities of elementary reactions S_{3p} and S_{4p}, which affect relative concentrations of radicals R_1 and R_2:

$$B = \begin{pmatrix} -k_{p12}[A_2] & k_{p12}[A_1] \\ k_{p21}[A_2] & -k_{p12}[A_1] \end{pmatrix} \quad (3)$$

At long chains approximation the condition of stationarity on active radicals assumes that their input v_{in} and output v_t rates are equal: $v_{in} = v_t$. This condition according to (2) is defined by expression

$$v_t = \sum_{i,j} k_{tij}[A_i][R_j] = L\sum_{i,j} k_{tij}[A_i]x_j, \tag{4}$$

in which $x_i = [R_i]/\sum_i [R_i] = L_i / L$ is relative radicals concentration of radicals of given type. It is determined from the condition $Bx = 0$, the solution of which is the expression

$$x_i = M_i / \sum_i M_i \tag{5}$$

Using (5) in the condition of stationarity $v_{in} = v_t$. and taking into account (4), we obtain

$$L = v_{in}\sum_i M_i / \sum_{i,j} k_{tij}[A_i]M_j. \tag{6}$$

Thus, kinetic model of the route (1) becomes as follows

$$r(v,w,u) = v_{in}M(v)\Delta(v,w,u) / \sum_{i,j} k_{tij}[A_i]M_j. \tag{7}$$

According to kinetic matrix B on (3), additional single minors M_i are determined by expressions:

$$M_1 = k_{p12}[A_1], \qquad M_2 = k_{p21}[A_2]. \tag{8}$$

In the graph $G = \{R, S\}$ we have 3 routes, two of them are represented on one elementary reaction and form cycles at the tops R_1 and R_2, the third route is represented by the sequence of elementary reactions S_{3p} and S_{4p}. Thus, according to abovementioned algorithm, we have

$$\begin{aligned} &u_1 = S_1,\ \Delta_1 = k_{p11}[A_1],\ M(1) = M_1 = k_{p12}[A_1],\\ &u_2 = S_2,\ \Delta_2 = k_{p22}[A_2],\ M(2) = M_2 = k_{p21}[A_2],\\ &u_3 = S_3 + S_4,\ \Delta_3 = k_{p12}k_{p21}[A_1][A_2],\ M(3) = M(1,2) = 1. \end{aligned} \tag{9}$$

From here the expression for the rates on the routes follows:

$$\begin{aligned} r_1 &= k_{p11}k_{p12}[A_1]^2 v_{in} / \sum_{i,j} k_{tij}[A_i]M_j,\\ r_2 &= k_{p22}k_{p21}[A_2]^2 v_{in} / \sum_{i,j} k_{tij}[A_i]M_j,\\ r_3 &= k_{p12}k_{p21}[A_1][A_2]v_{in} / \sum_{i,j} k_{tij}[A_i]M_j, \end{aligned} \tag{10}$$

where

$$\sum_{i,j} k_{tij}[A_i]M_j = k_{t11}k_{p12}[A_1]^2 + (k_{t12}k_{p21} + k_{t21}k_{p12})[A_1][A_2] + k_{t22}k_{p21}[A_2]^2 \quad (11)$$

Taking into account the stoichiometry of the routes, the rates of monomers A_1 and A_2 consumption are equal

$$- d[A_1] / dt = r_1 + r_3,$$
$$- d[A_2] / dt = r_2 + r_3. \quad (12)$$

Respectively, the total rate of copolymerization is equal

$$- d([A_1] + [A_2]) / dt = r_1 + r_2 + 2r_3 \quad (13)$$

Instantaneous composition of copolymer is defined by expression

$$d[A_1] / d[A_2] = (r_1 + r_3) / (r_2 + r_3) . \quad (14)$$

CONCLUSION

Derivation of kinetic model of copolymerization at linear chain termination, which is considered as the act of self-burial of active radical in the reaction of chain propagation, has been suggested. The obtained equations will be used at the analysis of kinetics of photoinitiated copolymerization of mono- and di(meth)acrylates to high conversions.

REFERENCES

[1] Dyachkov A. I., Efimov A. L., Efimov L. I. et al. About chain termination mechanism at radical polymerization of methymethacrylate to high conversions. *Vysokomol. soyed.* V. 25(A), 10, 1983, P. 2176-2181.

[2] Zhu S., Tian Y., Hamieles A. E. Radical trapping and termination in free-radical polymerization of methylmethacrylate. *Macromolecules,* 10, 1990, P. 1144-1150.

[3] Kurdikar D. L., Peppas N. A. A kinetic study of diacrylate photopolymerizations. *Polymer.* V. 35, 5, 1994. P. 1004-1011.

[4] Guzeeva E. V., Efimov A. L., Dyachkov A. I. et al. Chain propagation limitation mechanism in radical polymerization of styrene to high degrees of conversion. *Vysokomol. soyed.* V. 28(A), 8, 1986, P. 587-588.

[5] Brun E. B., Ivanov V. A., Kaminsky V. A. Kinetics of postpolymerization at high degrees of conversion. Proceedings of AS of USSR. V. 291, 3, 1986, P. 618-622.

[6] Medvedevskikh Yu. G., Zagladko E. A., Turovsky A. A., Zaikov G. E. Kinetics of the three-dimentional thermopolymerization. *Intern. J. Polym. Mater.* 43, 1999. P. 157-168.

[7] Medvedevskikh Yu., Zahladko E., Turovsky A., Zaikov G. Kinetic model of stationary and non–stationary process of three–dimensional block polymerization. *Russian Polymer News.* V. 4, 3, 1999. P. 33–39.
[8] Berlin A. A., Korolyov G. V., Kefeli T. Ya. et al. Acrylic oligomers and materials on their base. M.: *Khimiya,* 1983. 232P.
[9] Medvedevskikh Yu., Bratus A., Hafijchuk G, Zaichenko A., Kytsya A., Turovsky A., Zaikov G. Kinetic model of block photopolymerization of glycidyl methacrylate to high conversion. *Journal of Applied Polymer Science.* V. 86, 2002. P. 3556–3569.
[10] Medvedevskikh Yu. G., Kytsya A. R., Bazylyak L. I., Turovsky A. A., Zaikov G. E. Kinetic models of stationary and non–stationary photoinitiated polymerization of dimethacrylates. *Oxidation Communications.* V. 28, 1, 2005. P. 173–207.
[11] Medvedevskikh Yu. G., Kytsya A. R., Bazylyak L. I., Turovsky A. A. Zaikov G. E. Stationary and nonstationary kinetics of the photoinitiated polymerization. 2004. VSP. Utrecht. Boston. P. 313.
[12] Medvedevskik Yu. G., Kucher R. V. Routes and kinetic models of linear catalytic reactions i complicated systems. Proceedings of AS of Ukrainian SSR (B), 1, 1984. P. 42-46.
[13] Berlin A. A., Medvedevskikh Yu. G., Kucher R. V. Kinetic model of the routes of unbranched radical-chain processes. Proceedings of AS of Ukrainian SSR (B), 6, 1986. P. 32-36.

In: Progress in Chemistry and Biochemistry
Editors: E. M. Pearce and G. E. Zaikov

ISBN: 978-1-60692-344-3

Chapter 5

THE REACTION CESSATION IN POLYCONDENSATION PROCESS: FRACTAL ANALYSIS

G.V. Kozlov[1], G.B. Shustov[1] and G.E.Zaikov[2]
[1] Kabardino-Balkarian State University,
360004, Russia, KBR, Nal'chik, Chernyshevsky st., 173
[2] Institute of Biochemical Physics, Russian Academy of Sciences,
119991, Russia, Moscow, Kosygin st., 4

ABSTRACT

The main conclusion of the present paper is that, the reaction cessation in a polycondensation process is limited by a purely physical factor, namely, the achievement of macromolecular coil density of solution density in which the reaction proceeds. The possibility of such treatment is due fully to macromolecular coil fractality. The indicated processes are simulated within the framework of irreversible aggregation models over mechanism cluster-cluster. The critical values of molecular weight and reaction duration, above which a synthesis reaction ceases, exists.

Keywords: Reaction cessation, polycondensation, fractal dimension, structure, irreversible aggregation, mechanism cluster-cluster.

INTRODUCTION

It is well known [1, 2] that, depending on the reactionary medium and synthesis carrying out conditions, polycondensation processes can be ceased at different temporal scales of the experiment that in the long run results in a considerable variation of the main parameters characterizing this process: conversion degree Q and molecular weight MW. Let's note, that the indicated reaction cessation is not defined by the exhaustion of the available reactive groups. At present there are quite enough partial explanations of this effect, applicable to individual combinations of reagents and solvents [1, 2], that does not allow to develop a general conception of this important effect both theoretically and practically. Therefore, for the description of polycondensation process cessation, we used much more general notions, namely, irreversible aggregation models. The application of the mentioned models is due to the fact that any polymerization process in the most general treatment presents unification of small molecules (particles) in a much larger macromolecule (cluster).

Irreversible aggregation models of different universality classes (particle-cluster, cluster-cluster) were developed for the description frequently occurring in nature aggregation processes like flocculation, coagulation, polymerization and so on [3, 4]. In the last years one could see the applicability of these models for the description of all kinds of polymerization processes within the framework of the very general physical notations [5-7]. The main feature of the indicated models is, that processes of aggregation described by them generate fractal objects as final aggregates [3, 4]. As it is known [8], macromolecular coils in solution are fractals, which fractal dimension D varies within the range ~ 1.5-2.2. This circumstance allows one to use for the description of polymerization processes in solutions irreversible diffusion-limited aggregation models cluster-cluster [3]. At present several types of such models were developed, allowing one to take into consideration a number of specific features of polymerization processes, namely, simultaneous occurrence of aggregation and disaggregation [9], different probability of cluster joining [10] and so on. The purpose of the present paper is irreversible aggregation model used for the description of some features of different polymers polycondensation: aromatic copolyethersulfoneformals (APESF), diblockcopolyether of oligoformal 2,2-di-(4-oxiphenyl)-propane and oligosulfone of phenolphthalein (CP-OF-10/OS-10) and also polyarylate on the base of dichloroanhydride of terephthalic acid and phenolphthalein (F-2).

EXPERIMENTAL

APESF have the following chemical structure:

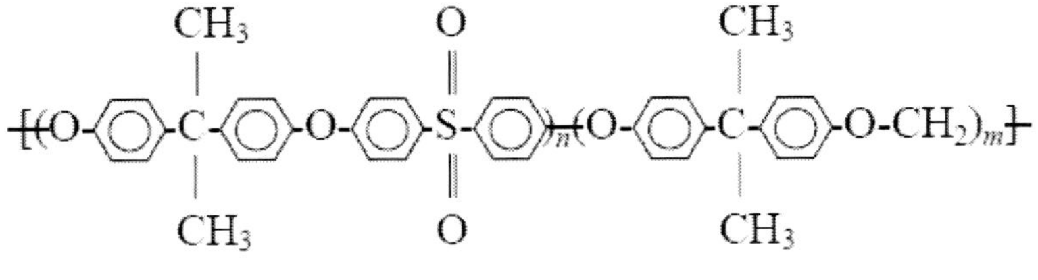

The APESF synthesis was carried out according to the following technique. In a three-neck retort of 750 ml capacity supplied with a stirrer, Din-Stark trap and cleaned by nitrogen, 40 ml of dimethyl sulfoxide and 7.8953 g (0.09285 mole) of sodium hydroxide in a state of concentrated aqueous solution, 10.5984 g (0.04642 mole) of 2,2-di-(4-oxiphenyl)-propane and 40 ml of toluene are placed. At stirring temperature was gradually raised and azeotropic mixture of water and toluene were distilled off. After cessation of water separation temperature was raised to 443 K and toluene was distilled off. To the solution of bis-phenolate cooled to 413 K 9.3322 g (0.032497 mole) of 4,4'-dichlorodiphenysulfone were added. The temperature was raised to 433 K and at this temperature the mixture was sustained for an hour. Then the mixture was cooled up to 303 K, Din-Stark trap was replaced by the opposite refrigerator and to 0.9 ml (0.01393 mole) of methylene chloride were added to the reactionary mixture. The temperature was gradually being raised to 423 K and the mixture was being sustained at this temperature for 2 hours. The obtained viscous solution was cooled off, diluted with dimethylsulfonoxide and poured into distilled water stirred intensively and oxidized by oxalic acid. After separation, purification and drying 16.3 g (92 %) copoether, contaiting 30 mol. % of methylene chloride residuals, were received. The reduced viscosity was equal to 0.78 dl/g (0.5 %-th solution in chloroform at 298 K). The other APESF were synthesized by the similar way.

Diblockcopolymers CP-OF-10/OS-10 have the following chemical structure:

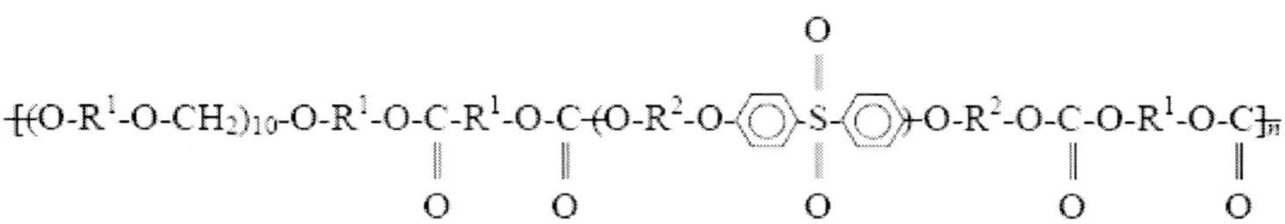

where 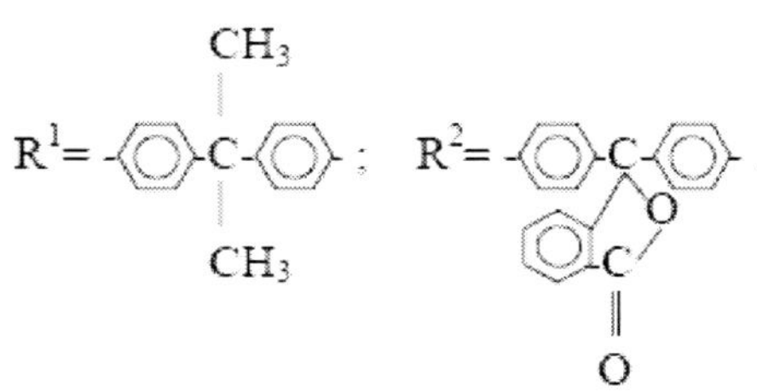

The synthesis of diblockcopolymers was made according to the following technique. In a three-neck retort of 250 ml capacity supplied with mechanical stirrer 1.365 g (0.000525 mole) of OF-10, 2.94 g (0.000525 mole) of OS-10 and 30 ml of methylene chloride are placed. After solution of the oligomers 0.3 ml (0.0021 mole) of triethylamine, then 0.3709 g (0.00105 mole) of bischloroformiate 2,2-di-(4 oxyphenyl)-propane were added. The reactionary mixture was being stirred during 2 hours at room temperature, after that the viscous solution was diluted by methylene chloride and it was poured into intensively stirred iso-propanol spirit. After separation, purification and drying 4.215 g (92 %) of diblockcopolymer containing 50 mol. % of OF-10 with reduced viscosity 0.52 dl/g (0.5 %-th solution in chloroform at 298 K) were received. The other diblockcopolymers were synthesized by the similar way.

The glass transition temperatures T_g of APESF and CP-OF-10/OS-10 were determined by thermomechanical technique. They are within the range 353-493 K. The T_g value for F-2 was accepted equal to 503 K [11]. The reduced viscosity of copolymers in simm-tetrachloroethane (good solvent) and 1,4-dioxane (θ-solvent) was measured with the aid of Ubbelode viscosimeter at temperature 298 K. For the intrinsic viscosity [η] calculation the following relationship was used [12]:

$$[\eta]=\frac{\eta_{red}/c_p}{1+K_\eta \eta_{red}}, \quad (1)$$

where η_{red} is reduced viscosity, c_p is polymer concentration in solution, K_η is coefficient, equal to 0.28.

The values of reduced viscosity η_{red} in various solvents for F-2 were accepted according to the data of paper [1]. The initial concentration c_0 is determined as general mass of loading reagents, reduced to solvent volume of 30 ml.

RESULTS AND DISCUSSION

Methodics of the Macromolecular Coil Fractal Dimension Determination

As it is known [13], linear polymeric macromolecules can be found in solution depending on *MM*, solvent quality, temperature, concentration and other factors in various conformational and/or phase states. The most trivial of them are statistically twisted coil in ideal (θ) solvent, nontransparent coil in good solvent and transparent coil. In all listed states the macromolecular coil in solution is fractal, i.e. self-similar object with dimension *D*, which does not coincide with its topological dimension d_t. The dimension *D* is named the fractal (Hausdorf) one and it characterizes the volume distribution of coil elements in space [3]. Value *D* determination is the first stage of macromolecular coils study within the framework of fractal analysis and similar estimations are usually made by the exponents measurement in the equations type of Mark-Kuhn-Houwink, connecting [η], translational diffusion coefficient D_0 or high-speed sedimentation coefficient S_0 with *MW* of polymers [14]:

$$[\eta]\sim MW^{a_\eta}, \quad (2)$$

$$D_0 \sim MW^{-a_D}, \quad (3)$$

$$S_0 \sim MW^{a_S}. \quad (4)$$

Then *D* value can be calculated according to the relationship [14]:

$$D=\frac{3}{a_\eta+1}, \quad (5)$$

$$D=\frac{1}{a_D}, \quad (6)$$

$$D=\frac{1}{1-a_S}. \quad (7)$$

All considered methods require complicated enough and laborious measurements [14-16]. The simplest and not requiring complicated devices from the listed methods is [η] measurement, which can be practically fulfilled in any laboratory. Therefore we will consider here a simple method of the value *D* estimation based on the same principles that were used at the equations (5)-(7) derivation.

As it is known [17], the coefficient of swelling of a macromolecular coil α is determined as follows:

$$\alpha = \left(\frac{\left\langle \overline{h}^2 \right\rangle}{\left\langle \overline{h}_\theta^2 \right\rangle} \right)^{1/2}, \qquad (8)$$

where $\left\langle \overline{h}^2 \right\rangle$ and $\left\langle \overline{h}_\theta^2 \right\rangle$ are the root-mean-square distances between the ends of the macromolecule in an arbitrary and ideal (θ) solvents, respectively.

In its turn, the value α is related with intrinsic viscosities [η] and $[\eta]_\theta$ of the polymer in arbitrary and ideal solvents, respectively, according to the equation [17]:

$$\alpha^3 = \frac{[\eta]}{[\eta]_\theta}. \qquad (9)$$

The parameter ε of the bulk interactions (which cause a deviation of the coil shape from the ideal Gaussian one) is determined according to the relationship [17]:

$$\varepsilon = \frac{d \ln \alpha^2}{d \ln MM} = \frac{\alpha^2 - 1}{5\alpha^2 - 3}. \qquad (10)$$

In its turn, both ε, and coil fractal dimension *D* depend on the exponent a_η value (further simply *a*) in Mark-Kuhn-Houwink equation (2) [16]:

$$\varepsilon = \frac{2a - 1}{3}, \qquad (11)$$

and the dependence *D* on *a* is given by the equation (5).

The combination of the equations (5) and (8)-(11) allows to obtain the following relationship for *D* determination only from intrinsic viscosities values [η] and $[\eta]_\theta$:

$$D = \frac{5\left([\eta]/[\eta]_\theta\right)^{2/3} - 3}{3\left([\eta]/[\eta]_\theta\right)^{2/3} - 2}. \qquad (12)$$

The value $[\eta]_\theta$ can be estimated either directly from the experiment, as it is made in the present paper for APESF and CP-OF-10/OS-10, or from the relationship (2) under the condition a=0.5, which is true for the θ-point, if the constant in this equation is known, as it is made for F-2.

Molecular Weight Estimation

MW estimation for all studied polymers was carried out with the aid of Mark-Kuhn-Houwink equation. For F-2 solution in simm-tetrachloroethane this equation has the form [18]:

$$[\eta] = 0.266 \times 10^{-4} MW^{0.938}. \tag{13}$$

For APESF and CP-OF-10/OS-10 the coefficient K in Mark-Kuhn-Houwink equation can be calculated as follows [18]:

$$K = \frac{21}{m_e}\left(\frac{1}{2500 m_e}\right)^a, \tag{14}$$

where m_e is an average weight of polymer elementary link (without side substituents).

The value a was determined according to the equation (5). The values m_e for APESF homopolymers were calculated from their molecular weight MW_e with the aid of the relationship [19]:

$$m_e = \frac{MW_e}{N_A}, \tag{15}$$

where N_A is Avogadro number.

For CP-OF-10/OS-10 in a homopolymers case molecular weight of one of 10 fragments of oligomer in brackets was calculated and the tenth of molecular weights sum of oligomer link and chain lengthener remaing part were added to it. From total MW_e obtained in this way the value m_e was calculated according to the equation (15). The values m_e for copolymers were calculated in supposition of their additive change with composition.

The combination of the equations (2), (5) and (14) allows to obtain Mark-Kuhn-Houwink equation fractal variant:

$$[\eta] = \frac{21}{m_e}\left[MW\left(\frac{1}{2500 m_e}\right)\right]^{(3-D)/D}. \tag{16}$$

Theoretical Principles of Irreversible Aggregation Models Application

Let's consider irreversible aggregation models application for polycondensation process cessation criterion obtaining. The dependence of aggregation processes on aggregating particles concentration c (with initial concentration c_0) in some closed space has the direct relation to the considered problem. If during the growth of cluster (macromolecular coil) its density is reduced up to the density of the environment (solution with concentration c at the given moment), then the transition between universality classes diffusion-limited aggregation (DLA) a particle-cluster and Eden happens and further macromolecular coil ceases its growth, achieving a critical gyration radius R_c [3]. Let's note, that earlier one more transition occurs within the framework of DLA between universality classes cluster-cluster and particle-cluster. This strongly decreases the reaction rate, but, unlike the transition indicated earlier, does not cease it [20]. The scale of transition from a particle-cluster aggregation to Eden model depends on the initial concentration of particles [3]:

$$R_c \sim c_0^{-\frac{1}{d^D}}, \tag{17}$$

where d is dimension of Euclidean space, in which fractal object was considered (obviously, in our case d=3).

It is obvious also, that the value c_0 can be expressed through weights of monomers loading during the reaction m_m and the particle (molecule of mono- or oligomer calculated according to the equation (15)) weight m_0:

$$c_0 = \frac{m_m}{m_0}. \tag{18}$$

In its turn, value R_c is related to the largest possible degree of polymerization N_c as follows [3]:

$$R_c \sim N_c^{1/D}. \tag{19}$$

The combination of the equations (17)-(19) allows to obtain the interrelation between MW and molecular weight of the aggregating particles MW_0 (accounting equality MW=$N_c MW_0$) in the following form:

$$MW \sim \frac{MW_0^{d/(d-D)}}{m_m^{D/(d-D)}}. \tag{20}$$

Comparison of Theoretical and Experimental Results for Copolymers

In figure 1 the comparison of the calculated according to Mark-Kuhn-Houwink equation and obtained within the framework of irreversible aggregation model values *MW* (the equation (20), the last ones were fitted up to the best correspondence with the first one) for copolymers APESF and CP-OF-10/OS-10 is shown. Let's note, that for APESF theoretical data are arranged systematically lower than experimental dependence of *MW* on formal blocks contents C_{form} and for diblockcopolymers CP-OF-10/OS-10 the good correspondence of theory and experiment up to $C_{form}\approx 30$ mol. % is observed and then (at $C_{form}\approx 50$ and 70 mol. %) the calculated data are essentially lower than the experimental ones. To explain and remove this discrepancy let's consider the dependences $T_g(C_{form})$ for the indicated copolymers, shown in figure 2. From the adduced plots nonadditive dependence $T_g(C_{form})$ follows for both copolymers: for APESF T_g exceeds additive glass transition temperature T_g^{ad} at all compositions and for CP-OF-10/OS-10 for small C_{form} the value $T_g<T_g^{ad}$, at $C_{form}\approx 30$ mol. % $T_g\approx T_g^{ad}$ and at $C_{form}>30$ mol. % T_g of copolymers exceeds T_g^{ad}. This means, that for APESF at any composition the same type of copolymer is formed, whereas at CP-OF-10/OS-10 synthesis depending on the composition different copolymer types are formed. As it was known [21], at the analysis of copolymers composition influence on their T_g Gordon-Tailor-Wood equation is often used:

$$T_g = K_m\left[\left(T_{g_2} - T_g\right)\left(\frac{W}{1-W}\right)\right] + T_{g_1}, \tag{21}$$

where T_{g_1} and T_{g_2} are the glass transition temperatures of homopolymers, K_m is copolymer constant, which serves as quantitative characteristic of distribution sequence of links in copolymer chain [22], *W* is comonomer molar fraction.

The value K_m, which can be calculated according to the equation (21), characterizes copolymer type [22]. For purely statistical copolymers K_m=1, for copolymers, forming long sequences of the same links, K_m=0.5 and for strictly alternating blocks K_m=2.0. This classification assumes that in the APESF case copolymers with more or less regular alternation of links are synthesized irrespective of the composition and for CP-OF-10/OS-10 at small C_{form} copolymers with long sequences of the same blocks are synthesized, at $C_{form}\approx 30$ mol. % the statistical copolymer was obtained and at $C_{form}>30$ mol. % copolymers with more or less regular blocks alternation is formed.

We assume that in the last case, i.e. at $K_m>1$, the intermediate aggregation process occurs in a polycondensation process the formation of regular sequences from the even number of oligomer fragments or from the even number of monomers, which being in such aspect are inserted into the growing macromolecule. For copolymer with strictly alternating blocks K_m=2.0 [22] that means, that all 100 % of monomer (oligomer) molecules intermediate aggregates. We obtain the values K_m=1.3 and 1.8 for CP-OF-10/OS-10 with C_{form}=50 and 70 mol. %, respectively, and this assumes, that in them about 65 and 90 % oligomer molecules respectively were formed by intermediate aggregates. The similar estimations were made for APESF. It is obvious, that in this treatment at invariable initial concentration of aggregating particles c_0 the value MW_0 increases proportionally to the intermediate aggregates fraction and that's why the effective value MW_0^{ef} can be expressed as follows:

$$MW_0^{ef} = K_m MW_0 . \tag{22}$$

It is easy to see, that in the relationship (22) only values $K_m>1$ can be used, as $K_m=1$ dois not change the value MW_0^{ef} and $K_m<1$ means oligomer (monomer) molecules destruction, which is not confirmed experimentally.

The replacement in the relationship (20) value MW_0 on MW_0^{ef} allows to obtain full correspondence of theory and experiment, as follows from figure 1 data. This confirms the offered above formation mechanism of copolymers APESF and CP-OF-10/OS-10.

Simulation of Parameters of Polyarylate F-2 Low-Temperature Polycondensation

In this section we will consider the simulation of final characteristics (conversion degree and molecular weight), obtained at polyarylate F-2 synthesis in different solvents [1]. In figure 3 the comparison of experimental MW and estimated according to the relationship (20) MW^T molecular weight for F-2 is shown. As follows from the data of this figure, relation between MW and MW^T is approximated by a straight line, passing through the origin, that confirms the offered model correctness.

Let's note, that polymers synthesis reactions simulation within the framework of the offered model is possible in virtue of macromolecular coil fractality only, whose density ρ depends on its gyration radius R_g and, hence, on molecular weight [3]:

$$\rho \sim R_g^{D-d} . \tag{23}$$

It is obvious, that in case of Euclidean object $D=d$ and ρ=const.

The gyration radius R_g of macromolecular coil (aggregate) in cluster-cluster model can be described by the following relationship [23]:

$$R_g \sim \left(\frac{4c_0 kT}{3\eta_0 m_0} \right)^{1/D} t^{1/D} , \tag{24}$$

where k is Boltzmann constant, T is testing temperature, η_0 is initial viscosity of reactionary medium, t is reaction duration.

If as the first approximation all parameters in the right part of the relationship (30), except for t, accept by constant, the under conditions $R_g=R_c$ and $t=t_c$ (t_c is time of the reaction cessation), we shall receive:

$$t_c \sim R_c^D. \quad (25)$$

As for polyarylates F-2, synthesized in different solvents, m_0=const, then we shall finally receive from the combination of the relationships (19) and (25) under conditions of $R_g=R_c$ and $MW=N_c m_0$:

$$t_c \sim MW. \quad (26)$$

Further it is possible to estimate theoretically conversion degree Q (Q^T) according to the following relationship [7]:

$$Q^T \sim m_m \eta_0 t_c^{(3-D)/2}. \quad (27)$$

If to suppose m_m=const and η_0=const, it is possible to receive Q^T from the more simple relationship (with allowance for the relationship (26)):

$$Q^T \sim MW^{(3-D)/2}. \quad (28)$$

In figure 4 the comparison of experimental values Q [1] and Q^T for polyarylates F-2 is shown. As one can see, though the variation tendencies Q and Q^T are identical (the shaded line in figure 4), but the quantitative correspondence is poor. It is explained by the acceptance of the condition η_0=const in the relationship (27). As the experimental values η_0 are absent, we shall make rough, but simple estimation of this parameter from the following considerations. From the relationship (27) at $Q^T=Q$, m_m=const and t_c=const=60 min. (representative reaction duration [2]) we'll estimate η_0 and then calculate Q^T according to the same relationship with variable η_0. The results of comparison of Q and estimated by indicated method Q^T are shown, in figure 5 from which much better correspondence of theory and experiment follows in comparison with figure 4.

Let's note in conclusion, that the treatment of polycondensation process within the framework of an irreversible aggregation models does not except conclusions made in paper [1]. Let's explain it on one example. In paper [1] it was shown, that the destruction F-2 in N,N-dimethylformamide is observed. Within the framework of an irreversible aggregation model it means, that two processes in parallel occur: aggregation and disaggregation. As it was shown in paper [9], such combination results to the increase D of final aggregates. Really, the value D for a macromolecular coil F-2 synthesized in N,N-dimethylformamide is equal to 2.03, that will well agree with theoretical value D for such aggregation type [9].

CONCLUSIONS

The main conclusion of the present paper is that, what reaction cessation in polycondensation process is limited by purely physical factor, namely, the achievement of macromolecular coil density of solution density in which reaction proceeds. The possibility of such treatment is due fully to macromolecular coil fractality. The indicated processes are simulated within the framework of an irreversible aggregation models over mechanism cluster-cluster. The critical values of molecular weight and reaction duration, above which a synthesis reaction ceases, exist.

REFERENCES

[1] Korshak V.V., Vinogradova S.V., Vasnev V.A. *Vysokomol. Soed. A,* 1968, v. 10, № 6, p. 1329-1335.
[2] Korshak V.V., Vinogradova S.V. Nonequilibrium Polycondensation. Moscow, *Nauka,* 1972, 695 p.
[3] Kokorevich A.G., Gravitis Ya.A., Ozol-Kalnin V.G. *Khimiya Drevesiny,* 1989, № 1, p. 3-24.
[4] Feder F. Fractals. New York, Plenum Press, 1989, 248 p.
[5] Kaufman J.H., Melroy O.R., Abraham F.F., Nazzal A.I. *Solid State Commun.,* 1986, v. 60, № 9, p. 757-761.
[6] Shiyan A.A. *Vysokomol. Soed. B,* 1995, v. 37, № 9, p. 1578-1580.
[7] Shogenov V.N., Temiraev K.B., Kozlov G.V. In book: *Physics and Chemistry of Perspective Materials.* Nal'chik, KBSU, 1997, p. 80-84.
[8] Vilgis T.A. *Physica A,* 1988, v. 153, № 2, p. 341-354.
[9] Botet R., Jullien R. *Phys. Rev. Lett.,* 1985, v. 55, № 19, p. 1943-1946.
[10] Kolb M., Jullien R. *J. Phys. Lett.* (Paris), 1984, v. 45, № 10, p. L977-L981.
[11] Büller K.-U. Heat- and Thermostable Polymers. Moscow, *Khimiya,* 1984, 1056 p.
[12] Brow D., Sherdron G., Kern V. The Practical Guidance on Synthesis and Polymers Properties Study. Ed. Zubkov V.P. Moscow, *Khimiya,* 1976, 256 p.
[13] Baranov V.G., Frenkel S.Ya., Brestkin Yu.V. Doklady AN SSSR, 1986, v. 290, № 2, p. 369-372.
[14] Karmanov A.P., Monakov Yu.B. *Vysokomol. Soed.* B, 1995, v. 37, № 2, p. 378-331.
[15] Pavlov G.M., Korneeva E.V., Mikhailova N.A., Anan'eva E.P. *Biofizika,* 1992, v. 37, № 6, p. 1035-1040.
[16] Pavlov G.M., Korneeva E.V. *Biofizika*, 1995, v. 40, № 6, p. 1227-1233.
[17] Ptytsin O.B., Eizner Yu.E. *Zhurnal Technicheskoi Fiziki,* 1959, v. 29, № 9, p. 1117-1134.
[18] Askadskii A.A. Physics-Chemistry of Polyarylates. Moscow, *Khimiya,* 1968, 216 p.
[19] Kozlov G.V., Shogenov V.N., Kharaev A.M., Mikitaev A.K. *Vysokomol. Soed.* B, 1987, v. 29, № 4, p. 311-314.
[20] Kolb M. *Physica A,* 1986, v. 140, № 2, p. 416-420.
[21] Aliguliev R.M., Oganyan V.A., Yurkhanov V.B., Ibragimov Kh.D. *Vysokomol. Soed.* A, 1987, v. 29, № 3, p. 611-615.

[22] Encyclopaedia of Polymers, v. 1, Ed. Kargin V.A. Moscow, *Sovetskaya Entsiklopediya,* 1972, 1223 p.
[23] Weitz D.A., Huang J.S., Lin M.Y., *Sung J. Phys. Rev. Lett.,* 1984, v. 53, № 17, p. 1657-1660.

In: Progress in Chemistry and Biochemistry
Editors: E. M. Pearce and G. E. Zaikov

ISBN: 978-1-60692-344-3

Chapter 6

HYBRID ANTIOXIDANTS

E.B. Burlakova, E.M. Molochkina and G.A. Nikiforov
N.M. Emanuel Institute of Biochemical Physics,
Russian Academy of Sciences,
4 Kosygin str., Moscow 119334, Russia
tel. (+7)(495) 137-31-27; fax(+7) (495) 137-41-01;
e-mail: *seren@sky.chph.ras.ru*

ABSTRACT

This article is a review of data on the synthesis and physiological activity of hybrid antioxidants. The introduction offers an explanation why, in some cases, it is necessary to add drug molecules with fragments responsible for various properties and aimed at various targets. A large group of hybrid antioxidants comprise stable nitroxyl radicals that behave as antioxidants in free-radical reactions of oxidation. Compounds of this type were synthesized extensively to form a group of antitumor agents. As a rule, the specific (antitumor) activity retained or even increased as compared with the initial compounds (without nitroxyl radicals); the toxicity decreased 5 to 10 times, which made it possible to apply the drug in considerably higher concentrations. There are reported data on nitroxyl derivatives of anthracycline antibiotics, antimetabolites, alkylating agents, and the recent results on platinum complexes with nitroxyl fragments. Much attention is given to hindered phenols with 'buoyancy' properties, particularly, to biochemical effects making them promising agents to treat Altzheimer's disease.

Keywords: antioxidants, synthesis, physiological activity, hybrid, Altzheimer's disease, 'buoyancy' properties, oxidation.

INTRODUCTION

In 1954, the book by Professor B.N. Tarusov "Principles of Biological Effects of Radioactive Emissions" was published [1]; the book made a great impression on me. The author, an outstanding Soviet biophysicist, Head of the Biophysics Department at the Faculty of Biology of the Moscow State University, put forward a hypothesis that the development of radiation-induced disease is associated with the induction of ramified chain reaction of oxidation of fats of cellular shells (membranes), the oxidation products are very toxic for the cell.

I had dreamed of studying the mechanism of radiation-induced disease but nobody at the Faculty of Chemistry dealt with these subjects then. Now two areas of my scientific interests have converged – chain reactions that had always been the main subject of investigation at the Chemical Kinetics Department of the Faculty of Chemistry (headed by N.N. Semenov, Nobel Prize Laureate) at the Moscow State University, where I wrote my diploma work, and radiation-induced disease, that was of interest at the Biophysics Department. I was permitted to write my diploma work on biology under the guidance of Academician N.M. Emanuel and Professors B.G. Dzantiev and G.B. Sergeev. The aim of my work was to find out what biologically toxic products are formed upon irradiation of lipids. A model substrate chosen for radiolytic oxidation of fat was natural cod-liver oil. The oil was exposed to radiation, its composition was investigated, and then the oil was oxidized. The investigation results showed that even after high-dose exposures no detectable quantities of specifically new products are formed and the toxicity of the exposed oil depended only on a degree of oxidation. The higher the oxidation degree, the more the deep-oxidation products (aldehydes, ketones, peracids, etc.) and the higher the oil toxicity. At equal oxidation degrees, the toxicity of exposed and unexposed oil was equal [2-4]. The main irradiation effect was reduced to decomposition of natural antioxidants in oil. Generally speaking, this result (a decrease in the quantity of antioxidants upon exposure) was not unexpected: the effect had been determined previously in in vitro experiments with irradiation of various fats.

However, this result interested us in view of its significance with respect to radiation-involved reactions. A simultaneous investigation of toxic effects of irradiation on plant and animal fats made it possible to conclude that toxicity of irradiated fat is not associated with the formation of new products of oxidation but rather with acceleration of oxidation of exposed fat because of decomposition of natural antioxidants in fat. Hence, an unambiguous practical conclusion was drawn: if animals could be administered with antioxidants before exposure, we could slow down processes associated with acceleration of oxidation of lipids and formation of oxidized toxic products. Therefore, we should introduce compounds that could fulfill functions of antioxidants decomposed upon exposure. We started with introduction of natural antioxidant – tocopherol and found out that we can increase the average life-span of irradiated animals. Then, by analogy with works on preventing fats from oxidative decay, we introduced nontoxic synthetic antioxidants used in the food industry [4]. This decision was important not only because we were the first who introduced antioxidants to animals to protect them from irradiation (although that was a new word in radiobiology), but also because *synthetic* antioxidants were introduced to animals. Previously, we were sure that irrespective of a particular structure of an antioxidant (synthetic or natural), its main characteristic was its ability to react with free radicals. Therefore, we believed that this ability that we had shown in model experiments can persist and manifest itself after introduction of an oxidant into lipids of animal organs. These experiments confirmed, to some extent, the Tarusov's concept about the great role of chain (free-radical) reactions in lipids of exposed animals in the development of radiation-induced damages. *(E.B. Burlakova)*

REVIEW

In 1960, N.M. Emanuel put forward the hypothesis that not only radicals of lipids but also radicals from other biochemical components of cell (DNA, protein, polysaccharides, etc.) that are alien to normal cell and are formed upon exposure may cause multiple damages and death of the cell [5].

It was shown that radicals formed upon exposures of DNA and proteins may, like lipid radicals, enter into exchange reactions with antioxidants; as a result, the free valence passes from biopolymer radical to the antioxidant molecule and forms inactive radical from the antioxidant [InH] [6-8.

$$R^{\cdot}_{biol.} + HIn \rightarrow RH + In^{\cdot}$$

In 1961, a supposition was made about the great role of free radicals alien to normal cell in the development of some other diseases and about the feasibility of inhibiting the free-radical reactions by applying synthetic inhibitors to achieve a curing effect [9-11]. This supposition could be made only by physicochemists, first of all, by specialists in kinetics of the Semenov–Emanuel school, who understood the importance of not only (and not so much of) a change in the composition of the reaction components but also of their physicochemical properties, i.e., when the same results may be obtained with different (in composition) components but with the one common physicochemical property – in this case, ability to react with free radicals. Therefore, synthetic compounds of the structure other than that of natural antioxidants may be used instead of (substitute) the latter ones in reactions with free radicals.

In the 1960s, the Emanuel Institute of Biochemical Physics, Russian Academy of Sciences, initiated studies in the new field – chemistry and biology of antioxidants. The scientists of the Institute had to solve an important task – to find out whether the biological activity of antioxidants as inhibitors of radical reactions depends on their properties. For this purpose, nontoxic different-structure antioxidants were synthesized: derivatives of hindered phenols and heterocyclic hydrocarbon hydroxy compounds [12, 13]. The existence of homologous arrays of antioxidant derivatives made it possible to determine the structure–activity dependence and select the most efficient and least toxic compounds.

In in vivo experiments, the correlation between the radioprotecting activity and antiradical properties of synthetic antioxidants was determined [14, 15]. Kinetic studies on *natural* antioxidants – vitamins were carried out; their constants as inhibitors of radical processes were determined [16, 17]. In the works by Khrapova et al., the chemiluminescence method adapted to studies on bioantioxidants in lipids was used; with this method, the problems of synergism and anthagonism of synthetic and natural antioxidants were studied and the antioxidant system in membrane lipids was characterized as a whole [18].

The above works were concerned mainly with investigating antioxidants in lipid components of cells. However, specially developed photochemiluminescence models were used to assess the antiradical activity of water-soluble natural and synthetic inhibitors; exchange reactions of this type of antioxidants with UV-induced peptide radicals were studied [19-20]. Great attention was given to extending research concerning the role of bioantioxidants in the development of some or other diseases and the feasibility of using antioxidants for prophylactic or curing purposes.

It is necessary to dwell on an important property of antioxidants – the dose–effect dependence of introduced antioxidant. After introduction of antioxidant, the antioxidant activity (AOA) increases, then it returns to normal, and then, after a short-time AOA increasing, it decreases drastically below the normal. Therefore, antioxidants may produce a curing effect by decreasing (at low doses) or increasing (at high doses) the rate of free-radical reactions.

The changes in the antioxidant activity of organs and tissues lipids in the process of carcinogenesis were studied in [21]. The staged mechanism of changes in the antioxidant activity in the process of carcinogenesis caused by different cancinogens: benzopyrrene, orthoaminoazotoluene, and γ-irradiation was established. At the initial stage of the carcinogen toxic effect and unduction of tumor cells, AOA decreases, then it increases, reaches the normal, and then increases above the normal at the stage of transition from diffusional to localized hyperplasia. The efficiency of synthetic antioxidants depends on their concentration and time of intriduction [22]. At the first stage of carcinogenesis, the protective effect is caused by doses that increase AOA; at the late stages, these doses may accelerate the development of carcinogenesis and increase the number of tumors induced. At the late stages, it is necessary to introduce larger antioxidants quantities, which can cause an opposite effect – to decrease AOA [23, 24] and inhibit the process of carcinogenesis.

N.M. Emanuel and O.S. Frankfurt were the first who discovered the anticancer effect of the antioxidant dibutyloxytoluene [25].

A great number of works were devoted to studying free-radical processes associated with the tumor growth and the antitumor effects of antioxidants [26-28]. It was found out that tumor growth is associated, as a rule, with an increased level of antioxidants and only high doses of antioxidants produce the antitumor effect. In this case, antioxidants do not increase the AOA of organs and tissues but, on the contrary, decrease it and act as prooxidants. It should be noted that there is a general trend not only for antioxidants but also for various antitumor agents: their efficiency is the higher, the stronger they decrease AOA [29].

Many specialists at IBCP RAS studied antioxidants with respect to radiation-induced disease. The radioprotective effect of the compounds was in conformity with their AOA [15]. Similar data were obtained not only in experiments in animals but also in model experiments with exposed solutions of DNA, proteins, and lipids [30-32].

A novel field of science was commenced in the 1970s – applications of antioxidants in gerontology. Kinetic studies on model reactions of aging, investigation of age-related changes in antioxidants, theoretical concepts of aging, particular experimental studies of antioxidants as geroprotectors showed that this science field is of both theoretical and practical importance [33-37]. It is very strange that the idea of using antioxidants in gerontology is now put forward as a new one and the authorship is ascribed to people other than those who have been working in this field for 40 years already.

Along with studies on using antioxidants for some particular diseases, extensive studies on the role of antioxidants in normal physiological processes were commenced. Palmina et al., studied the role of antioxidants in cell proliferation and showed that the factors that increase the antioxidant activity accelerate the proliferation; those that decrease it, inhibit [38-40].

Alesenko et al., studied the effect of antioxidants on the genetic apparatus activity [41-43]. The authors showed that bioantioxidants are able to affect the cell lipids composition and change the activity of lipid-dependent enzymes of synthesis and reparation of DNA and affect the activity of chromatin.

The end of 1970s was marked by extensive studies on the role of antioxidants in the normal metabolism of cell. There was drawn the conclusion that AO are universal modifiers of composition, structure, and functional activity of membranes and that many of their effects on cell metabolism may be interpreted from these posotions [44, 45]. There was discovered the physicochemical system of regulation of cell metabolism by membranes based on interrelation between membranes lipid peroxidation (LPO), on the one hand, and changes in the composition of membrane lipids and their oxidizability, on the other hand [46, 47].

Proceeding from the parameters of this system, it is possible to use antioxidant to convert a cell, organ, and organism from one metabolic state to another.

In 1970s, antioxidants found wide use in cardiology, oncology, and treatment of neurodegenerative and other classes of diseases [48-50]. Extensive studies on antioxidants were commenced in the field of plant growing and farming as plant growth stimulators and for preventive and curing treatment of cattle and poultry [51-55].

The main conclusions made in the works by Russian scientists in 1970s are as follows:

1. Non-toxic inhibitors of radical processes – antioxidants exhibit a wide gamut of biological activity.
2. The biological effectiveness of antioxidants correlates with their antioxidizing properties.
3. Depending on dose, antioxidants may either increase (at low doses) the antioxidizing activity or decrease it .
4. The activity of antioxidants for curing any particular disease depends on the time of introduction in the course of medical treatment because the development of the disease may be accompanied by stages of changing the antioxidizing activity. The compound may be efficient only if it is introduced at a low dose at the stage of reduced AOA or at a high dose at the stage of AOA elevation.

It is natural, and it could not be otherwise, that the pioneering works on antioxidants and free-radical reactions occurring in living systems were attacked furiously by opponents – scientists of various profiles: biologists, physicians, and even some of chemists and physicists. In spite of all arguments, vitalistic tendencies were strong: nobody could even dare to think that synthetic antioxidants may substitute natural ones. "Neither of free radical reactions can develop in a living organism", the opponents claimed, "because these reactions are not controlled but strict regulation is essential for a living organism". In addition, according to the concepts prevailing at that time, the membrane structure was such that it excluded the valence transfer from one lipid molecule to another since a lipoprotein model of membrane showed that lipids were separated by protein molecules. The opponents considered the absence of specific enzymes governing these reactions a strong argument against inhibitors of radical reactions and radical processes as such in organisms. The wide-spread opinion was that the antioxidant function, even that of tocopherol, was a side effect of its activity and important only for in vitro processes but did not play any role in bioobjects life. This opinion was supported by the fact that the deficiency of tocopherol (E-avitaminosis) can not be cured completely by applying synthetic antioxidants [56, 57]. Finally, it was considered doubtful that works, in which peroxides were detected in lipids isolated from organs and tissues, dealt with true amounts of products of free radical processes in vivo but not with the amounts of products formed during the process of isolation.

All the objections and skepticism have been rejected in due time. Antioxidant enzymes were discovered, the model of membrane was revised. The development of biochemistry and biophysics held the course in the direction of verification of this concept [58, 59].

These several pages of history should be concluded by the acknowledgment that the works carried out by Soviet scientists in the field of free-radical biology were pioneering and many if not all data obtained at that time remain valuable until the present time.

In spite of the fact that antioxidants are much spoken about because of their extensive application for various purposes, we should return to the definition of bioantioxidants.

Bioantioxidants are substances that exhibit the properties of inhibitors in model free-radical reactions of oxidation and retain these properties when introduced into a living organism (cell culture etc.). The absence of even one of these qualities does not permit calling a substance a bioantioxidant (BAO). Although the antioxidizing activity of lipids can be increased by applying substances that are synergists to natural antioxidants or those transformed into antioxidants in the course of metabolism, bioantioxidants, by our definition, should necessarily possess the ability to inhibit an oxidizing free-radical process in model reactions. This property of BAO makes it possible to predict the spectrum of their biological effects and perform a targeted synthesis of compounds.

At present, the following pathways of antioxidants effects on the cell metabolism are considered:

1. Interaction of BAO with free radicals of different nature.
2. Incorporation of BAO into the membrane structure and changes in the membrane functional activity due to changes in the membrane viscous properties (fluidity).
3. Effect of BAO on the activity of membrane proteins – enzymes and receptors.
4. Effect of BAO on the cell genetic apparatus including gene expression.
5. Effect of BAO on the regulatory systems of cell and, indirectly, on its metabolism as a whole.

Note that the reaction rate constants of the same inhibitors with different radicals differ considerably (by several orders of magnitude) [60, 61] (see table 1). There are cases when antioxidants that are active for some radicals can not compete in interacting with others and we can not protect cell components from the effects of these radicals because the affinity of radicals to them will be higher than that of inhibitors introduced.

Another obstacle to effective using of antioxidants is their extreme concentration–effect dependence. As noted above, antioxidants applied in high concentrations produce an opposite effect and do not inhibit but accelerate free radical reactions. The phenomenon may be attributed either to a high activity of radicals accumulated from inhibitors or to the prevailing consumption of natural antioxidants as compared with synthetic ones introduced. Many of these effects depend on the initial characteristics of free radical processes and the initial level of antioxidants.

Thus, because of the versatility of antioxidants properties and feasibility of affecting various normal and pathological states, we are obliged to know exactly the nature of radicals responsible for pathological changes, the time of AO introduction, concentration, and elementary constants of inhibitors. A negligence in or an erroneous approach to the antioxidant therapy may lead to negative results.

Table 1. Rate constants of interaction of inhibitors-antioxidants with biologically active radicals

Inhibitors	Radicals (rate constants, l/m sec			
	$OH^{\bullet}$	$R.^{\cdot}$ proteins	$RO_2^{\cdot}$ lipids	$0_2^{-\cdot}$ superoxide anion radical
2-ethyl-6-methyl-3-hydroxypyridine, hydrochloride	3.3×10^{1o}	1.9×10^{6}	9.0×10^{5}	26-4
N 3.5 ditert-butyl-4-hydroxyphenyl propionic acid (phenozan)	4.4×10^{1o}	1.2×10^{6}	$2..2 \times 10^{4}$	10-2
5,7,8-Trimethyltocopherol (α -tocopherol)	8×10^{1o}		3.4×10^{6}	47×10^{4} (soluble form)

A study on the mechanism of BAO effects showed that there is a whole system of relations between separate indices of cell metabolism that vary under the action of antioxidants.

Investigation of the physicochemical regulatory system maintaining the level of free-radical reactions in lipids, on the one hand, and regulating the metabolism of membrane lipids and the rate of consumption of antioxidants in lipids, on the other hand, has been further developed [46, 47].

The components of this system are antioxidants, free radicals, peroxidation products, composition and oxidizability of lipids, and the rate of consumption of antioxidants.

It was shown that enhancement of the antioxidant level is associated with reduction of the lipid peroxidation (LPO) rate, reduction of oxidation products concentration, reduction of the rate of lipids exit from membranes and enrichment of membranes with unsaturated lipids, and enhancement of lipids oxidizability. The latter effect results, in turn, in an increase in the rate of consumption of the antioxidant activity and, correspondingly, in returning of the AOA and LPO rate to the normal. An opposite situation is observed with decreasing in the AOA concentration, increasing the LPO rate, etc. Similar systems of regulation were discovered almost for all intracellular and cellular membranes of animals, plants, and microorganisms. Note that changes in the composition and oxidizability degree of lipids are associated with changes in the viscosity of various layers of membranes. The above parameters affect the activity and kinetic characteristics of membrane-bound proteins – enzymes and receptors; changes in the LPO rate may result in changing of not only the structure but also of the functional activity of membranes. In normal membranes, we observe identical relationships between the parameters; the difference concerns the system relaxation time (from several minutes to several days).

Exposure of organism to any damaging factor is associated with changes in this system of regulation. Long-term changes may result from (i) the action of a chronic factor that does not cause breaking bonds in the system of regulation; the system can return to the normal after cessation of the action; (ii) there may occur situations when exposure to a damaging factor results in transition to a new level of regulation; and (iii) there may occur breaking bonds in the system of regulation, which prevents the system to return to the normal. In the latter case, antioxidants may be efficient as a component of complex therapy. Such conclusions were drawn both in experimental studies and in clinical tests.

At present, scientists-pharmacologists must answer the question: whether all pharmacologically active compounds should be multitargeted drugs, i.e., should be aimed at several targets but not at one specific target. The answer to this question is not unambiguous, although the scientists who pose it cite the data that are evidence for the fact that most diseases are associated with changes and defects of various pathways of metabolism and drugs should not be aimed at one critical target but at all affected by the disease.

Meanwhile, an alternative to this approach is complex therapy that uses several compounds aimed eact at its specific target; the compounds may be applied at various concentrations (not only at those specified by the compound structure), at various times of application, in various solvents, etc.

In our opinion, both views have the right to exist and different approaches should be taken in each specific cases for the benefit of patients.

As was noted above, breaking bonds in the system of regulation points to impossibility of returning the system to the initial state by applying AO.

Investigation of the regulation system as a whole but not only of separate changes in the system makes it possible to decide when the AO monotherapy is sufficient and when a complex therapy is needed, in which, apart from AO, other biologically active substances aimed at other targets are needed. To some extent, this may be accomplished by synthesis of hybrid molecules.

The term of hybrid antioxidants implies molecules whose structure contains parts responsible for antioxidant properties and fragments of molecules responsible for other specific functions.

In most cases, synthesis of hybrid molecules does not yield a new polyfunctional structure but cross-linked or integrated molecules that produce a high AO effect and are aimed at other targets specific of a certain disease.

Very often, when designing a molecule, it is necessary to retain the specific activity of one of the hybrid components and, at the same time, to reduce side-effects, e.g., toxicity of the compound. One of the promising ways is incorporating nitroxyl radicals in the molecule structure.

Nitroxyl derivatives of biologically active substances are the most numerous and earliest-synthesized hybrid antioxidants. Previously, it was shown that the nitroxyl radical exhibits the AO properties in model reactions pf oxidation and in vitro and in vivo experiments. One of the pioneers in organic chemistry who synthesized nitroxyl derivatives of BAS was a Soviet scientist – A.B. Shapiro [62]. Since then, nitroxylation of BAS has been put into practice in pharmacology. Most extensively, antitumor compounds are nitroxylated. Konovalova N.P. [63-68] who had gained a great experience in the field of nitroxyl antitumor compounds made up a list of synthesized and well-studied antitumor agents referred to the class of nitroxyl-containing antibiotics, antimetabolites, and alkylating agents (table 2).

Table 2. Nitroxyl derivatives of antitumor agents

Initial components	LD_{50} mg/кg	Nitroxyl derivatives	LD_{50} mg/кg
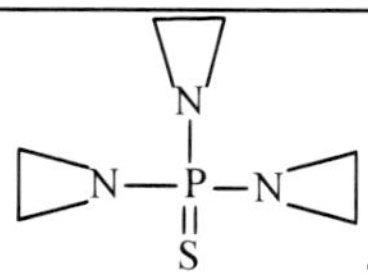 Thio TEF	18	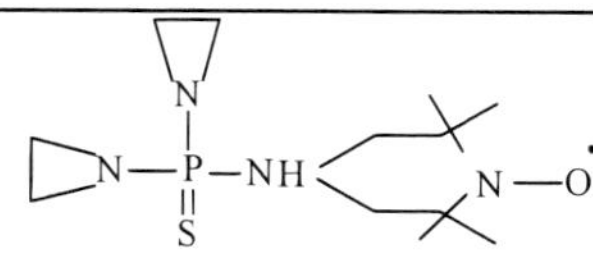	187
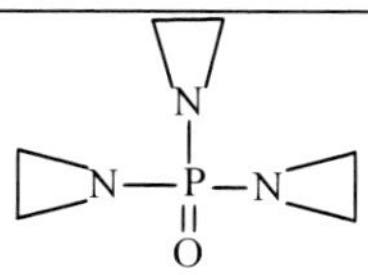 TEF	15	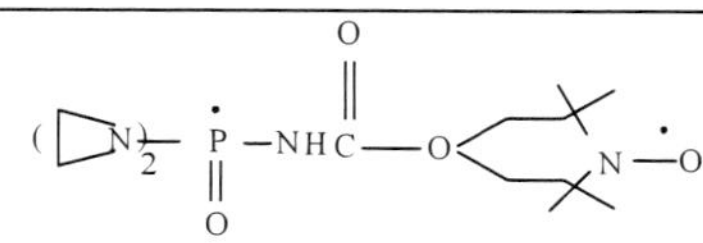	150
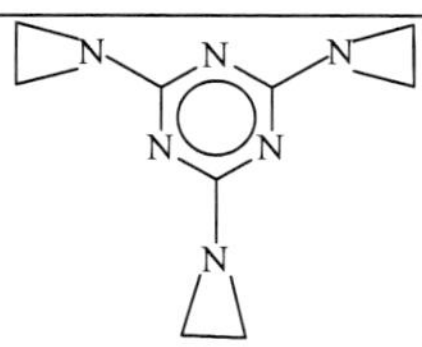 TEM	1,5	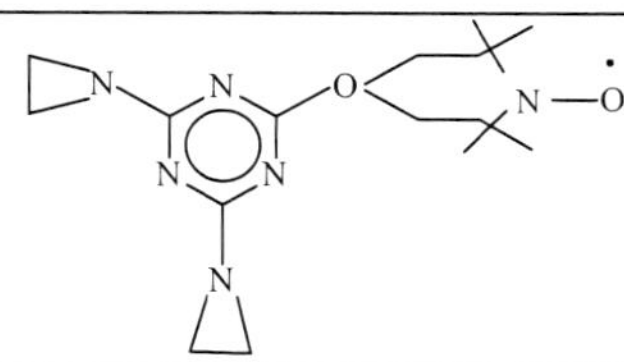	15
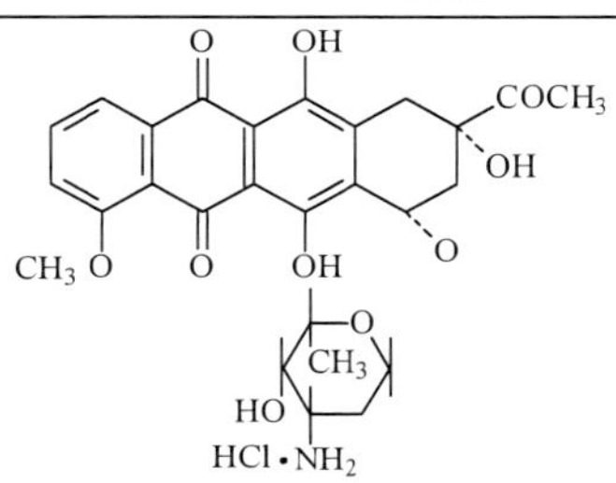 Rubomycin	5,6	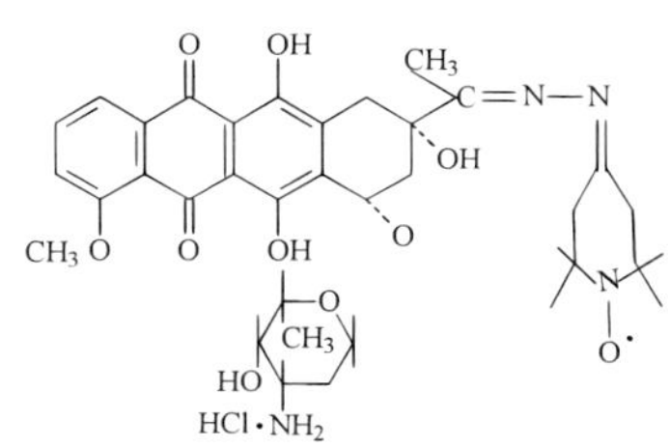 Ruboxyl	50
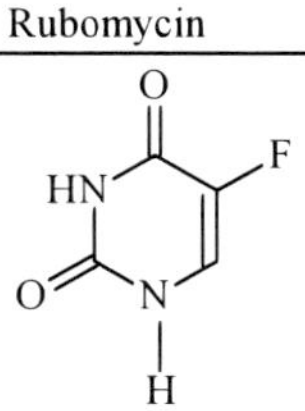 5-Fluorouracil	100	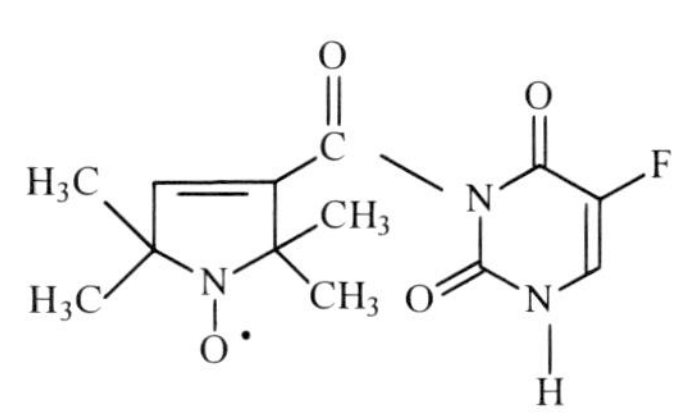 Magnicyl	510
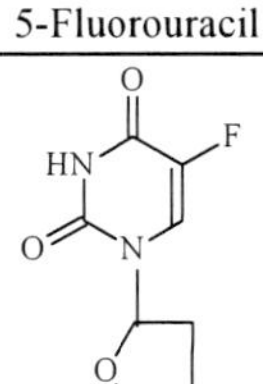 Fluoroaphur	750		

Table 2. (Continued).

Initial components	LD$_{50}$ mg/кg	Nitroxyl derivatives	LD$_{50}$ mg/кg
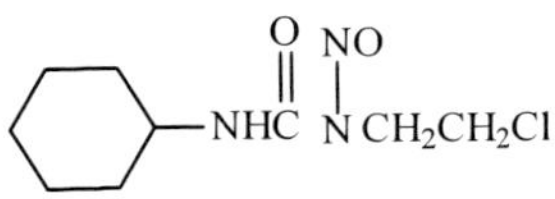 Cyclohexyl - Nitrosourea	47	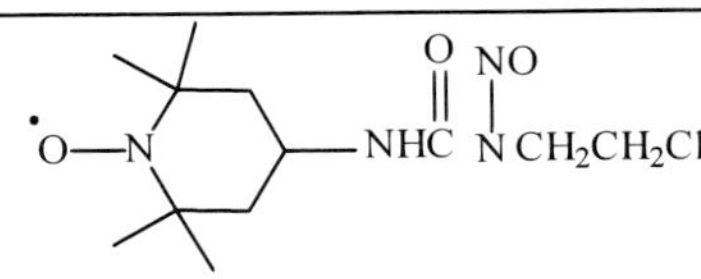	60
		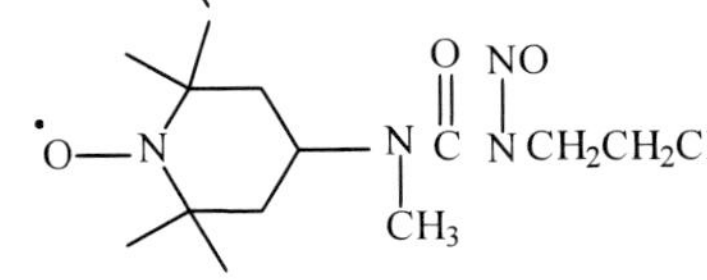	150

As is seen from table 2, nitroxyl derivatives are 5 to 10 times less toxic than the initial compounds.

At present, derivatives of platinum compounds have found wide use in chemotherapy of malignant tumors. It should be noted that, along with their high antitumor effect, these compounds exhibit a high toxicity. Supposedly, their cytotoxic effect and other side-effects (nephro- and ototoxicity, nausea, etc.) are associated with intensification of free-radical processes and formation of active oxygen species [$\bar{O_2^{\cdot}}$, OH$^{\cdot}$] induced in cell by cisplatinum. In fact, the antitumor effect is associated, apart from the interaction with the DNA molecule, with enhancement of free-radical reactions. As a rule, reduction of toxicity results in decreasing the antitumor effect. Development of hybrid molecules based on platinum derivatives and nitroxyl radical makes it possible to reduce the toxicity and retain the activity of the agent.

At present, extensive studies are being carried out on the synthesis, structure, and biological activity of mixed-ligand complexes of platinum (II) with aminonitroxyl radicals [69].

The largest group of antitumor compounds includes nitroxyl derivatives of anthracycline antibiotics. One of the major achievements of the chemistry and biochemistry of hybride antioxidants of this class is the development of emoxypin (ruboxyl) – a nitroxyl derivative of the anthracycline antibiotic – rubomycin.

This compounds have a great advantage of hybride molecules, namely, in the background of high antitumor effect (higher than that of the parental compound – rubomycin), its toxicity decreased by 40% and cardiotoxicity and depilative properties vanished almost completely. The compound acquired some novel properties that are not characteristic of rubomycin and nitroxyl radical. It is of interest that there appeared no cross-resistance with rubomycin. The development of this compound is the greatest practical achievement in this field of investigation. I has passed the second stage of clinical tests.

Other promising hybrid compounds are antitumor agents prepared on the basis of mixed-ligand platinum II and platinum IV complexes containing antioxidant fragments of the array of aminonitroxyl radicals. The feasibility of preparing such mixed complexes relies on the fact that cisplatinum ($[NH_3]_2PtCl_2$) transforms readilty into a $Na[(NH_3)PtCl_3]$ complex, which interacts with primary amines on the background of sodium iodide and transforms into a $(RNH_2)(NH_3)PtCl$ complex. Having been treated with silver nitrate the latter complex exchanges readily halogen anions for the mobile NO_3^- anion; as a result, they can be substituted for other anions – Cl^- and anions of dicarboxylic acids.

The above complexes were characterized in various tumor models; it was shown that the DNA binding rate for these new complexes is comparable with that of cisplatinum but their oxidizing effect is opposite: the initial cisplatinum accelerates oxidation in model radical reaction; the platinum–nitroxyl complexes, on the contrary, inhibit oxidation. From this point of view, it is easy to explain the reduction of the toxicity of the synthesized complexes.

In the recent time, a great interest was aroused for platinum IV compounds. The scientists-organochemists of the Institute of Theoretical Problems of Chemical Physics synthesized complexes of platinum IV with aminonitroxyl radicals. These compounds were tested on the experimental model of P-388 leukemia. In the experiments, the complexes exhibited a strong effect on mice–leukemia-carriers: sometimes to complete recovery; the toxicity reduced two- to four-fold [70].

The most impressive results were obtained for low doses of cisplatinum and complexes of platinum IV with amino nitroxyl radicals applied in combination for treatment of P-388 leukemia. The survival rate for leukemia mice was 100%.

As was noted above, nitroxylation of antitumor compounds was successful. There are some other examples of application of hybrid compounds containing nitroxyl radicals in their molecules.

At the Institute of Problems of Chemical Physics, novel biologically active compounds were synthesized –nitroxyl derivatives of azidothymidine of the common formula

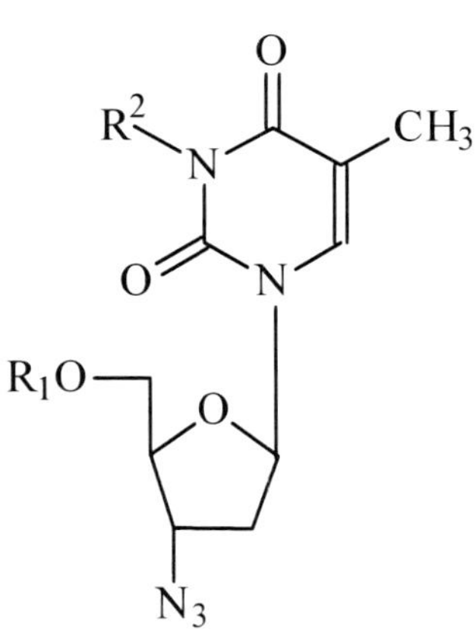

where R^1 is the radical containing a nitroxyl group > N – O and $R^2 = R^1$ or H.

These compounds produce the antiviral effect against RNA-containing viruses (human immunodeficiency virus and vesicular stomatitis virus) and the DNA-containing virus – cytomegalovirus [71].

It should be noted that hybrid molecules containing azidothymidine and antioxidant fragments in their structure inhibit the reproduction of cytomegalovirus; other azidothymidine derivatives do not possess this property. Considering that death of immunodeficiency patients is caused by cytomegalovirus-provoked diseases, this property of antioxidant- and azidothymidine-based hybrid molecules should be emphasized particularly.

One of the most promising and important practical areas of the modern chemistry related to phenol bioantioxidants is the synthesis of hybrid compounds that combine the antioxidant activity and the capacity for structural interactions with a biosystem. This type of compounds includes so-called 'buoy' compounds synthesized from quaternized derivatives of dialkylaminoalkylsubstituted 2.4-2.6-di-tert-butylphenols and dialkylaminoalkyl ethers of phenozan acid. For these compounds, a wide spectrum of biological activity was discovered – the antimicrobial, antiviral, analgetic, etc. activities. Also, it was discovered that gradual (step-by-step) redox and solvolytic conversions result in the formation of a cascade of intermediates with different inherent activities – antioxidizing, chelating, capacity for incorporating into the charge transfer chain, etc. In the cascade mechanism of the effect of 2.4-2.6-di-tert-butylphenol derivatives, the tendency to the formation of heterocyclic compounds, which make a contribution to the total biological activity, plays a great role.

Another important currently developing research area is the synthesis of hybrids of functional di- and tert-butylphenols and biocompatible macromolecules. In this area, it is possible to achieve the highest values of antioxidizing activity of hybrid compounds of a wide variety of hydrophobic–hydrophilic relationships and particular structural properties in solutions [72].

The presence of the positively-charged nitrogen atom in the hybrid molecule provides for the antioxidant adherence to the surface of a cell membrane and its fixation in a certain place by means of the lipophilic long-chained alkyl fragment R_1. Such a structure ensures the targeted use of the antioxidant and favors the inhibition of pathological processes in cell, e.g. intensification of LPO and disorders of cell membranes functions.

Hence, a hybrid molecule – an analog of acetylcholine [$CH_3COOCH_2CH_2N^+(CH_3)_3OH^-$] was constructed. In this molecule, instead of acetic acid, the ester bond is formed by carboxylic acids that contain a 2.6-di-tert.butyl-4-hydrophenyl fragment. We called these hybrid antioxidants ichfans.

It might be expected that the above structure should give rise to a bioantioxidant with an anticholineesterase activity. Indeed, anticholineesterase compounds are the most efficient up-to-date therapeutic agents applied for the Altzheimer's disease (AD); they make it possible to maintain the level of acetylcholine (responsible for memory and cognitive functions) in the disease-affected sections of brain. On the other hand, the oxidative stress, i.e., promotion of LPO in cell membranes of brain and in cells of peripheral systems and organs plays an important role in the development of AD [73].

The above-described properties and interactions of the AOs with biological systems, e.g., cell membranes, made it possible to suggest using these compounds as drugs for the therapy of the Altzheimer's disease.

In studies performed with the use of various oxidation models, antioxidizing properties of ichfans were determined and assessed quantitatively. It was found out that new hybrid AOs – ichfans –possess the antioxidant activity (revealed in the model of oxidation of homogenate lipids) and anticholineesterase activity that exceed the corresponding indices of the initial substances. The addition of the molecule with alkyl substituents with different lengths of the alifatic chain on the nitrogen atom promotes the AOA and inhibiting properties. With increasing the length of the carbon chain in the alkyl substituent, the capacity of the compounds to inhibit AChE increases. The type of inhibition depends on an alkyl substituent. Although, the inhibiting power of the substances under study for a membrane-bound enzyme is by an order of magnitude lower than that for soluble AChE, the same relationships between the structure and inhibiting activity of the compounds were detected for the membrane-bound and soluble enzyme [81-84].

A strict direct correlation between the anticholineesterase and antioxidative properties of ichfans was detected; the correlation is also of the same character for membrane-bound and soluble AChE.

On the basis of results of in vitro experiments, by the criteria of the anticholineesterase and antioxidative properties, the optimum compound was chosen for further in vivo studies. That was a hybrid with the radical $R_1 = C_{10}H_{21}$; hereinafter, it will be referred to as ichfan. An additional lengthening of the tail resulted in undesirable perturbating effect on membranes; the effect manifested itself by the erythrocytes increasing sensitivity to hemolysis. In accordance with the data published, the compound with $R_1 = C_{10}H_{21}$ is sufficiently hydrophobic to permeate through the hematoencephalic barrier.

With regard to the AChE and antioxidizing effects and feasibility of permeation through the HEB, ichfan is of considerable interest in view of using it as a drug for treatment of AD.

It should be noted that the oxidative stress as an important factor of AD may be not only a source of free radicals damaging cell structures and macromolecules but also a symptom of a disorder in the operation of the system of homeostasis of lipid peroxidation (LPO) in biological membranes. This system plays an important role in the regulation of cell metabolism; it controls the structure and structure-related functions of various cellular membranes.

An analysis of data [85-97] on changes in the lipid metabolism, composition and structure of the membrane lipid phase, which plays a great role in transmission, processing, and storing the information in cell, showed that the development of AD is associated, along with enhancement of LPO, with enrichment of lipids with unsaturated compounds and increasing the fluidity (decreasing the viscosity) of the lipid phase. In other words, on the background of enhancement of LPO, the lipid bilayer fluidity increases; the effect favors the increasing of oxidation and, hence, the development of the pathological process. This property of membranes in AD makes it impossible to prevents from the oxidative stress by means of traditional phenol antioxidants, which promote the fluidity increasing and complicate the disease development.

We suggested that addition of a saturated fatty-acid tail (buoy) that incorporates in the antioxidant molecule membrane may make the membrane more rigid and thus contribute to the therapeutic effect of ichfan. As is seen from table 1, after introduction of the compound to mice, the microviscosity of the membrane near-surface sites studied by the method of EPR spin probes either changes insignificantly or increases; the latter is a desirable effect. It should be emphasized particularly for membranes isolated from a coarse fraction of synaptosomes because AD is associated mostly with damages of nerve fibers.

According to many of the researchers, one of the AD risks is an elevated cholesterol level [98-100]. It should be noted that the cholesterol content in rat brain tissues decreased by 40% within 2 h after introduction of 15 mg/kg of the compound to the animals. The cholesterol content in a cytoplasmic fraction isolated from mice brain decreased almost two-fold within 2.5 h after introduction of 6 mg/kg of ichfan.

Thus, in addition to the ability of ichfan to inhibit the cholineesterase activity, it can inhibit the oxidative stress (LPO) and, in contrast to traditional phenol-type antioxidant, rigidize the structure of the membrane lipid bilayer or at least prevent from increasing its fluidity. The combination of these properties may be beneficial for the therapy of AD through the correction of the AD-damaged system of regulation of lipid peroxidation homeostasis that participates in controlling of cell metabolism. A certain contribution to the therapeutic effect of ichfan may be provided by its lowering effect on the level of cholesterol.

REFERENCES

[1] Tarusov B.N., Mechanisms of biological effects of radioactive emissions, M., *Medgiz,* 1954, 140 pp. (in Russian)

[2] Burlakova E.B., Dzantiev B.G., Sergeev G.B., Emanuel N.M., *Izv. vuzov,* 1959, vol. 2, no. 4, p. 533. (in Russian)

[3] Burlakova E.B., Dzantiev B.G., Sergeev G.B., Emanuel N.M., *Nauchn. dokl. vyssh.shk.,* 1960, vol. 1, p. 145. (in Russian)

[4] Burlakova E.B., Dzantiev B.G., Sergeev G.B., Emanuel N.M., in Proceedings of Conference "Biochemical and physicochemical principles of biological effects of radiation, M., *Izd. MGU,* 1957, p. 10. (in Russian)

[5] Emanuel N.M., in Proceedings of the Moscow society of nature researchers, M., *Nauka,* 1963, p. 73. (in Russian)

[6] Sapezhinskii I.I., Emanuel N.M., *Dokl. AN SSSR,* 1965, vol. 163, no. 4, p. 845. (in Russian)

[7] Zhizhina G.P., Kruglyakova K.E, Emanuel N.M., *Dokl. AN SSSR,* 1965, vol. 163, no. 4, p. 931. (in Russian)

[8] Emanuel N.M., Burlakova E.B., Kruglyakova K.E., Sapezhinskii I.I., *Izv. akad. nauk AN SSSR, Ser. biol.,* 1965, no. 2, p. 183. (in Russian)

[9] Burlakova E., *Moderne Medizin,* 1972, vol. 2, no. 4, p. 215.

[10] Emanuel N.M., Lipchina L.P., *Dokl. AN SSSR,* 1958, vol. 121, no. /, p. 141. (in Russian)

[11] Emanuel N.M., Pathways of synthesis and investigation of antitumor drugs, M., *Nauka,* 1962, 22 pp. (in Russian)

[12] Ershov V.V., Nikiforov G.A., Volodkin A.A., Spatially-hindered phenols, M., *Khimiya,* 1972, 351 pp. (in Russian)

[13] Smirnov L.D., Sholina S.I., et al., Izv. *AN SSSR, Ser. khim.,* 1963, no. 5, p. 890. (in Russian)

[14] Burlakova E.B., Alesenko A.V., Molochkina E.M., Palmina N.P., Khrapova N.G., Bioantioxidants for radiation damage and malignant growth, M., *Nauka,* 1975, 214 pp. (in Russian)

[15] Burlakova E.B., Gaintseva V.D., Slepukhina L.V., Khrapova N.G., Emanuel N.M., *Dokl. AN SSSR,* 1964, vol. 155, p. 1398. (in Russian)

[16] Burlakova E.B., Khrapova N.G., Yadykin G.I., *Biofizika,* 1971, vol. 16, no. 2, p. 201. (in Russian)

[17] Aristarkhova S.A., Burlakova E.B., Khrapova N.G., *Izv. akad. nauk, Ser. khim.,* 1971, no. 2, p. 2714. (in Russian)

[18] Burlakova E.B., Burobina S.A., Khrapova N.G., Yadykin G.I., *Biofizika,* 1971, vol. 16, no. 1, p. 39. (in Russian)

[19] Sapezhinskii I.I., Biopolymers: Kinetics of radiation-induced and photochemical conversions, M., *Nauka,* 1988, 214 pp. (in Russian)
[20] Lozovskaya E.L., Sapezhinskii I.I., *Radiation biology. Ecology,* 1993, vol. 33, no. 1(4), p. 601. (in Russian)
[21] Burlakova E.B., Molochkina E. M., *Biofizika,* 1968, vol. 13, p. 443. (in Russian)
[22] Burlakova E.B., Burobina S.A., Khrapova N.G., *Dokl. akad. nauk SSSR,* 1971, vol. 200, p. 461. (in Russian)
[23] Burlakova E.B., Molochkina E. M., *Vopr. onkol.,* 19 74, vol. 20, p. 62. (in Russian)
[24] Burlakova E.B., Palmina N.P., *Vesti AMN SSSR,* 1982, no. 3, p. 74. (in Russian)
[25] Frankfurt O.S., Lipchina L.P., Bunto T.V., Emanuel N.M., *Bull. experim. biol. med.,* 1967, vol. 8, p. 86. (in Russian)
[26] Emanuel N.M., Konovalova N.P., Dronova L.M., *Dokl. akad. nauk SSSR,* 1962, vol. 143, no. 3, p. 737. (in Russian)
[27] Emanuel N.M., Dronova L.M., Konovalova N.P., *Dokl. akad. nauk SSSR,* 1963, vol. 152, no. 2, p. 481. (in Russian)
[28] Burlakova E.B., Palmina N.P., *Vopr. onkol.,* 1990, no. 10, p. 1155. (in Russian)
[29] Palmina N.P., Burlakova E.B., *Vesti AMN SSSR,* 1985, no. 1, p. 91. (in Russian)
[30] Emanuel N.M., Kruglyakova R.E., Nikolaeva N.V., Zakharova N.A., *Dokl. akad. nauk SSSR,* 1964, vol. 157, no. 4, p. 935. (in Russian)
[31] Sapezhinskii I.I., Silaev Yu.V., Emanuel N.M., *Dokl. akad. nauk SSSR,* 1964, vol. 159, no. 6, p. 1378. (in Russian)
[32] Sharpatyi V.A., Radiation chemistry of biopolymers, M., *Energoizdat,* 1981, 168 pp. (in Russian)
[33] Obukhova L.K., *Uspekhi khimii,* 1975, vol. 44, no. 10, p. 1914. (in Russian)
[34] Emanuel N.M., Obukhova L.K. *Exp. Gerontol.,* 1979, vol. 13, p. 25.
[35] Obukhova L.K., Nakaidze N.Sh., Serebrjany A.M., Smirnov L.D. Akifiev A.P. *Exp. Gerontol.,* 1979, vol. 14, p. 335.
[36] Sadovnikova I.P., Obukhiva L.K., Bunto T.V., Smirnov L.D., *Izv. AN SSSR,* 1984, Ser. biol., no. 4, p. 543. (in Russian)
[37] 37. Emanuel N.M., Obukhova, Bunto T.V., Dyakova V.V., *Izv. AN SSSR, Ser. biol.,* 1997, no. 1, p. 32. (in Russian)
[38] Burlakova E.B., *Biofizika,* 1967, vol. 12, no. 1, p. 82. (in Russian)
[39] Burlakova E.B., Palmina N.P., Ruzhinskaya N.L., *Izv. AN SSSR,* 1971, no. 1, p. 134. (in Russian)
[40] Alesenko A.V., Sokolova I.S., Kukushkina G.V., Burlakova E.B., Gorbacheva L.B., *Dokl. akad. nauk SSSR,* 1989, vol. 254, no. 6, p. 1472. (in Russian)
[41] Alesenko A.V., Burlakova E.B., *Dokl. akad. nauk SSSR,* 1972, vol. 207, p. 1471. (in Russian)
[42] Alesenko A.V., Burlakova E.B., Vainson A.A., *Dokl. akad. nauk SSS,* 1972, vol. 202, p. 208. (in Russian)
[43] Alessenko A. V., Burlakova E. B.//*Bioetectrochemistry.* - 2002. - V. 58. - P. 13.
[44] Burlakova E.B., Goloshchapov A.N., *Biofizika,* 1975, vol. 10, no. 5, p. 816. (in Russian)
[45] Burlakova E.B., Dzhalyabova M.I., Molochkina E.M., in "Structure, biosynthesis and conversions of lipids in animal and human organisms", 1975, Frunze, Izd. *FAN,* 1975, p. 70. (in Russian)

[46] Aristarkhova S.A., Arkhipova G.V., Burlakova E.B., et al., *Dokl. AN SSSR,* 1976, vol. 228, p. 215. (in Russian)
[47] Burlakova E. B., Dzhalyabova M. I., Molochkina E. M., Khokhlov A. P., *Biophysical and Biochemical Information Transfer in Recognition and Aging,* 1979.- P. 1583.
[48] Burlakova E.B., *Kardiologiya,* 1980, vol. 20, p. 48.
[49] Arkhipova G.V., Burlakova E.B., Semiokhina A.F., Fedotova I.B., Krushinskii L.V., *Dokl. AN SSSR,* 1981, vol. 256, no. 3, p. 746. (in Russian)
[50] Disvetova V.V., Genieva E.I., et al., *Klinich. med.,* 1968, no. 3, p. 126. (in Russian)
[51] Zoz N.N., Lemanova I.B., Akhmedov S.A., Suleimanov D.S., Serebryanyi A.M., Morozova I.S., Sultanova O.D., S.-*khoz. biol.,* 1985, no. 4, p. 71. (in Russian)
[52] Lipsits D.D., Kruglyakova K.E., Postnikova M.S., *Dokl. AN SSSR,* 1962, vol. 145, p. 212. (in Russian)
[53] Sadykov A.S., Kruglyakova K.E., et al., Chemistry of phyto substances, Tashkent, *FAN,* 1968, vol. 3, p. 86. (in Russian)
[54] Burlakova E.B., Laricheva E.P., *VNIIZh,* 1973, p. 15. (in Russian)
[55] L.D. Smirnov, Yu.V. Kuznetsov, L.M. Apasheva, K.D. Poltorak, A.L. Grinchenko, K.M. Dyumaev, N.M. Emanuel, Author's certificate 1098934, Feb. 23, 1983. (in Russian)
[56] Krashakov S.A., Burlakova E.B., Khrapova N.G., *Biol. membr.,* 1998, vol. 12, no. 2, p.173. (in Russian)
[57] Burlakova E.B., Krashakov S.A., Khrapova N.G., Kinetic characteristics of tocopherols as antioxidants, M., *Nauka,* 1988, 247 pp. (in Russian)
[58] Burlakova E.B., *Uspekhi chimii,* 1975, vol. 44, no. 10, p. 1871. (in Russian)
[59] Burlakova E.B., Khrapova N.G., *Uspekhi khimii,* 1985, vol. 54, p. 1540. (in Russian)
[60] Pobedimskii D. G., Burlakova E.B. in Mechanism of Antioxidant Action in Living Organisms, in Atmospheric Oxidation and Antioxidants, Ed.J. Scott, v. 3, no.9, p. 223, 1993.
[61] Burlakova E.B.,Khrapova N.G., Shtolko V.N., Emanuel N.M., *Dokl. AN SSSR*, vol. 169, no. 3, p. 688. (in Russian)
[62] Shapiro A.B., Kropacheva A.A., Suskina V.I., et al., *Mass-spectroscopic study of paramagnetic derivatives of ethylenephosphoramides,* 1971, vol. 4(1), p. 864 (in Russian)
[63] Konovalova N.P., Nitroxyl radicals as modifiers of biological effects of antitumor compounds, *Khimfizika,* 1991, vol. 10(6), pp. 861-868 (in Russian)
[64] Emanuel N.M., Konovalova N.P., Povarov L.S., et al., Author's certificate 977462 SSSR, //BI, 1980.
[65] Emanuel N.M., Konovalova N.P., Diatcbkovkaja RF. *Potential anticancer agents--nitroxyl derivatives of Rubomycin.*Neoplasma 1985,v.32, p.285-92.
[66] Emanuel N.M., Zhdanov R.I., Konovalova N.P., et al., Paramagnetic diethyleneimides of urethan phosphoric acids as antitumor substances, *Voprosy oncol.,* 1980, vol. 36, pp. 54-58. (in Russian)
[67] Seminara P., Franchi F., Konovalova N. et al. Activity of a nitroxylated analog of daunirubomocin, ruboxyl in B- lymphoproliferative disorders. *Acta Haematol.* 2001, 105: 77-82.
[68] Emanuel N.M., Rozenberg A.V., Golubev V.A., et al., *Urgent problems of experimental chemotherapy of tumors,* 1987, vol. 5. (in Russian)

[69] Sen' V.D., Rukina N.A., Tkachev V.V., et al., *Synthesis, structure and biological activity of platinum (II) complexes,* 2000, vol. 9, pp. 1624-1630. (in Russian)

[70] Sen' V.D., Tkachev V.V., Volkova L.M., et al., Synthesis, structure and antitumor properties of platinum (IV) complexes with aminonitroxide, *Izv AN, Ser. khim.,* 2003, vol. 2, pp. 403-408. (in Russian)

[71] Sen' V.D., Golubev, V.A., Rukina, N.A., Orlova, T.G., Burlakova, E.B. *Nitroxyl derivatives of 3'-azido-2'3'-dideoxythymidine with antiviral activity.* Patent RU 2103274 C1 1998.

[72] Behl C. Alzheimer Disease and oxidative stress: implications for novel therapeutic approaches. *J. Biol. Chem.* 1996, 271: 26482-26489.

[73] Nitsh R., Blusztajns J. K. Pittas A. G. et al. Evidence for a membrane defect in Alzheimer disease brain. *Proc. Natl. Acad. Sci. USA.* 1992, 89: 1671-1675.

[74] Zafrilla P., Mulero J., Xandri J.M. et al. Oxidative stress in Alzheimer patients in different stages of the disease. *Curr. Med. Chem.* 2006,13:1075-1083.

[75] Frank B., Gupta S. A review of antioxidants and Alzheimer's disease. *Ann. Clin. Psychiatry.* 2005, 17: 269-286.

[76] Moreira P.I., Honda K., Liu Q. et al. Oxidative stress: the old enemy in Alzheimer's disease pathophysiology. *Curr. Alzheimer Res.* 2005, 2: 403-408.

[77] Mariani E., Polidori M.C., Cherubini A. et al. Oxidative stress in brain aging, neurodegenerative and vascular diseases: an overview. *J. Chromatogr. B Analyt. Technol. Biomed. Life Sci.* 2005, 827: 65-75.

[78] Moreira P.I., Smith M.A., Zhu X. et al. Oxidative damage and Alzheimer's disease: are antioxidant therapies useful? *Drug News Perspect.* 2005, 18:13-19.

[79] Irizarry M.C. Biomarkers of Alzheimer disease in plasma. *NeuroRx.* 2004,1:226-234.

[80] Perevozkina M.G., Storozhok N.M., Nikiforov G.A., Correlation between chemical structure and inhibiting properties of sterically-hindered phenols of the ichfan group, *Biomed. khim.,* 2005, vol. 51, pp. 413-423. (in Russian)

[81] Braginskaya F.I., Zorina O.M., Molochkina E.M., et al., Synthetic bioantioxidants as inhibitors of the AChE activity, Izv. AN SSSR, 1992, vol. 5, pp. 690-698. (in Russian)

[82] Braginskaya F. I., Molochkina E. M., Zorina O. M. et al. *New synthetic bioantioxidants - acetylcholinesterase (AChE) inhibitors.* in: Alzheimer Disease: From Molecular Biology to Therapy. R. Becker & E. Giacobini (eds.). Birkhauser-Boston. 1996, 337-342.

[83] Ozerova I.B., Molochkina E.M., Burlakova E.B. Ichfan – new potential drug for treatment of Alzheimer's disease. *Advances in Gerontology*. 2001, 6: 30.

[84] Molotchkina E. M., Ozerova I. B., Burlakova E. B. «ICHFAN» - new antioxidant drug for the treatment of Alzheimer's disease. *Free Rad. Biol. Med.* 2002, 33: 229- 230.

[85] Buchet R. and Piku S. Alzheimer's disease: Its origin at the membrane, evidence and questions. *Acta Biochimica Polonica.* 2000, 47 : 725–733.

[86] Prasad M.R., Lovell M.A., Yatin M., Dhillon,H. & Markesbery, W.R. Regional membrane phospholipid alterations in Alzheimer's disease. *Neurochem. Res.* 1998, 23: 81–88.

[87] Wells K., Farookui A.A., Liss L. & Horrocks L.A. Neural membrane phospholipids in Alzheimer disease. *Neurochem. Res.* 1995, 20: 1329–1333.

[88] Soderberg M., Edlund C., Kristensson K. & Dallner G. Fatty acid composition of brain phospholipids in aging and in Alzheimer's disease. *Lipids.* 1991 26:421-425.

[89] Pettegrew J.W., Panchalingam K., Hamilton R.L. et al. Brain membrane phospholipid alterations in Alzheimer's disease. *Neurochem. Res.* 2001, 7:771-782.

[90] Soderberg M., Edlund C., Alafuzoff I. et al. Lipid composition in different regions of the brain in Alzheimer's disease/senile dementia of Alzheimer's type. *J. Neurochem.* 1992, 5:1646-1653.

[91] Nitsch R.M., Blusztajn J.K, Pittas A.G. et al. Evidence for a membrane defect in Alzheimer disease brain. *Proc. Natl. Acad. Sci. U S A.* 1992, 89: 1671-1675.

[92] Wells, K., Farookui, A.A., Liss, L. & Horrocks, L.A. Neural membrane phospholipids in Alzheimer disease. *Neurochem. Res.* 1995, 20: 1329–1333.

[93] Markesbery W.R., Leung P.K., Butterfield D.A. Spin label and biochemical studies of erythrocyte membranes in Alzheimer's disease. *J. Neurol. Sci.* 1980, 45:323-330.

[94] Zubenko G.S., Kopp U., Seto T et al. Platelet membrane fluidity individuals at risk for Alzheimer's disease: a comparison of results from fluorescence spectroscopy and electron spin resonance spectroscopy. *Psychopharmacology.* (Berl) 1999,145:175-180.

[95] Kozubski W., Swiderek M., Kloszewska I. et al. [Platelet membrane fluidity and receptor exposition in patients with Alzheimer's disease]. *Neurol. Neurochir. Pol.* 1999, 33:1275-1284. (in Polish)

[96] Muller W.E., Kirsch C., Eckert G.P. Membrane-disordering effects of beta-amyloid peptides. *Biochem. Soc. Trans.* 2001, 29:617-623.

[97] Braginskaya F.I., Zorina O.M., Pal'mina N.P. et al. Some blood biochemistry parameters during the cholinergic treatment of Alzheimer's disease. *Neurosci. Behav. Physiol.* 2001 31: 457-461.

[98] Simons M., Keller P., De Strooper B et al. Cholesterol depletion inhibits the generation of beta- amyloid in hippocampal neurons. *Proc. Natl. Acad. Sci. U.S.A.* 1998, 95: 6460–6464.

[99] Eckert G.P., Kirsch C., Muller W.E. Brain-membrane cholesterol in Alzheimer's disease. *J. Nutr. Health Aging.* 2003,7:18-23.

[100] Raffaï R. L. and Weisgraber K. H. Cholesterol: from heart attacks to Alzheimer's disease. *J. Lipid Res.* 2003, 44: 1423–1430.

In: Progress in Chemistry and Biochemistry
Editors: E. M. Pearce and G. E. Zaikov

ISBN: 978-1-60692-344-3

Chapter 7

MECHANISM OF ACTION AND EFFECTIVENESS OF ESTER THIOLS AS THERMAL STABILIZERS FOR POLY(VINYL CHLORIDE)

Xianlong Ge,[1] Elizabeth K. Culyba,[2] Cole L. Grinnell,[2] Alexander G. Zestos[2] and William H. Starnes, Jr.*[1,2]

Department of Applied Science[1] and Department of Chemistry,[2]
College of William and Mary, P. O. Box 8795,
Williamsburg, Virginia 23187-8795

ABSTRACT

Organic thiols containing at least one carboxylate ester group (ester thiols) are excellent thermal stabilizers for both rigid and plasticized poly(vinyl chloride) (PVC). Their mechanism of action is shown to involve the deactivation of unstable structural defects by nucleophilic chloride displacement, the retardation and removal of coloration through thiol additions to polyene double bonds, and the prevention of autoacceleration during thermal dehydrochlorination through polyene shortening reactions and the scavenging of free radicals formed from polyenes and HCl. An unusually facile displacement of labile chloride that is favored by thiol acidity can account, at least in part, for the relatively high effectiveness of dipentaerythritol hexakis(mercaptoacetate) as a stabilizer.

INTRODUCTION

Previous work by this research group has shown that organic thiols containing one or more carboxylate ester functions (ester thiols) are remarkably effective as thermal stabilizers for poly(vinyl chloride) (PVC). Some prototypical examples of these thiols are compounds 1–9, where R = 2-ethylhexyl and R' = 3-heptyl [1–10].

* Correspondence to: William H. Starnes, Jr.; e-mail: whstar@wm.edu

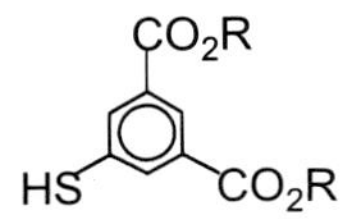
1

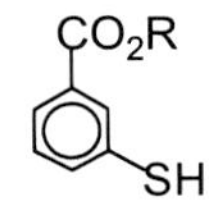
2

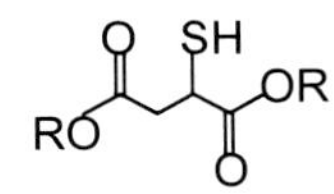
3

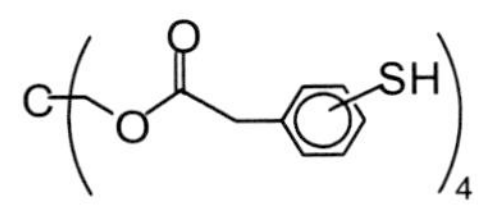
4

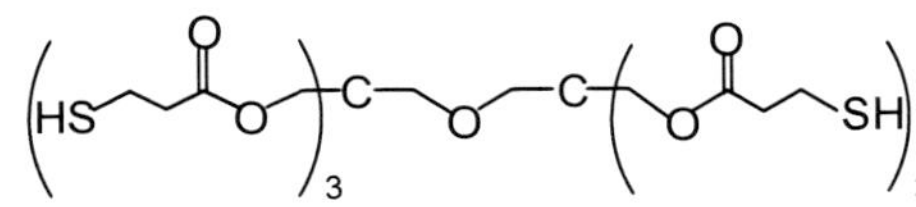
5

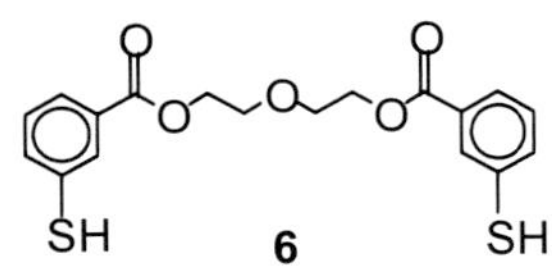
6

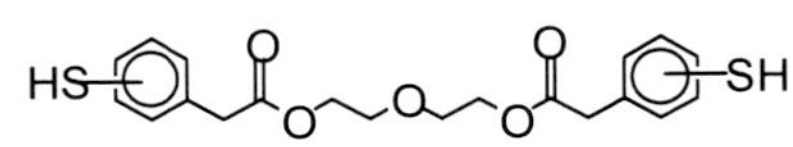
7

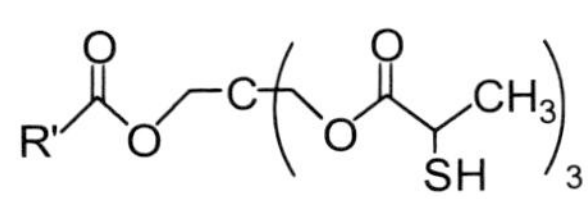
8

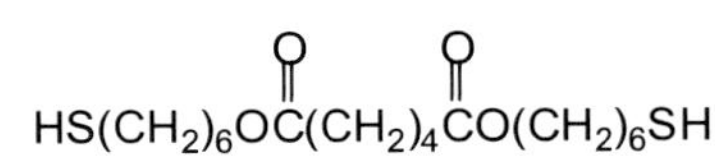
9

Major advantages of these additives are the following [1–10]:

1. When they are introduced at typical stabilizer loading levels, the ester thiols confer high static and dynamic stability to both rigid and plasticized PVC.
2. When they are used at much higher loading levels (e. g., 30–40 parts by weight), some of the ester thiols are excellent primary plasticizers as well.
3. The ester thiols do not require the presence of costabilizers (including those containing metals) in order to be highly effective. Even so, in certain cases, their use in conjunction with well-known HCl scavengers [for example, epoxidized soybean oil (ESO)] has given even better results.
4. Because of their low volatilities resulting from their high molecular weights, the ester thiols do not have unpleasant odors when they are pure.
5. With few exceptions, the ester thiols can be synthesized quickly and easily in high yields (frequently in a single step) from starting materials that are available commercially, sometimes in bulk quantities.
6. A part of the ester thiol stabilization technology was licensed for commercialization in 2005.

A detailed study of the mechanism of action of the ester thiols has not been reported heretofore. We now describe the results of experiments that have a bearing on this question. They will be discussed within the context of the mechanism for the thermal dehydrochlorination of PVC, which is generally agreed to involve the following steps [11, 12]: (*a*) rapid losses of HCl from thermally labile structural defects that contain allylic or tertiary chloride and are not at the ends of the polymer chains and (*b*) subsequent growth of conjugated polyene chromophores through very fast sequential losses of HCl from adjacent monomer units. Both step *a* (polyene initiation) and step *b* (polyene propagation) have been shown to occur by either an ion-pair pathway or a concerted four-center mechanism with a highly polar transition state [11, 12].

RESULTS AND DISCUSSION

Mechanistic Polymer Studies

When blended with PVC, some ester thiols cause moderate decreases in the initial rate of thermal dehydrochlorination, while others cause moderate rate enhancements [1]. These divergent results support the removal of labile chloride by nucleophilic displacement (Eq. 1). The displacement process

$$\mathrm{RSH} + \mathrm{R'Cl} \longrightarrow \mathrm{RSR'} + \mathrm{HCl} \qquad (1)$$

itself produces HCl, but this effect is opposed kinetically by the ability of the displacement to prevent the subsequent stepwise HCl losses that would form a polyene sequence.

All of the ester thiols retard the development of color upon heating. Their ability to do so also is consistent with the prevention of polyene production via reaction 1, where R'Cl is either a growing polyenyl chloride [i. e., $-(CH{=}CH)_nCHCl-$] or an unstable structural defect in the undegraded polymer.

Thiols, in general, are well-known to undergo additions to alkene double bonds by a free-radical mechanism or in an ionic process that is promoted by strong acids such as HCl. During PVC thermolysis, such additions of ester thiols would tend to inhibit coloration by shortening polyene lengths. They also would tend to inhibit polyene growth if they occurred, for example, as in Eq. 2, which removes an unstable allylic chloride function at the end of a polyene sequence.

$$\mathrm{RSH} + -(\mathrm{CH{=}CH})_n\underset{\displaystyle\mathrm{Cl}}{\overset{|}{\mathrm{CH}}}- \longrightarrow -\overset{\displaystyle\mathrm{SR}}{\underset{}{\overset{|}{\mathrm{C}}}}\mathrm{H}(\mathrm{CH{=}CH})_{n-1}\mathrm{CH_2}\overset{\displaystyle\mathrm{Cl}}{\overset{|}{\mathrm{C}}}\mathrm{H}- \qquad (2)$$

A twofold amount by weight of ester thiol 2 was combined with a reddish-brown sample of PVC that had been subjected to thermal degradation for 4 h at 180°C under argon. After an additional 6 h of heating under the same conditions, the color of the sample actually became much lighter rather than darker! This observation strongly supports the addition of 2 to polyene double bonds formed during the initial heating period.

In a related kinetics experiment, summarized in figure 1, PVC (0.1 g) was degraded under argon at 180°C for 2.3 h and then treated with an excessive amount (0.4 g) of 2 in an attempt to maximize contact of the ester thiol with the now heavily crosslinked and intractable polymer. As figure 1 reveals, this addition immediately caused a great increase in the rate of evolution of HCl. Yet, after another hour of heating, the rate decreased to about half of its value prior to thiol inclusion. The initial rate increase produced by 2 can be attributed to the formation of HCl in reaction 1, while the subsequent rate deceleration is consistent with the ability of that reaction to deactivate labile structures, such as –

$(CH{=}CH)_nCHCl$–, that would lead to continued polyene growth. (Their concurrent deactivation via reaction 2 may have occurred as well.)

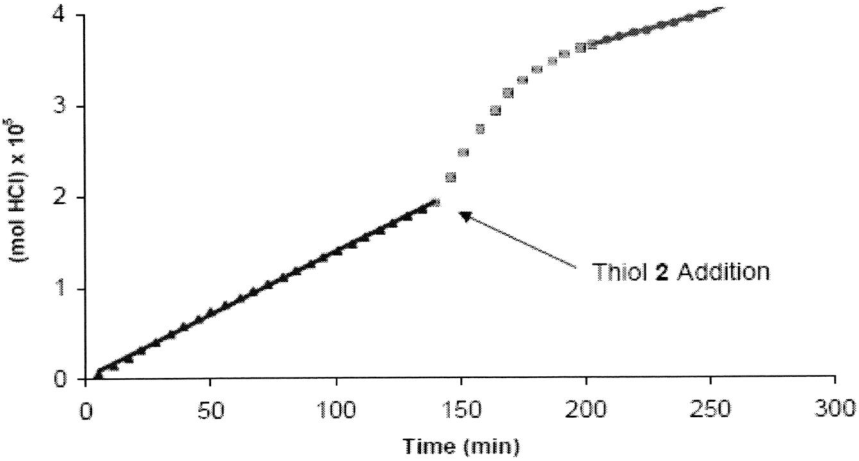

Figure 1. Effect of the addition of a fourfold excess (by weight) of ester thiol **2** to PVC experiencing degradation at 180°C under argon. [Color figure can be viewed in the online issue, which is available at www.interscience.wiley.com].

A 1:1 (w/w) blend of PVC and ester thiol 2 was heated under argon at 190°C for 4 h, and the polymer was recovered and purified by dissolution in tetrahydrofuran, precipitation into methanol, and Soxhlet extraction with methanol for 24 h. Its dehydrochlorination rate at 180°C under argon was then found to be less than that of the starting material by about a factor of 2. This result confirms the destruction, by the thiol, of labile sites that were present originally. However, it also suggests that the thiol had retarded the formation of new unstable defects during the first heating period. Further evidence for the latter effect was provided, in fact, by all of our kinetic runs in which PVC was being heated with an ester thiol. In those experiments, the HCl evolution rate always exhibited autodeceleration, rather than the autoacceleration that occurred when no thiol was present.

Autoacceleration of dehydrochlorination is now known to involve free-radical intermediates [13]. Figure 2 is instructive in this regard. It shows the temporal profiles of some PVC dehydrochlorinations that were carried out under flowing argon and in the presence or absence of BHT (a very effective free-radical scavenger) and β-carotene (a model for PVC polyene sequences). In these experiments, the HCl

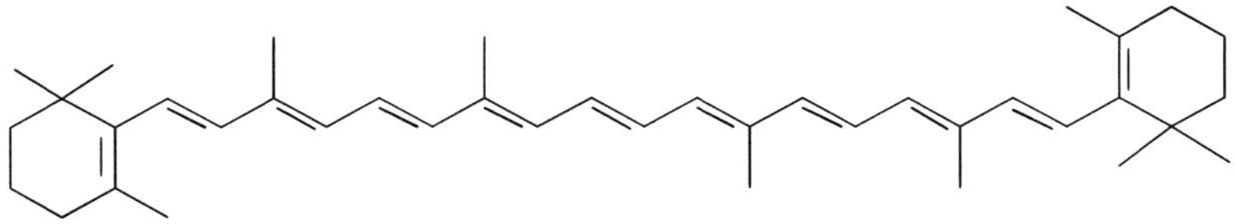

β-carotene

concentration in and around the degrading polymer was varied from low to high by using argon flow rates of 80 and 10 mL/min, respectively.

A

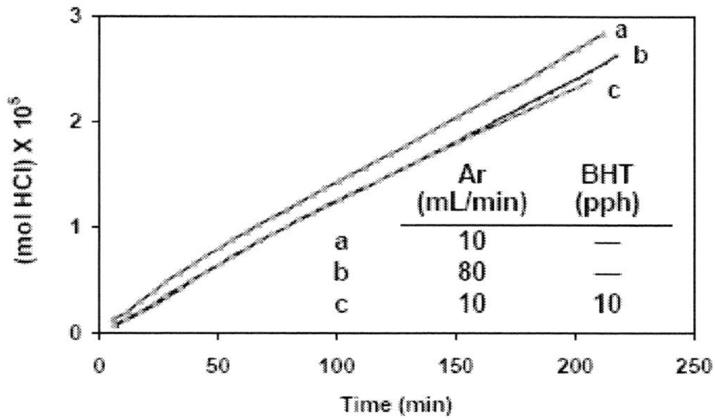

B

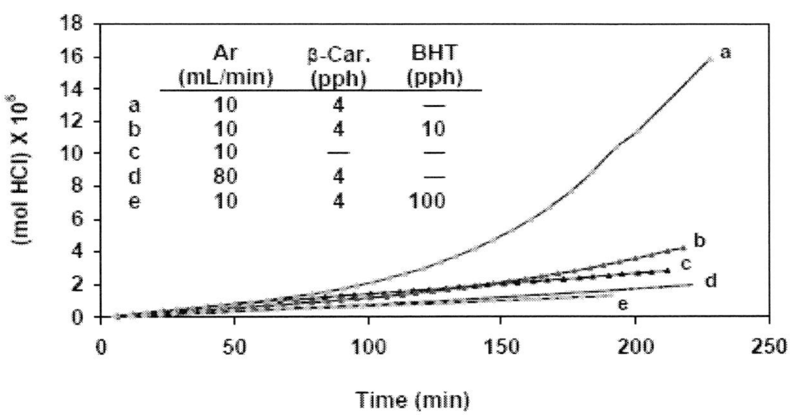

Figure 2. Typical rate curves for the thermal dehydrochlorination of PVC at 180°C under argon, where "BHT" is 2,6-di-*t*-butyl-*p*-cresol; "β-Car." is β-carotene; and "pph" is parts by weight per hundred parts of polymer. Curve a in part A and curve c in part B are for the same run. [Color figure can be viewed in the online issue, which is available at www.interscience.wiley.com].

Part A of figure 2 shows that in the early stages of degradation, where polyene formation is not extensive, the rate is not affected to a large extent by a change in the HCl concentration or by the presence of BHT. On the other hand, part B of the figure reveals the occurrence of a dramatic autoacceleration when the HCl concentration is high and β-carotene is present (cf. curves a and c). β-Carotene does not cause the autoacceleration, however, when the HCl concentration is low (cf. curve d of part B). Moreover, curves b and e of part B show that the autoacceleration is inhibited very strongly by BHT. Similar results were obtained when triphenylmethane and metallic mercury (a highly specific chlorine-atom scavenger) were used as radical traps [13].

As argued elsewhere in detail [13], the foregoing observations support a mechanism for autoacceleration that involves the reaction of HCl with polyenes to generate free radicals. These radicals, or radicals formed from them, then abstract methylene hydrogen from the ordinary monomer units of PVC, as shown in Eq. 3 (wherein the nature of $X^{\bullet}$ is not yet specified), and the resultant carbon-centered radicals experience β-scission in order to give free chlorine atoms and allyl chloride defect structures (Eq. 4). These structures initiate the growth of new polyene sequences in the usual (non-radical) way, and the chlorine atoms may function, in turn, as $X^{\bullet}$ in Eq. 3.

$$X^{\bullet} + \text{–CHClCH}_2\text{CHCl–} \longrightarrow \text{HX} + \text{–CHCl}\dot{\text{C}}\text{HCHCl–} \tag{3}$$

$$\text{–CHCl}\dot{\text{C}}\text{HCHCl–} \longrightarrow \text{Cl}^{\bullet} + \text{–CH=CHCHCl–} \tag{4}$$

What, specifically, are the first-formed $X^{\bullet}$ intermediates? Their identity is not entirely clear. Nevertheless, some reasonable possibilities are polyenyl cation monoradicals, polyenyl cation diradicals, and chlorine atoms formed by the β-scission of one or both of those species [13]. The most likely possibility, at present, is the latter one, because metallic mercury completely prevents the autoacceleration [13]. In any event, it now appears that the inhibition of autoacceleration by the ester thiols is likely to result, at least in part, from free-radical scavenging. Thiols, in general, are well-known to react very readily with a large variety of free radicals via Eq. 5, and the resultant $RS^{\bullet}$ species are

$$Y^{\bullet} + \text{RSH} \longrightarrow \text{HY} + \text{RS}^{\bullet} \tag{5}$$

known to be quite unreactive in hydrogen-atom abstractions such as reaction 3. Such radicals are likely to terminate quickly by coupling instead.

The presence of ester groups causes the ester thiols to be compatible with PVC, a property that is absolutely indispensable for effective stabilization. When a sample of PVC was heated with the essentially incompatible additive, 1-dodecanethiol (0.044 mol per mol of monomer units, 1.5 h under argon at 170°C), the thiol gave no color stabilization whatsoever [1].

Model-Compound Studies

Table 1 shows the yields of the sulfides (RSR') obtained from reactions of three thiols with some model compounds. Compound 10 is a rather crude model for thermally labile tertiary chloride groups in PVC. Compound 12 is a model for another unstable defect, while compound 13 is the product of its dehydrochlorination. Compound 11 models a long-chain end that also occurs in the polymer but does not contribute significantly to its thermal instability [11, 12, 14]. In nucleophilic chloride displacements, the mechanisms for 10, 11, and 12 are expected to be predominantly SN1, SN2, and (probably) a mixture of SN1 and SN2, respectively [15]. Under the conditions used for the tabulated experiments performed at 160°C, ester thiol 2 did not react with *trans*-7-tetradecene or 6-chloroundecane, a rudimentary model for the ordinary monomer units of the polymer.

Table 1. Sulfide percentage yields from reactions of thiols with model compounds (1:1 mole ratio, 4 h under N_2)

	ArSH[a]		$Me(CO_2R)CHSH$[b]		RSH[c]	
	Temperature (°C)					
Model	160	80	160	80	160	80
$MeBu_2CCl$ (**10**)	19	1	20	1	9	0
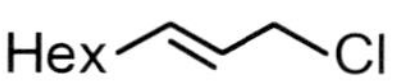 (**11**)	100	4	24	1	14	<1
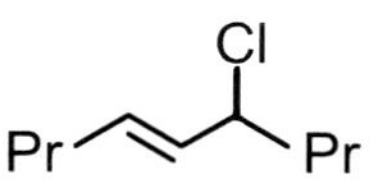 (**12**)	88	33	6	<1	9	0
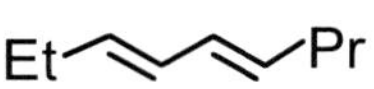 (**13**)	51	14	5	<1	8	<1

[a] 2-Ethylhexyl 3-mercaptobenzoate (2). [b] 2-Ethylhexyl 2-mercaptopropionate.
[c] 1-Dodecanethiol.

The most striking results in table 1 are the high yields of sulfides obtained from the reactions of thiol 2 with chlorides 11 and 12. In contrast, the yields derived from the three thiols and chloride 10 are lower and differ from one another by no more than a factor of 2. From inductive, steric, and resonance effects, the expected order of thiol nucleophilicities is RSH > $Me(CO_2R)CHSH$ > ArSH, and the nucleophilicities of the corresponding thiolate anions should vary in the same way. Surprisingly, however, none of the chlorides gave sulfide yields that conform to this trend. On the other hand, the expected order of thiolate anion concentrations is ArS^- > $Me(CO_2R)CHS^-$ > RS^-, judging from the following pK_a values, which were determined in water at 25°C: thiophenol, 6.50; EtO_2CCH_2SH, 7.95; $CH_3CH_2CH_2SH$, 10.86 [16]. At 160°C, the sulfide yields from 11 fall in this same order, as do the yields from 12, within the likely limits of error. These observations strongly suggest that in the reactions of these two substrates, the effective nucleophiles were, in fact, the thiolate anions, rather than the thiols themselves. Electrophilic catalysis, by the thiols, of C-Cl heterolysis would lead to the same yield order but is currently thought to be of lesser importance, because the thiol acidities (and thus their probable catalytic activities) are so low. On the other hand, comparison with the yields obtained from the thiols and diene 13 at both reaction temperatures shows that some of the sulfides ensuing from 12 actually could have resulted from additions of the ester thiols to 13 formed *in situ*. These additions might have been promoted by trace amounts of acid or O_2 (which reacts with thiols to generate thiyl radicals), but at present, their mechanism is unclear.

More information obviously would be needed in order to achieve full rationalization of the yield data in table 1. Nevertheless, the information now in hand shows that thiols can destroy both tertiary chloride and allylic chloride structures by nucleophilic displacement, and that at least in the latter instance, thiolate anion concentration (as determined by thiol acidity) may be more important than thiol or thiolate anion nucleophilicity. The results obtained with diene 13 confirm the addition of thiols to conjugated double bonds and show that the reactivity of the aromatic ester thiol, 2, was superior to those of the other thiols in this respect, as well.

Stabilizing Effectiveness

The foregoing observations imply that in the absence of major complications, such as large differences in the compatibility of the ester thiols with PVC, the effectiveness of these compounds as thermal stabilizers should increase with increasing acidity of the SH group(s). Some results we have obtained recently with thiols 5 and 14 [17] relate to this hypothesis. Blends of PVC powder (100 parts by weight) with 3 parts of each of these thiols, either with or without 5 parts of ESO, were heated at 180°C under air while the color changes caused by degradation were observed as a function of time. Visual inspection showed that under these conditions, compound 14 was significantly better than compound 5 as a color stabilizer, both in the presence and the absence of ESO. This conclusion was supported by quantitative color measurements made according to ASTM D2244. They revealed, for example, that after 1 h of heating with ESO absent, the sums of the absolute values of *dL* and *dE* were 173.2, 103.4, and 80.0 for pure PVC, PVC with 5, and PVC with 14, respectively.

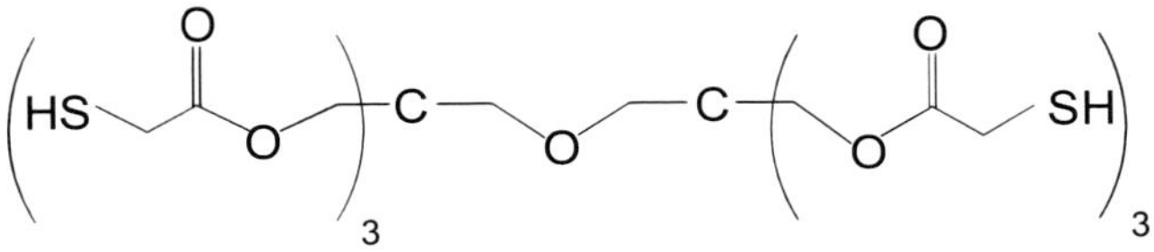

14

The thiol groups of 14 should be significantly more acidic than those of 5, owing to their closer proximity to the electron-withdrawing ester groups. Thus our observed color changes correlate with thiol acidity in the expected way. It should be noted, however, that owing to the different molecular weights of the two ester thiols, the relative effectiveness of 14 also should be enhanced by the presence of an additional 12% of SH groups when the thiols are compared at the same loading level.

The superiority of 14 to 5 is especially noteworthy because 5 is one of the most effective thiol stabilizers identified in previous studies [1–10]. Another advantage of 14 is its ease of preparation in a single step (an acid-catalyzed esterification) from relatively inexpensive starting materials (dipentaerythritol and mercaptoacetic acid) that are available commercially in bulk quantities [17].

Experimental

The PVC was a powdered commercial product containing no additives. Full details for the synthesis of the ester thiols are reported elsewhere [5–10, 17]. The materials used in the model-compound experiments were either obtained from commercial sources or prepared by standard methods. Reactions of the models with thiols were carried out in sealed tubes that were filled in a drybox under dry nitrogen, and the yields of the resulting sulfides were determined by GC/MS analysis with a Hewlett-Packard HP 5890A/7988A apparatus. The tabulated yields are averages obtained from at least two runs; average deviations generally were no greater than ±5–10% of the values reported. Dehydrochlorination kinetics experiments were performed by a standard procedure that used a Metrohm 702 SM Titrino apparatus for acid-base titrimetry [13].

CONCLUSIONS

The studies reported here show that ester thiols can thermally stabilize PVC in the following ways:

1. *They destroy thermally labile structural defects.* These defects are tertiary chloride, internal allylic chloride, and polyenyl chloride groups. Their destruction involves the nucleophilic displacement of chloride and possibly the addition of the thiols to polyene double bonds (as in reaction 2, for example). The displacement is facilitated by increasing thiol acidity.
2. *They inhibit coloration by shortening polyene sequences.* Even if the addition of thiols to polyenes does not stop their growth, it will tend to make them less colored by decreasing their lengths. Long, intensely colored polyenes may actually be converted into colorless structures by multiple thiol additions.
3. *They scavenge free radicals that lead to the formation of new unstable groups.* The evidence for this scavenging process is indirect, but the presence of the radicals has been established conclusively, and thiols, in general, are well-known to be superlative radical traps.

Of these three mechanisms, the first is the most important, because it tends to prevent the formation of conjugated polyene segments. If such segments are never produced, the degrading polymer will not acquire color, and the polyenes will not be available to react with HCl in order to form free radicals. In other words, if mechanism 1 is sufficiently effective, the other mechanisms are not required.

We hope that the mechanistic information disclosed in this paper will aid in the further development and application of the ester thiol technology for the stabilization of PVC.

ACKNOWLEDGMENTS

This work was supported in part by the National Science Foundation (Grant No. DMR-9610361) and by the Edison Polymer Innovation Corporation. We are indebted to Soungkyoo Kim for the synthesis of ester thiol 2 and to Wai-Kwong Ho of the Sherwin-Williams Company for quantitative color measurements on degraded PVC specimens.

An abbreviated version of this paper received the annual Best Student Paper Award of the SPE Vinyl Plastics Division and was presented by awardee Xianlong Ge at SPE ANTEC 2007, Cincinnati, Ohio, May 7, 2007.

Reprinted with permission from the *Journal of Vinyl and Additive Technology*, 13, 170-175 (2007).

References

[1] W. H. Starnes, Jr., B. Du, and V. G. Zaikov, *J. Vinyl Addit. Technol.*, 7, 250 (2001).
[2] W. H. Starnes, Jr., B. Du, and V. G. Zaikov, *Polym. Prepr. (Am. Chem. Soc., Div. Polym. Chem.)*, 42 (2), 881 (2001).
[3] T. C. Jennings and W. H. Starnes, Jr., in *PVC Handbook*, C. E. Wilkes, J. W. Summers, and C. A. Daniels, eds., Hanser, Munich, Chapter 4 (2005).
[4] W. H. Starnes, Jr., B. Du, S. Kim, V. G. Zaikov, X. Ge, and E. K. Culyba, *Thermochim. Acta*, 442, 78 (2006).
[5] W. H. Starnes, Jr., and B. Du, U. S. Patent 6,667,357 (2003).
[6] W. H. Starnes, Jr., U. S. Patent 6,747,081 (2004).
[7] W. H. Starnes, Jr., and B. Du, U. S. Patent 6,762,231 (2004).
[8] W. H. Starnes, Jr., and S. Kim, U. S. Patent 6,927,247 (2005).
[9] W. H. Starnes, Jr., and B. Du, U. S. Patent 7,250,457 (2007).
[10] W. H. Starnes, Jr., B. Du, and S. Kim, U. S. and foreign patents pending.
[11] W. H. Starnes, Jr., *Prog. Polym. Sci.*, 27, 2133 (2002) (see also references therein).
[12] W. H. Starnes, Jr., *J. Polym. Sci., Part A: Polym. Chem.*, 43, 2451 (2005) (see also references therein).
[13] W. H. Starnes, Jr., and X. Ge, *Macromolecules*, 37, 352 (2004).
[14] W. H. Starnes, Jr., and S. Girois, *Polym. Yearb.*, 12, 105 (1995) (see also references therein).
[15] J. March, *Advanced Organic Chemistry*, 4th ed., McGraw-Hill, New York, Chapter 10 (1992).
[16] J. A. Dean, *Lange's Handbook of Chemistry*, 14th ed., McGraw-Hill, New York, Section 8 (1992).
[17] W. H. Starnes, Jr., U. S. patent pending.

In: Progress in Chemistry and Biochemistry
Editors: E. M. Pearce and G. E. Zaikov
ISBN: 978-1-60692-344-3

Chapter 8

KINETICS AND MOLECULAR MODELLING OF 1-HYDROXYCYCLOHEXYL HYDROPEROXIDE DECOMPOSITION ACTIVATED BY ALK$_4$NBR

N.A. Turovskij[1], E.V. Raksha[1], O.I. Gevus[2], I.A. Opeida[3] and G.E. Zaikov[4]

[1] Donetsk National University, Donetsk, Universitetskaya str. 24, 83055, Ukraine E-mail: turov@dongu.donetsk.ua; Ukrainian-American Laboratory of Computational Chemistry, Kharkov, Ukraine, Jackson, USA
[2] National University «Lvivska Politechnika», S. Bandera 12, 79013 Lviv, Ukraine
[3] Institute of Physico-Organic and Coal Chemistry of the National Academy of Sciences of Ukraine, R. Luxemburg 70, 83114 Donetsk, Ukraine
[4] Institute of Biochemical Physics, Russian Academy of Sciences, 4 Kosygina, 117334 Moscow, Russian Federation, E-mail: chembio@sky.chph.ras.ru

ABSTRACT

Kinetics of the chemically activated 1-hydroxycyclohexyl hydroperoxide decomposition in the presence of quaternary ammonium salts (Et_4NBr, Pr_4NBr, Bu_4NBr, and Hex_4NBr) has been investigated. Kinetic parameters of the hydroperoxide – catalyst complex decomposition has been determined. The correlation between reactivity and structural characteristics of ammonium cations was found. Scheme for the chemically activated hydroperoxide decomposition in the presence of ammonium salts is proposed.

Keywords: hydroperoxides, ammonium salts, chemical activation, complexation, molecular modeling.

INTRODUCTION

Organic peroxides are widely used as initiators as well as components of binary initiating systems in the processes of vinyl monomers polymerization, polymers modification [1-3], oxidation of organic substances by molecular oxygen [4, 5]. Initiating systems on the base of quaternary ammonium bromides were found to be the most effective in the case of the liquid phase oxidation of isopropyl benzene by O_2 [5]. 1-hydroxycyclohexyl hydroperoxide - Et_4NBr system initiates the reaction of liquid phase isopropyl benzene oxidation at 308-340 K [6] while the hydrocarbon oxidation in the presence of only one component of the binary system did not occur. The present work presents the study of 1-hydroxycyclohexyl hydroperoxide decomposition activated by the tetraalkylammonium bromides to investigate the role of ammonium salt cation in the process of chemical activation of peroxide bond.

EXPERIMENTAL

1-hydroxycyclohexyl hydroperoxide has been prepared from cyclohexanone and H_2O_2 in anhydrous ether according with [7]. Tetraalkylammonium bromides (Et_4NBr, Pr_4NBr, and Bu_4NBr) were recrystallized from the saturated acetonitrile solution by addition of diethyl ether excess. The solvent (acetonitrile) purity was controlled by electrical conductivity value, which was within $(8.5 \pm 0.2)\cdot 10^{-6}$ Om^{-1} sm^{-1} at 303 K. Reactions were carried out in the glass-soldered ampoules in argon atmosphere. Hydroperoxide kinetic concentration was controlled by the iodometric titration with potentiometric fixation of the equivalent point.

Quantum chemical calculations of hydroperoxides molecules and corresponding radicals were carried out by AM1 semiempirical method implemented in MOPAC package [8]. The RHF method was applied to the calculation of the wave function. Optimization of structure parameters of hydroperoxide and hydroperoxide complexes was carried out by Eigenvector Following procedure. The molecular geometry parameters were calculated with boundary gradient norm 0.01. Solvent effect was considered in COSMO approximation [9].

RESULTS AND DISCUSSION

Kinetic Features of the Activated Hydroperoxide Decomposition

Reaction of 1-hydroxycyclohexyl hydroperoxide (ROOH) decomposition activated by the tetraalkylammonium bromides (Alk_4NBr) has been investigated at 323 – 353 K under conditions of ammonium salts abundance ($[ROOH]_0 = 5\cdot 10^{-3}$ mol·dm^{-3}, $[Alk_4NBr]_0 = 2.5\cdot 10^{-2} - 1\cdot 10^{-1}$ mol·dm^{-3}). Kinetics of ROOH activated decomposition on these conditions could be described by the first order reaction proceedings low. The reaction was carried out up to 50 % hydroperoxide conversion and the products did not effect on the reaction proceeding as kinetic curves anamorphouses are linear in the corresponding first order coordinates. The effective rate constant (k_{ef}, sec^{-1}) was found to be independent from the hydroperoxide initial concentration within $[ROOH]_0 = 2.5\cdot 10^{-3} - 1\cdot 10^{-2}$ mol·dm^{-3} and constant amount of Alk_4NBr ($5\cdot 10^{-2}$ mol·dm^{-3}). Nonlinear character of the dependence of reaction rate effective constants from the ammonium salt initial concentration (figure 1) in the present conditions points out onto the occurrence of complexation stage between the ROOH and Alk_4NBr. These facts are in an agreement with reaction scheme for the isopropyl benzene hydroperoxide activated decomposition recently proposed [10]. Thermolysis contribution into the total reaction rate the ROOH activated decomposition was found to be negligibly small because the thermal decomposition rate constant [6] was two orders lower then corresponding k_{ef} values.

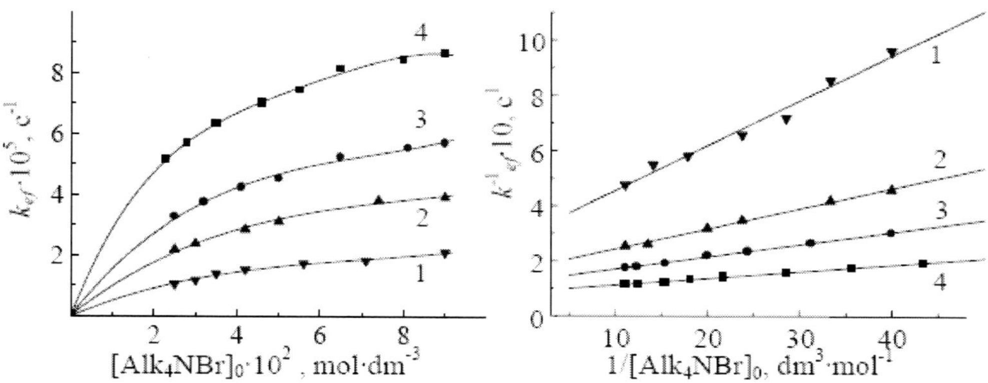

Figure 1. Dependence of k_{ef} from the catalyst initial concentration in the direct (a) and inverse (b) coordinates. $[ROOH]_0 = 5.0\cdot10^{-3}$ mol·dm^{-3}, 333 K, 1 – Hex_4N^+, 2 – Bu_4N^+, 3 – Pr_4N^+, 4 – Et_4N^+.

Some experiments were carried out under conditions of ROOH excess as compared with salts concentration ($[Alk_4NBr]_0 = 5\cdot10^{-3}$ mol·dm^{-3}, $[ROOH]_0 = 5\cdot10^{-2}$ mol·dm^{-3}) at 323 K. Even at present concentration ratio of ROOH and Alk_4NBr the rate of hydroperoxide activated decomposition is several orders higher then ROOH thermolysis rate. The Br^- concentration during reaction proceeding is unaltered. These facts point out onto catalytic character of Alk_4NBr action in the process. Catalytic scheme of the hydroperoxide decomposition is supported also in works [11, 12] in which it was noted that O-O bond cleavage in the hydroperoxide – catalyst complex proceeds homolytically, catalyst is not consumed and deactivated in the system.

The kinetic scheme is proposed for the chemically activated ROOH decomposition. It includes the complexation stage between ROOH and salts ions as well as stage of the complex-bonded hydroperoxide decomposition with catalyst regeneration:

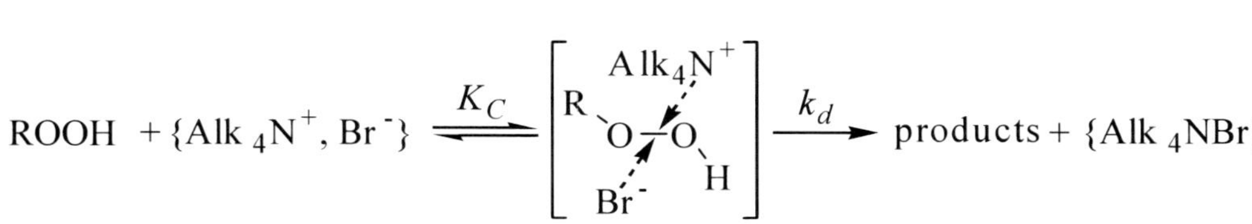

k_{ef} dependence on the salt concentration can be expressed by the relationship (1):

$$\frac{1}{k_{ef}} = \frac{K_C k_d[Alk_4NBr]}{1+K_C[Alk_4NBr]} + \frac{1}{k_d} \qquad (1)$$

The dependence of k_{ef} on Alk_4NBr concentration is linear in double inverse coordinates (figure 1). It is in the agreement with the proposed kinetic scheme. The values of rate constants of complex-bonded peroxide decomposition (k_d) obtained for the investigated salts decrease in the following order: $Et_4N^+ > Pr_4N^+ > Bu_4N^+ > Hex_4N^+$. The values of equilibrium constants of complexation between ROOH and Alk_4NBr ions change similarly (table 1).

Table 1. Kinetic parameters of 1-hydroxycyclohexyl hydroperoxide decomposition in the presence of thetraalkylammonium bromides

T, K	$k_d \cdot 10^4$, sec^{-1}	K_C, dm^3 mol^{-1}	E_a, kJ mol^{-1}	lg$A$, (A, c^{-1})	ΔH_{com}, kJ mol^{-1}	ΔS_{com}, J mol^{-1} K^{-1}
ROOH + Et_4NBr						
333	1.14	36	73 ± 1	7.5 ± 9	-20 ± 1	-30 ± 4
343	2.5	29				
353	5.1	23				
ROOH + Pr_4NBr						
333	0.8	28	81 ± 2	8.6 ± 7	-16 ± 2	-21 ± 5
343	1.80	24				
353	4.20	20				
ROOH + Bu_4NBr						
333	0.59	23	87 ± 3	9.4 ± 7	-12 ± 2	-10 ± 4
343	1.51	20				
353	3.48	18				
ROOH + Hex_4NBr						
333	0.34	18	96 ± 2	10.5 ± 5	-9 ±1	-3 ± 3
343	0.97	16				
353	2.40	15				

Isokinetic relationship between complexation parameters in the system (figure 2) leads to the insignificant changes in free energy of complexation for different alkyl substituent in ammonium cation. Considering the intermolecular bonds energy the strongest complex is formed between hydroperoxide and Et_4NBr, the weakest - in the case of Hex_4NBr (see the corresponding ΔH_{com} values in table 1). The reactivity of complex-bonded ROOH also decreases with increasing of alkyl substituent size. Isokinetic relationship for the obtained activation parameters points out upon the unified mechanism of ROOH - Alk_4NBr complex decomposition (figure 2).

Changes of the peroxide bond activation in the complex in the case of different ammonium cations point out the role of steric factor at the stage of complex formation as well as at the formation of their decomposition transition state. In the simplest case the own volume of the investigated cations could describe the steric effect. A good correlation between activation parameters of the complex decomposition and calculated values of Van-der-Waals volumes of cations has been obtained (figure 3). Van-der-Waals volumes of the tetraalkylammonium cations were calculated in HyperChem package. Calculated volumes are in agreement with corresponding own volumes of cations listed in [13].

Summarizing, obtained experimental facts have shown that the salt cation participates both in complexation stage and in stage of complex ROOH - Alk_4NBr decomposition. Ammonium cation has the regulating action upon the catalytic reactivity of the halide-anion in the reaction of catalytic decomposition of the 1-hydroxycyclohexyl hydroperoxide in the presence of Alk_4NBr.

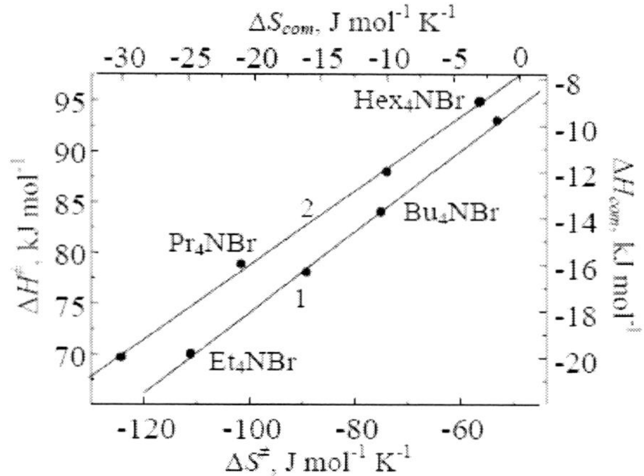

Figure 2. Isokinetic relationships between activation parameters of complex decomposition (1) and parameters of complex formation (2) in the reaction of activated decomposition of 1-hydroxycyclohexyl hydroperoxide.

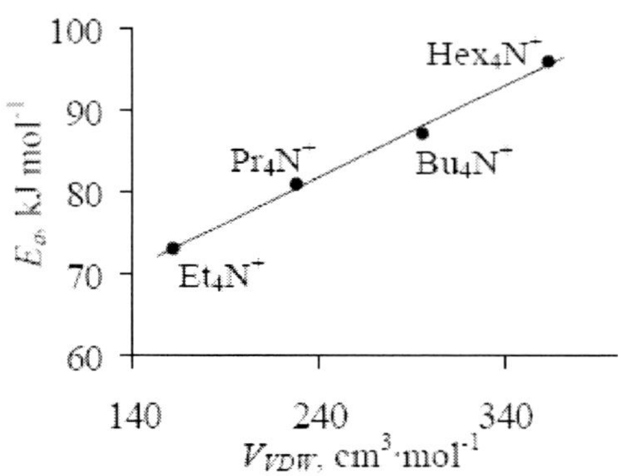

Figure 3. Relationship between activation parameters of the complex decomposition and Van-der-Waals volume of the tetraalkylammonium cations.

Molecular Modeling of the Activated Hydroperoxide Decomposition

To obtain the additional structure information the molecular modeling of the ROOH-Alk_4NBr ion-molecular complexes has been carried out for the case of Et_4NBr. In contrast to the aralkyl hydroperoxides (like isopropylbenzene hydroperoxide) the molecule of 1-hydroxycyclohexyl hydroperoxide does not contain the aromatic ring but there is the hydroxide group OH in their structure that also can participate in the intermolecular hydrogen bonds formation. As the kinetics parameters of the hydroperoxide decomposition are influenced by the nature of the salt anion as well as cation one can assume the model of complex formation with combined action of cation and anion (figure 4) such as previously proposed model of substrate separated ion pair (SubSIP) [14, 15]. Formation of such type associate is accompanied by considerable conformation changes of the hydroperoxide fragment. Association between hydroperoxide molecule and the salt ions occurs by the intermolecular hydrogen bonds formation. It is confirmed by charge increasing on the hydrogen atoms of the corresponding bonds C-H_3, O-H_1, and O-H_2 and elongation of these bonds from 1.11 Å to 1.17 Å and from 0.96 Å to 1.08 Å correspondingly. Calculations in COSMO approximation show that the electron density transfer from bromide-anion to the hydroperoxide fragment is lower in the case of solvent effect taken into account (table 2).

Stage determining the rate of the hydroperoxide decomposition is that of O-O bond cleavage. Thus the reaction activation energy will be determined by the energy of peroxide bond homolytic decomposition. Some possible elementary stages of the complex decomposition with the free radical formation are shown on scheme 1 and are listed in table 3.

On the figure 4 the enthalpies of complexation are shown. In the case of SubSIP we assume that the complex is formed according to the exchange mechanism (2) while possibility of the threemolecular reaction (3) is low.

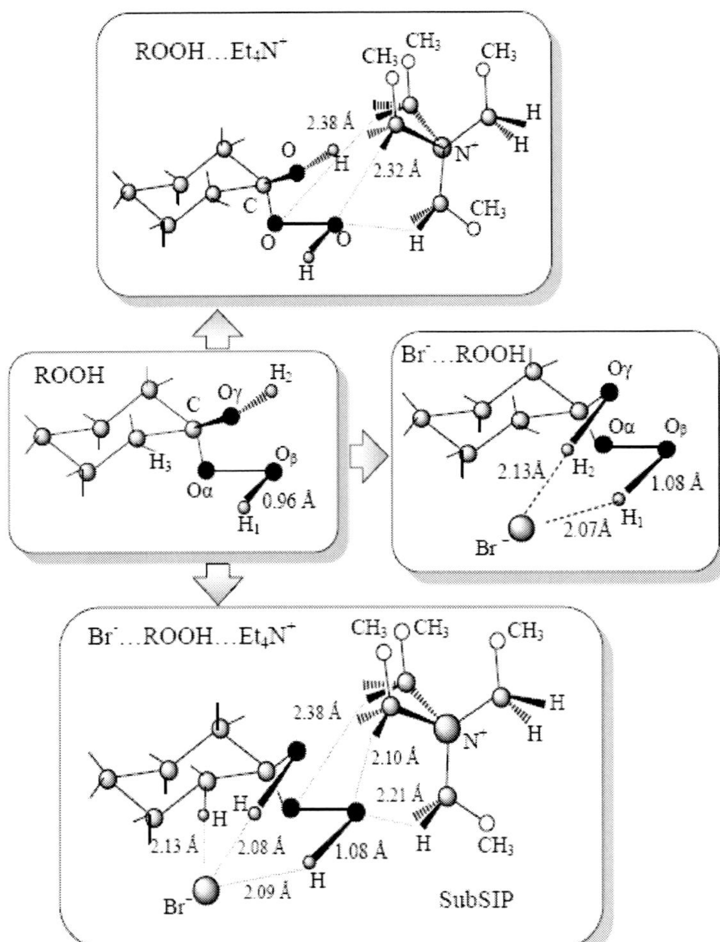

Figure 4. Associative models of 1-hydroxycyclohexyl hydroperoxide with tetraethylammonium bromide.

Table 2. Electron and stereochemical characteristics of COOH-fragment in ROOH – Et_4NBr complexes

	ROOH	Et_4N^+…ROOH…Br^-	ROOH…Br^-	ROOH…Et_4N^+
$q_{O\alpha}$, e	- 0.18	- 0.24	- 0.18	- 0.20
$q_{O\beta}$, e	- 0.20	- 0.26	- 0.29	- 0.23
q_{H1}, e	0.20	0.25	0.25	0.22
q_{H2}, e	0.22	0.27	0.27	0.04
q_{H3}, e	0.10	0.17	0.11	0.10
$\angle CO_\alpha O_\beta H$	90.8	33.4	68.3	77.5
q_{Br^-}	-	- 0.81	- 0.85	-

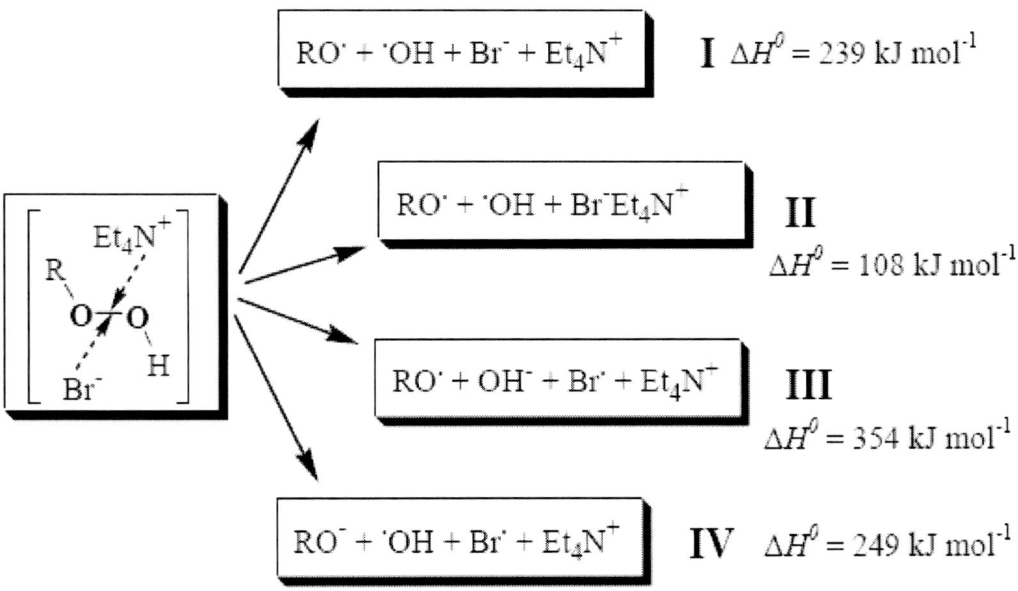

Scheme 1. Some elementary stages of ROOH – Et_4NBr complex decomposition ($\Delta H°$ values are listed for acetonitrile).

Table 3. Heats of some elementary stages of ROOH – Et_4NBr complex decomposition

Reaction	$\Delta H°$, kJ mol^{-1}	
	gas phase	acetonitrile
ROOH → RO˙ + ˙OH	148	151
ROOH…Et_4N^+ → RO˙ + ˙OH + Et_4N^+	192	179
ROOH…Br^- → RO˙ + ˙OH + Br^-	242	208
ROOH…Br^- → RO^- + ˙OH + Br˙	188	219
ROOH…Br^- → RO˙ + OH^- + Br˙	376	323

Solubility of the Et_4NBr in acetonitrile is sufficiently high, and reactivity of the ions and ion pairs in peroxide bond activation is similar. The threemolecular reaction (3) determines the second reaction order on salt while this value changes from 1 (at small salt concentrations) to 0 (at salt abundance) in experimental conditions. Thus the reaction (2) is rather preferable.

$$Et_4N^+...Solv...Br^- + ROOH \rightarrow Et_4N^+...ROOH...Br^- + Solv \quad (\Delta H_{r(2)} = -53.7\ kJ{\cdot}mol^{-1}) \tag{2}$$

$$Et_4N^+ + ROOH + Br^- \rightarrow Et_4N^+...ROOH...Br^- \quad (\Delta H_{r(3)} = -88.1\ kJ{\cdot}mol^{-1}) \tag{3}$$

Corresponding models like $Et_4N^+...ROOH...Br^-$ were obtained for tetraalkylammonium salts (Pr_4NBr, Bu_4NBr, и Hex_4NBr). Linear relationship has been observed between the complex heat of formation - $\Delta H°_f$ and experimental activation parameters - $\Delta H^{\neq}$ (figure 5).

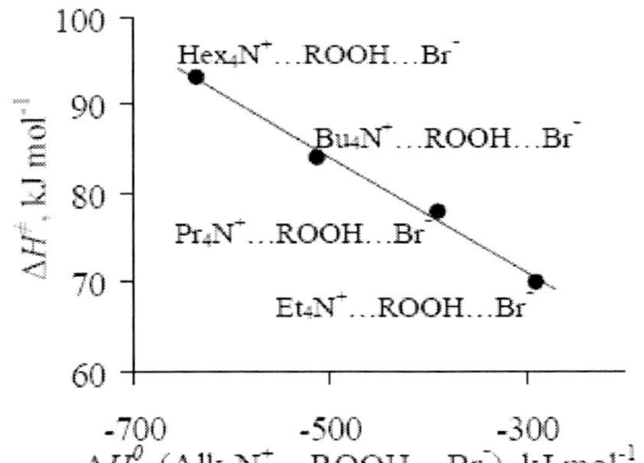

Figure 5. Relationship between the complex heat of formation ($\Delta H°_f$) and experimental activation parameters ($\Delta H^{\neq}$).

Changes in peroxide bond energy (i.e. total of electronic and nuclear energies of two-center term O-O atom pair) can be suggested as extent of the peroxide bond activation. Stable peroxide structure has the lowest O-O bond energy (-12.08 eV for free hydroperoxide) and less stable (or activated) peroxide conformation has the highest one (-11.78 eV for complex $Et_4N^+...ROOH...Br^-$). Difference between energy of stable and activated peroxide structure increases in the catalyst row: Hex_4NBr, Bu_4NBr, Pr_4NBr, and Et_4NBr indicating weakening of the O-O bond in the ROOH-Alk_4NBr complex. From this point of view the O-O bond activation in the case of $Et_4N^+...ROOH...Br^-$ is the highest and in complex with Hex_4NBr is the lowest. This suggestion is in correspondence with experimental data.

Thus the cation structure influences on the reactivity of the hydroperoxide complex and on the extent of peroxide bond activation. The molecular modeling of the hydroperoxide-catalyst reactive complex can be used to preliminarily predict the reactivity of the system.

Conclusions

Summarizing, obtained experimental facts have shown that the chemical activation of the peroxide bond is observed in the presence of Alk_4NBr. The kinetic parameters of the hydroperoxide catalytic decomposition have been obtained for the Et_4NBr, Pr_4NBr, Bu_4NBr, and Hex_4NBr. Catalysis of the 1-hydroxycyclohexyl hydroperoxide decomposition has been shown to occur through accompanied action of the ammonium salt cation and anion.

References

[1] E.T. Denisov, T.G. Denisova, T.S. Pokidova. *Handbook of Free Radical Initiators*. John Wiley & Sons Inc.: Hoboken, New Jersey, 2003. - 879 p.

[2] N.A. Turovskij, I.O. Opeida, O.V. Kush, E.L. Baranovskij. *J. Appl. Chem.* 77, 1887 (2004).

[3] C.J. Perez, E.M. Valles, M.D. Failla. *Polymer,* 46, 725 (2005).

[4] I.A. Opeida, N.M. Zalevskaya, E.N. Turovskaya. *Kinetika i Kataliz,* 45, 776 (2004).

[5] I.A. Opeida, N.M. Zalevskaya, E.N. Turovskaya, U.I. Sobka, *Petroleum Chemistry*, 42, № 6. 460 (2002).

[6] M.A. Turovskyj, I.A. Opeida, E.N. Turovskaya, O.V. Raksha, N.O. Kuznetsova, G.E. Zaikov. *Oxid. Commun.*, 29, 249 (2006).

[7] N.A. Milas, S.A. Harris, P.C. Panagiotacos. *J. Am. Chem. Soc.,* 61, No 9 2430 (1939).

[8] J.J.P. Stewart. *MOPAC 2000.00 Manual;* Fujitsu Limited: Tokyo: Japan, 1999.

[9] A. Klamt. J.Chem. Soc. Perkin Trans., №2 799 (1993).

[10] M.A. Turovskyj, A.M. Nikolayevskyj, I.A. Opeida, V.N. Shufletuk. *Ukrainian Chem. Bull.,* 8, 151 (2000).

[11] L.M. Pisarenko, O.T. Kasaikina. *Russ Chem Bull*, 3, 419 (2002).

[12] L.M. Pisarenko, V.G. Kondratovich, O.T. Kasaikina. *Russ Chem Bull*, 10, 2110 (2004).

[13] Y. Marcus. *Ion salvation*. Chichester etc.: Wiley. – 1985. – 306p.

[14] M.A. Turovskyj, I.O. Opeida O.M. Turovskaya, O.V. Raksha, N.O. Kuznetsova, and G.E. Zaikov. *Order and Disorder in Polymer Reactivity*. Howell New York: Nova Scince Publishers, Inc.-2006. - pp. 37-51.

[15] N.A. Turovskij, S.Yu. Tselinskij, Yu.E. Shapiro, A.R. Kal'uskij. *Teor. Eksp. Khim.*, 28, №4. 320 (1992).

In: Progress in Chemistry and Biochemistry
Editors: E. M. Pearce and G. E. Zaikov

ISBN: 978-1-60692-344-3

Chapter 9

INCIDENCE OF THE AGGRESSIVE ATMOSPHERIC AGENTS FROM THE CUBAN CLIMATE ON POLYMERIC ARTICLES AND MATERIALS

***Orlando Reinosa Espinosa*[1], *Francisco Corvo Perez*[2] *and G.G. Makarov*[3]**
[1] Central National Research Center of Cuba, Chemical Division, Habana, Cuba; Orlando.Reinosa@cnic.edu.cu
[2] Institute of Materials and Chemical Compounds for Electronics, Habana University, Cuba
[3]N.M. Emanuel Institute of Biochemical Physics, Russian Academy of Sciences
4 Kosygin str., 119334 Moscow, Russia
Makarov@sky.chph.ras.ru

ABSTRACT

The present research shows the high aggressiveness on the polymeric material and articles, and it especially shows the aggressive factors of the humid and tropical climate of Cuba on the rubbers-the different and more significative atmospheric factors, that is to say (temperature, solar radiation, ozone concentration), were supplied by the Instituto de Meteorología de la República de Cuba (The Institute of Meteorology of the Republic of Cuba). The samples of the tested rubbers, especially elaborated with no antioxidant protections were made using the coded rubber, used by the Rubber enterprise Union with the number 471.

The evaluation of the spouting and development of the ageing of the rubber samples mounted on the resting frames and under tension was carried out according to prior studies based on international standards.

The early apparition of cracks (days of exposition perpendicular to the direction they were mounted on the testing frames) is a demonstration of the high aggressiveness of the humid and tropical climate of Cuba.

The results obtained and presently shown in this work, show, on one side, to give a good antioxidant protection to the rubber and their articles to be used under the atmospheric condition under the humid, tropical climate of Cuba and also that the atmospheric testing stations used could be employed to select antioxidant protection on systems of high protective effectiveness.

Keywords: climate, materials, polymers, atmospheric agents, aggressive media.

Introduction

The Cuban climate favors the initiation and development of certain chemical and physical processes that reduce the useful properties of polymeric articles and materials, mainly those of massive use like rubbers, plastic and composites. These processes occur in very short periods of time in comparison with the durability of similar articles in some other countries with very different kinds of weather.

Although the significant impact that these processes cause on the Cuban economy is known, there are not specific dates to reflect them. No evaluation has been made up to the present moment.

Due to the easy appearance of this phenomenon in our country and its undesirable aspects not only from the technical point of view, but also from the economic one, research is demanded with the aim of knowing all the mechanisms that take part in the process and also the request for the appropriate procedures to inhibit its stabilization.

Though a particular research at a certain level is carried out, nowadays a deeper investigation is demanded. The purpose is to obtain a higher durability on our articles that will result in higher production, durability and better quality.

Aggressive Atmospheric Agents Present in the Cuban Climate

In our country there is a tropical climate with a high humidity and marine influence. Another significant aspect is the condition of the island and our location in a region where the influence of cold fronts (north) and tropical storms and hurricanes (south) is received. Among the determinant factors on the formation of the Cuban climate we may identify the amount of sun radiation that is received, the characteristic of the atmospheric circulation over Cuba and the influence of the different physical and geographical traits in our territory.

Effect of the Sun Radiation on Polymeric Materials

Sun radiation has a high impact on polymeric materials and its articles. The most aggressive components are the radiations from the UVA region that cause strong damage not only to materials but also to human beings.

Table 1. Comparison of bond energy of most common chemicals groups present on polymers and radiation energies

Bond	Bond energy (KJ / mol)	Radiation Energy. (KJ / mol)	Wavelength (nm)
C-C	955.0	1197.0	100
C≡N	877.8	-	-
C=O	730.8	798.0	150
C=C	520.8	598.5	200
C-F	499.8	-	-
OH	462.0	478.0	250
C-H	411.8	-	-
S_i-O	373.8	-	-
N-H	352.8	399.0	300
C-Cl	327.6	342.0	350
S_i-C	294.0	299.0	400

In Cuba the average annual temperature is 25 °C, highest temperatures are around 37 °C and the lowest between 1 °C-10 °C. The hottest region is the eastern one. In the following frame there are some figures reported by two stations in Havana city.

Table 2. Some atmospherics parameters from two different assay stations.

Factors	CNIC station	FLORES station
Humidity (%)	79	79
Solar Radiation mean (MJ / m^2)	17.9	17.9
Solar activity	236.3	--
Temperature mean (ºC)	28.0	28.0
Salinity (mg / m2 / day)	21.1	430.5
$S0_2$	21.1	430.5
Dust (mg / m^2 / day)	64.9	--

ACTION OF OZONE ON POLYMERIC MATERIALS AND RUBBER ARTICLES

Ozone is one of atmospheric factors highly significant on degradation of polymers. It has been demonstrated that low concentrations of ozone in the environment, ground 10-50 micrograms/m^3 in air, are enough to provoke the initiation and development of degradation on polymeric materials mainly on rubbers and as a result of it these rubbers are rejected since they are not good to be use.

In the following frame the differences between the two seasons regarding ozone's concentration have been stated. There is a dry season, which is (winter time), with low temperatures and not very much of rain, but a higher concentration of ozone and on the other hand we have the rainy season (summer period) with high temperatures and a lot of rain, but lower concentration of ozone and effects of sun radiation on polymeric materials.

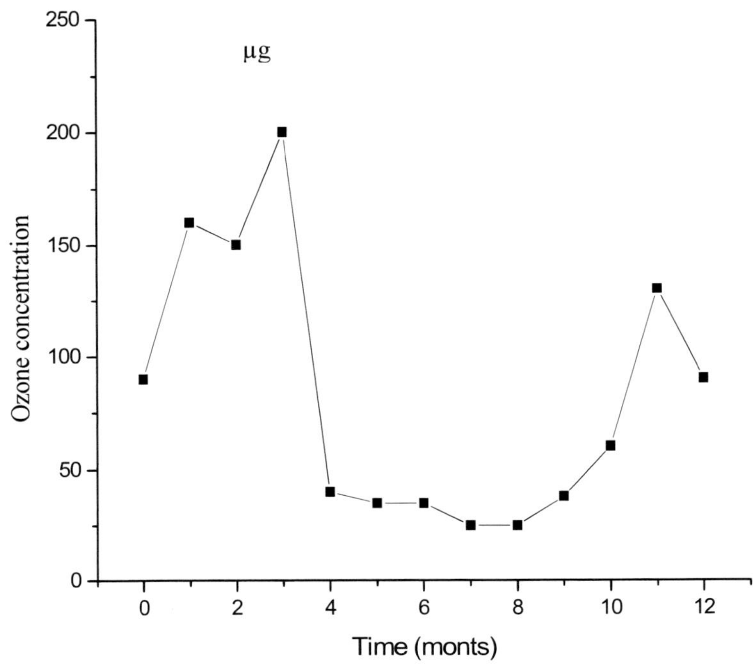

Casa Blanca station.

Figure 1. Annual Ozone concentration on Havana City, Cuba.

Effects of Sun Radiation on Polymeric Materials

Sun radiation is very significant on the behavior of polymeric materials and articles. The most aggressive components are undoubtedly, the radiations from UV region. These radiations are enough to provoke serious damages not only on materials but also on human beings.

In the following diagram, the different behaviors of three rubbers formulations are shown. Every formulation has a different ozone protection and they were exposed at two different regimes: shade and sun. The effect of temperature and sun radiation can be appreciated depending on the exposure conditions.

Diagrams reflecting the development of the aging process in rubbers wether they are exposed to shade or to the sun.

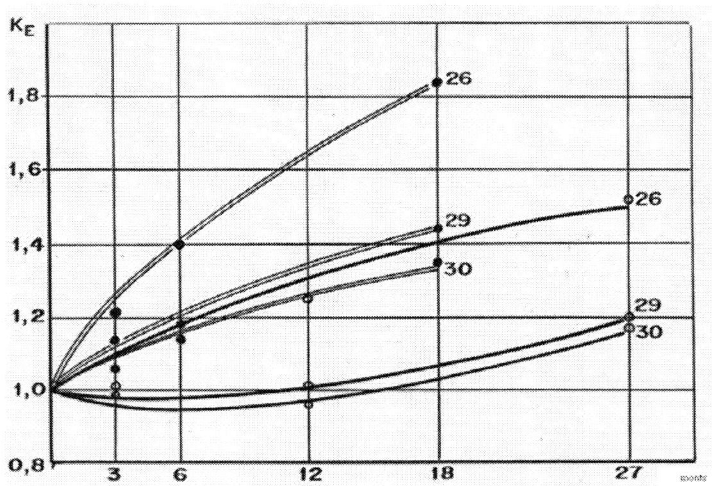

Figure 2. Dependence of weathering aging coefficients on exposition conditions; Legend: bold traces – not exposed to sun radiation, lightly traces – exposed to sun radiation.

Rubber samples, specially prepared without the required antioxidant protection show in a brief time the high aggressiveness of the environmental factors in different regions of Havana City.

Table 3. Components of rubber from national formulation number 471 without antioxidant protection

Ingredients	Parts by 100 parts of rubber (pphr)
Styrene butadiene - copolymer (E-1712)	50.00
Styrene butadiene - copolymer (E-1502)	20.00
Sorprene-293	30.00
Activator (Oxide of Zinc)	3.00
Stearic acid	2.00
Carbon Black (N-339)	60.00
Oil ICP-68	12.00
Vulcacit-CZ	1.10
TMTD	0.10
Vulcalent G	0.20
Sulphur	1.50
	$\sum$ 179.90

Table 4. Initial physico – chemical properties of experimental rubber formulations

Samples	Modulus at 100 %, MPa	Modulus at 300%, MPa	Tensile strength, MPa	Elongation, %
1	2.8	9.93	19.32	480
2	2.69	11.16	20.39	480
3	3.08	11.4	19.16	480
4	2.71	12.36	21.24	_
5	2.17	-	_	-
Xm	**2.7**	**11.2**	**20.0**	**480.0**
σ	**0.3**	**1.0**	**1.0**	**0.0**

The rubber samples specially prepared without the required antioxidant protection are show on the following photographs (figure 3). They expose in a short time of exposition to the atmospheric factors, the high aggressiveness present in two different regions of the city of Havana.

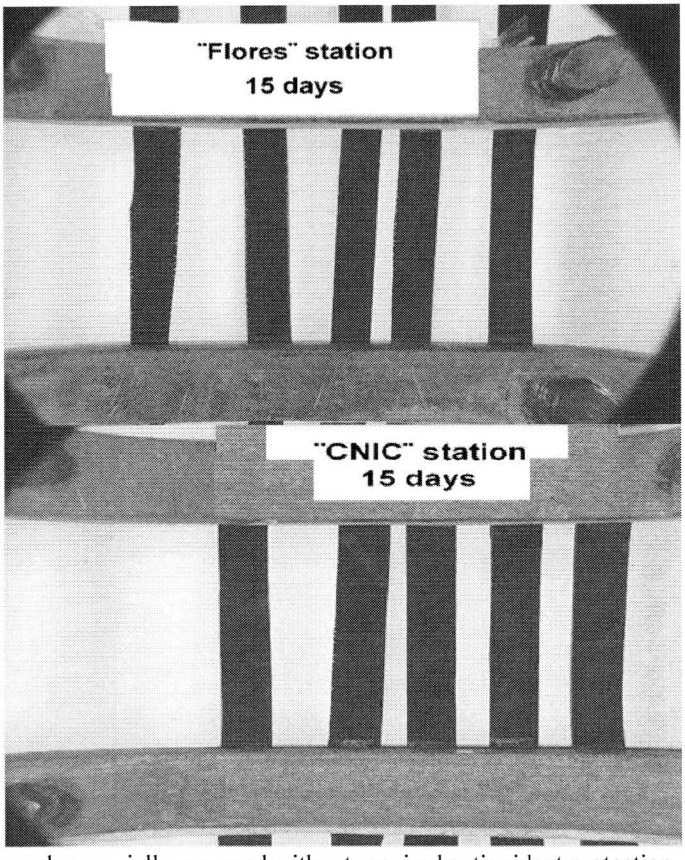

Figure 3. Rubber samples specially prepared without required antioxidant protection.

In the former photographs you may appreciate that all the changes on experimental rubbers occurred very fast. The incidence is higher in the station named "Flores" which is located only 20 meters from the northern coast in Havana city.

There are some other experiences related with the behavior of different rubber articles. Though these rubber articles were formerly protected with certain antioxidant levels, they also showed the aging symptoms in a short period of exploitation (months). These aging symptoms made them unuseful and as result they were rejected for use.

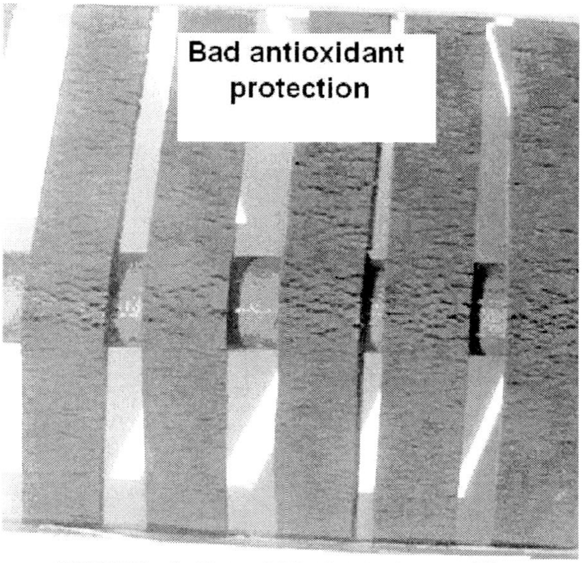

"CNIC" station 120 days exposition

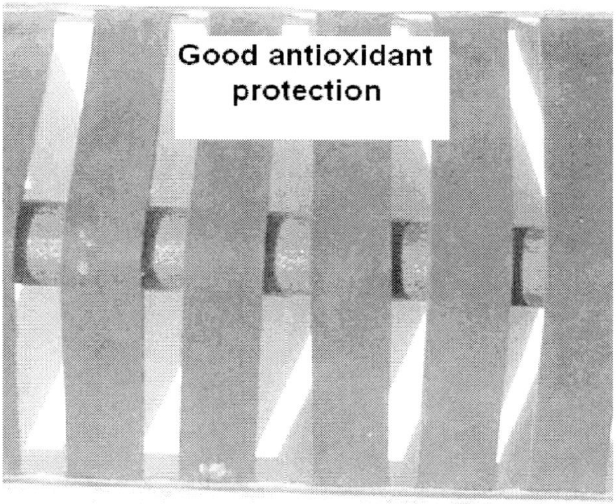

"CNIC" station 120 days of exposition

Figure 4. Different enhancement of rubber formulations with similar composition but with different antioxidant protection, during out door exposition and by strength.

The same has been observed in the case of rubber materials and articles under the action of the atmospheric factors from the Cuban environment, not only during their exploitation but also during the conservation. The following image shows what it has been formerly stated.

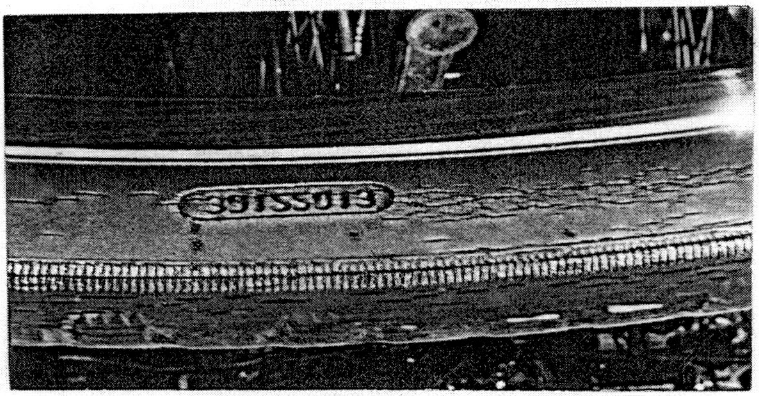

Conclusions

- The results show the high aggressiveness of the Cuban climate on polymeric materials and articles.
- The research stations allow the evaluation not only of the resistance of the different materials but also the effectiveness of the different protecting systems whether you are

using them or planning to use for the protection of the protection of the mentioned materials during their exploitation of and conservation in Cuba.

- The high aggressiveness of the different Cuban regions is very similar to the conditions given by "Xenotex" climatic chambers which are the ones in use during quick evaluations.

RECOMMENDATIONS

The atmospheric research stations on materials from different kinds, may be used to evaluate the behavior of previously conceived and elaborated materials when they are under the action of the aggressive agents from the Cuban climate. These stations are useful for the selection of new protecting systems to prevent the aging process on different materials and also to confirm the validity of the results obtained during the research recommended to be carried out.

REFERENCES

[1] F. Corvo, N. Betancourt, J.C. Díaz, C. Lariot, I. León, J. Pérez, O. Rodríguez, E. Bricuyet, F. Catalá, M. Castro, R. González, C. Echevarria, M. Lorente, M.E. Ladrón de Guevara, 2da variante del Mapa Regional de Agresividad Corrosiva de la Atmósfera de Cuba. Proceeding del PrimerTaller Internacional de Corrosión. CONACYT-CINVESTAV. Mérida. Yucatán. México. 23-28 marzo 1992.

[2] Razumovsky S.D, Podmasteryev V.V, Zaikov G.E Physico-Chemical Principles Governing the Ageing of Elastomers Under Action of Atmospheric Ozone *.Inter.J. Polymeric Mater.* 1990. Vol.13 pp 81-101

[3] Rakovski S.K, and Cherneva D.R Sínthesis of Antioozonants for the preparation of Rubber.*Intern. J. Polymeric mater.*1990 Vol.14, pp 21-40

[4] Popov A.A and Parfenov V.M. Efficiency of antiozonants under the conditions of static deformation of elastomers. *Intern. J.Polymeric Mater.* 1990, Vol. 13 ,pp 123-135

[5] Ramirez Almoguea Jesús. Trabajo de Tesis para optar por el Grado Científico de Dr. En Ciencias Quimicas . *Ciudad de la Habana.* 1989

[6] Ricard J.L Tesis presentada para optar por el grado científico de Dr. En Ciencias Técnicas.*Ciudad Habana.* 1994

[7] Bolivar S., Reinosa O. Envejecimiento y estabilización de la goma para la banda de rodamiento de neumaticos de bicicletas de producción nacional. *Tesis de maestría. Habana.*1999

[8] Reinosa O,Bolivar S., Ricard. J.L. Reforzamiento de la actividad antioxidante de una mezcla de aminas aromáticas empleadas en la protección de gomas al envejecimiento atmosférico. *Revista CNIC.Ciencias Quimicas Vol.* 29 pp 145-149 1998

[9] Emanuel N.M. Some Problems of Chemical Kinetics in Polymer aging and Stabilization. *Polymer Engineering and Science Vol.*20, pp 662-67

In: Progress in Chemistry and Biochemistry
Editors: E. M. Pearce and G. E. Zaikov
ISBN: 978-1-60692-344-3

Chapter 10

INTERFACIAL ADHESION BETWEEN POLY (ETHYLENE TEREPHTHALATE) FABRIC AND A THERMOPLASTIC POLYURETHANE MATRIX

***Arun Ghosh*[1], *Mohit Gupta and David A. Schiraldi*[2]**
Department of Macromolecular Science and Engineering
Case Western Reserve University
2100 Adelbert Road; Cleveland, OH 44106-7202

ABSTRACT

A method for the improvement of interfacial adhesion between a poly(ethylene terephthalate) (PET) fabric and a thermoplastic polyurethane (TPU) matrix is reported in this article. The PET fabric is treated sequentially with aqueous sodium hydroxide and 3-aminopropyl trimethoxy silane solution. The treatment process has improved the interfacial adhesion between PET fabric and TPU matrix dramatically. The increased interfacial adhesion between the PET fabric and TPU coating is accompanied by an overall improvement in the mechanical and dynamic mechanical properties of the PET fiber/TPU matrix composites.

Keywords: Poly(ethylene terephthalate), thermoplastic polyurethane, interfacial adhesion, fiber surface.

INTRODUCTION

Improved adhesion between fabric and coating is a key property in coated systems, which protects fabrics from a wide range of adverse environmental conditions and also allow the fabric materials to be handled and welded. The level of adhesion depends upon mechanical penetration of the coating polymer into pores or crevices in the fabric, as well as direct

---

[1] Present address: AgResearch Lincoln, Cnr Springs Rd & Gerald St, Christchurch 8140, New Zealand

chemical interactions between molecules of the coating and fabric materials [1,2]. The failure of adhesion between the coating and substrate can result from degradation of the coating, stresses resulting from manufacturing defects, and/or mechanical work applied to the finished good while in service [1,2].

Poly(ethylene terephthalate) (PET) fabric has emerged as a preferred candidate for textile, architectural and other load-bearing applications due to its high strength and modulus, low shrinkage, and chemical resistance to a variety of environmental insults [2,3]. High tenacity PET ropes are coated with poly(vinyl chloride) to increase abrasion resistance, UV resistance and water barrier properties. Poly (vinyl chloride) coated polyester fabrics are used in a range of applications including high strength belts, tarps, and awnings [3]. Polyurethanes constitute one of the most important classes of synthetic coatings which show numerous desirable properties such as toughness, flexibility, high resistance to abrasion, excellent electrical insulation, and good moisture and chemical resistance [4]. The previous communication reports a methodology of improved polyurethane coating on copper foil [5].

Delamination of the coating from the coated substrates limits their lifetimes and performance. In order to improve the performance of polymer-coated substrates, coupling agents based on titanate and silane chemistries, have been utilized [1]. Previous researchers have reported that adhesion is promoted by interdiffusion of the polymerized silane film and polymer, resulting in an interpenetrating polymer network (IPN) [1,6-12]. *Mittal* has described advances in understanding the chemistry of coupling agents/adhesion promoters in coated systems [12]. *Plueddemann*,[14,15] *Arkles*,[16] *Rosen*,[17] *Walker*[18] and *Boxall*[19] have studied various silanes and the mechanism of their functions as adhesion promoters. There are many reports on the application of silanes as adhesion promoters for urethane and epoxy coatings on metals, glass and other inorganic substrates [20-24]. Silane coupling agents are also used as adhesion promoters to improve adhesion of vinyl chloride and polypropylene on glass and metal surfaces [25-27]. The adhesion of the modified alkyd and acrylic polymers on glass or metallic substrates are also improved by using silane coupling agents [28-31].

The prime goal of the present work is to improve the interfacial adhesion of TPU coated PET fabric through chemical surface modification of the fabric. The aqueous alkali treatment creates activate surfaces on polyester surface [32-38]. In the present investigation an alkali treatment followed by silane treatment is used to enhance the coating matrix/fiber interfacial adhesion.

EXPERIMENTAL

Materials

Thermoplastic polyurethane (ESTANE® 58315 TPU) produced by Noveon was used as received. ESTANE® 58315 is a Shore 85A hardness aromatic polyether–based TPU. Sodium hydroxide pellets were purchased from Fischer Scientific and 3-aminopropyl trimethoxy silane was purchased from Sigma-Aldrich. High strength poly(ethylene terephthalate) webs (2.5 cm width) were supplied by the BioPlastics Company, North Ridgeville, OH.

[2] Corresponding author: David A. Schiraldi; david.schiraldi@case.edu

Chemical Treatment and Characterization of PET Fabric

In order to improve the adhesion between the fabric and the TPU matrix, the PET fabric was subjected to surface treatment with aqueous alkaline solution. The fabric was dipped into a NaOH aqueous solution (5, 15 and 25 wt %) for 15 second, heated in an oven at 180° C for 15 min, and then washed with distilled water. The alkali-treated fabrics were subsequently treated with aqueous solution of 3-aminopropyl trimethoxy silane (5, 10 and 15 wt % solution) by dipping into the solution for about 15 second, and then dried at room temperature for 24 h. The diameters of the PET fibers were measured using an optical microscope (Olympus BX 60) at magnification of 500×. The chemical natures of the surfaces of untreated and treated PET fabrics were analyzed by X-ray photoelectron spectroscopy (XPS). XPS analyses were carried out on VG MicroLab MKIII spectrometer at a base vacuum higher than 10- 9 Torr with non-monochromatized Mg Kα (1253.6 eV) radiation. The measurements were performed with a take-off angle of 60° relative to the surface normality. XPS spectra were an energy range from 20 to 110 eV for survey scans and the C (1s), O (1s) and Si (2p), Na (1s) and N (1s) regions of interest, respectively.

Preparation and Characterization of TPU Coated PET Webs

Films (~2 mm thickness) were initially prepared from TPU pellets using a Carver model C press at 170°C. Woven PET fabric web were then laminated with the TPU films using the same hot press at 170°C temperature, 7 MPa pressure and 1 min press time. The samples were pre-heated for approximately 5 min prior to application of pressure. A thin PTFE film tab (~0.13 mm) was inserted between the TPU and PET webs at one end to facilitate peel testing of the coated webs.

Measurement of interfacial adhesion was performed by examining peel strength of the TPU coated PET webs at 25°C according to ASTM Standard D903-2004 using an Instron model 5565 Universal Testing Machine at an extension of 100mm/min. Each coated web was tested 5 times with 5 prepared specimens; averaged results and standard deviations are reported.

Preparation and Characterization of PET Fiber/TPU Composites

PET webs were cut into small fibers (average length, ~1mm and average diameter, ~33 μm) and mixed with TPU matrix polymer using a DACA model 2000 twin-screw co-rotating micro-compounder. The blending of 10 wt % PET fiber (both chemically treated and untreated) with TPU was carried out at 100 rpm screw speed, blending for 5 minutes prior to extrusion. The extrusion temperature of the composite was 180°C. Samples were then compression molded into films for mechanical testing using a Carver model C press. The composites were first preheated at 180°C for 5 min and then rapidly compressed (15MPa) and released to remove trapped air bubbles. The samples were then compression molded under 15MPa pressure for 5 min and cooled down rapidly between two water-cooled aluminum cooling plates to produce the final composite films (0.3-0.4 mm in thickness) for testing.

Stress-strain properties were measured at room temperature according to ASTM D696 using an Instron model 5565 Universal Testing Machine at an extension of 50mm/min. Each composite was tested 5 times with 5 prepared specimens; averaged results and standard deviations are reported. A Tritec 2000 DMA (Triton Technology Co.) was used in tensile mode at an oscillatory frequency of 1 Hz with applied 1% strain and at a heating rate of 2°C/min for all samples. The temperature scan was performed in the range from -70° to 100°C for the composites. Sample dimensions were typically 4.7 mm long, 5-6 mm wide and 0.3 mm thick. SEM photomicrographs of the samples were taken at 20kV using XL 30 ESEM manufactured by PHILIPS. The 10 wt % PET fiber filled extruded samples were fractured cryogenically and the exposed surfaces were coated with palladium and analyzed by SEM.

RESULTS AND DISCUSSION

Characterization of PET Fabric

The effects of chemical treatment of fabrics on constituent fiber diameters, measured by optical microscopy, are shown in table 1. The average diameter of the untreated PET fiber is 33 μm. Upon treatment with aqueous NaOH solution, the alkaline agent attacks the PET surfaces eroding the fiber. Treatment with 25 wt% aqueous NaOH solution results in a reduction of PET fiber diameter to an average of 24 μm. The alkaline treatment followed by dipping into 10 wt% aqueous silane (3-aminopropyl trimethoxy silane) solution increases the PET fiber diameter back to an average value of 27μm; the apparent silane layer thickness is therefore approximately 3 μm. It is discussed later in this paper that the optimal adhesion is obtained with treatment at 10 wt% aqueous silane solution. Further increase in silane concentration decreases the interfacial adhesion, due to larger, less effective silane coatings.[6-8] The methoxy groups of the silane are hydrolyzed in the aqueous environment and enable condensation reactions with carboxylate groups produced on PET fabric surfaces by alkali treatment. The subsequent drying results in development of polymerized silane film on PET fiber. Previous researchers have reported that the adhesion promotion is due to interdiffusion of the polymerized silane film and polymer, and the resulting formation of an interpenetrating polymer network (IPN). [1,6-12] These authors have reported that silane must be applied in amounts sufficient to produce a silane film of finite thickness (multiple monolayers) that can interdiffuse with polymer and increase bond strength. Excessive application of silane, however, can lead to a larger and weaker boundary layer, and decrease interfacial adhesion. [6-8] *Teh* et al have reported comparative results on the morphology and chemical composition of the untreated and NaOH treated PET fibers examined by SEM and ToF-SIMS.[38] The highly oriented and stressed smooth PET fiber becomes progressively delustered on alkali treatment. PET fiber usually shows pitting when being treated with aqueous alkali solution.[39-41] It is known that conformational changes occur during the extension of PET yarns and films,[42,43] and that beyond the yield point structural rearrangement occur, giving more oriented structure. These conformation changes may modify the fiber surface, thus allowing diffusion of aqueous alkali into the fiber.

Table 1. Effect of chemical treatment of PET fibers on TPU composites

PET fiber type	Diameter (μm)	Mechanical properties of TPU composites with 10wt% PET fibers		
		Tensile strength (MPa)	Tensile modulus at 200 % strain (MPa)	DMA storage modulus at 25°C (MPa)
Untreated	33±0.2	14.4±0.5	6.3±0.1	18
25 wt% NaOH solution treated	24±1.0	16.3±0.2	6.8±0.1	26
25 wt% NaOH and 10 wt% silane solution treated	27±1.0	26.1±0.5	8.6±0.1	35

The elemental compositions on untreated and treated PET fabrics are summarized in table 2. The untreated PET fiber shows the C1s and O1s atomic ratio of about 9:2 as expected. The NaOH treated PET fiber exhibits significantly higher percentage of O1s atoms along with a considerable presence of Na1s atoms, which can be ascribed to the reaction of NaOH with the ester linkages of surface PET molecules and result in the formation of sodium carboxylate salts. Further treatment on the NaOH treated PET fiber with silane solution decreases the O1s and Na1s atomic concentration significantly, diluted by the addition of significant amounts of Si2p (5.5%) and N1s (7.1%) on the fiber surfaces. While the Si:N ratio detected in these treated fabrics does not exactly match theory, it clearly demonstrates a high level of surface treatment.

Table 2. Surface chemical composition on PET fibers examined by x-ray photoelectron spectroscopy

PET fiber type	Atomic (%)				
	C1s	O1s	Si2p	Na1s	N1s
Untreated	78.9	20.2	0.9	--	--
25 wt% NaOH solution treated	64.4	29.7	0.8	5.1	--
25wt% NaOH and 10wt% Silane solution treated	65.4	21.7	5.5	0.3	7.1

Adhesion Peel Test

The adhesion properties of the TPU coated woven PET webs were analyzed by comparing the peel strength values. Figure 1 shows the peel strength values of TPU coated PET laminates, where PET webs are treated under different conditions. The TPU coated untreated PET web laminate shows average peel strength of 2.8 kN/m. The aqueous silane solution (5, 10 and 15 wt%) treatment on PET webs moderately increases (ca. 50%) the peel strength value of TPU coated PET webs, while aqueous NaOH solution (5 to 25 wt%) treatment increases the peel strength value significantly (ca. 70 -100%). The alkali treatment followed by silane treatment on PET webs dramatically improves the adhesion peel strength. The laminates show about 150-170% higher peel strength as compared to untreated laminate, when PET webs are treated 25 wt% alkali solution followed by 10 or 15 wt % silane solution. The optimal adhesion is obtained with 10 wt % aqueous silane solution treatment; there is drop in peel strength beyond that level of treatment. With the present chemical treatment method, the TPU-coated PET webs maintain their original flexibility. The peel strength of TPU coated PET webs could be doubled using a concentrated silane solution (80-100%) or epoxy primer as adhesion promoters. These high peel strength laminates were very rigid ("boardy"), limiting their utility in applications where web flexibility is required.[44] The increase in peel strength of treated PET webs can apparently be attributed to improved interfacial adhesion achieved between fiber and TPU matrix that is due to coupling effect of the silane compounds.

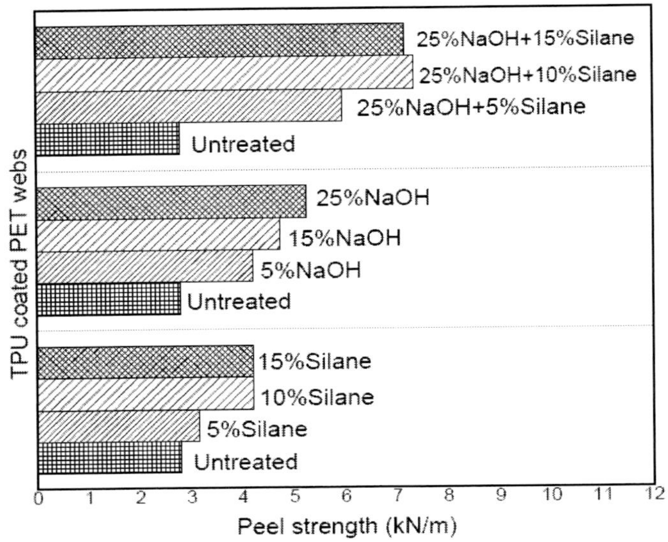

Figure 1. Effect of chemical treatment on peel strength of TPU coated PET webs.

Mechanical Properties of PET Fiber/TPU Composites

The TPU coated PET fabric work in this study was augmented by the production of loose PET fiber/TPU matrix composites. The mechanical properties of composite materials are greatly affected by the interface between the incorporated fiber and polymer matrix. With addition of PET fiber into TPU matrix, the tensile strength and breaking elongation of the resultant composites are reduced (figure 2). A moderate increase in the tensile modulus of these TPU composites is obtained with untreated and alkali treated PET fiber. The alkali treatment followed by silane treatment on PET fiber significantly enhances the tensile strength and modulus of PET fiber/TPU composites (table 1, figure 2). The mechanical properties of composites are determined by the characteristics of the polymer matrices themselves, together with reinforcements, and the adhesion fiber/matrix interface that mainly depends on voids and the bonding strength at the interface.[45] The objective of the present work is to fabricate high peel strength TPU coated PET webs, not to reinforce TPU composites with PET fiber. The possible drop of tensile strength and/or ultimate elongation of the neat TPU upon incorporation of PET fiber is therefore not necessarily germane to the current study. The prime objective is to improve the PET fiber/TPU interfacial adhesion which is reflected in increasing modulus and tensile strength of PET fiber/TPU composites after chemical treatment on PET fiber (figure 2).

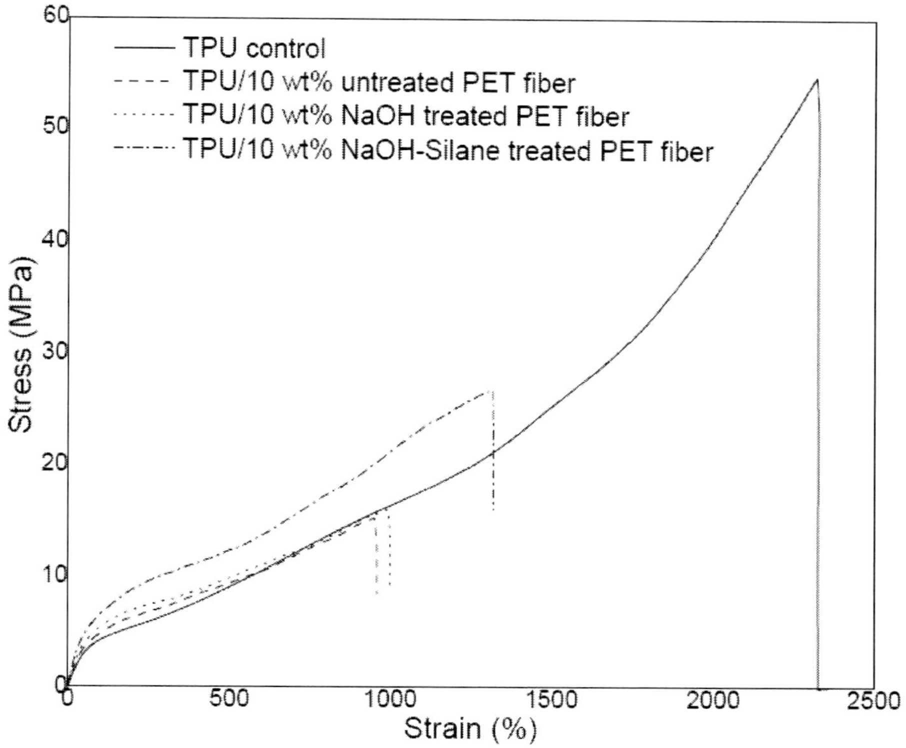

Figure 2. Stress-strain plots of unfilled and 10wt% PET fiber filled TPU composites.

To evaluate the effects of both unmodified and chemically modified fibers on mechanical properties of TPU, the storage modulus (*E'*) as a function of temperature is graphically represented in figure 3. As can be seen in the DMA spectra, the incorporation of PET fiber gives rise to a considerable increase of the TPU modulus. Moreover, the composite stiffness is increased as the PET fiber treated with NaOH and followed by silane coupling agent. The storage modulus at 25°C of the TPU composites is increased by 50% on addition of 10 wt% PET fiber treated with 25wt% alkali and there is a *ca.* 100% increase of modulus when alkali treated PET fiber is further treated with 10 wt% aqueous silane solution (table I). These results are in agreement with the micrographs taken in the scanning electron microscope as shown in figure 5. The position of the glass transition temperature (T_g, -23°C) of TPU composites remains unchanged in presence of PET fiber. The variation of tan δ values of TPU composites with temperature is presented in figure 4. It is found that the intensity and magnitude of the damping peak (tan δ) of the TPU composites sensibly are reduced in presence of PET fiber. These mechanical properties depend on the composition of the composites (i.e., type of PET fiber used) and affectivity of interaction between the fiber and the matrix. *Lopez-Manchado et al* have reported similar observation of dynamic mechanical behavior of PP composites with PET fiber treated with azide based non-silane coupling agent.[46]

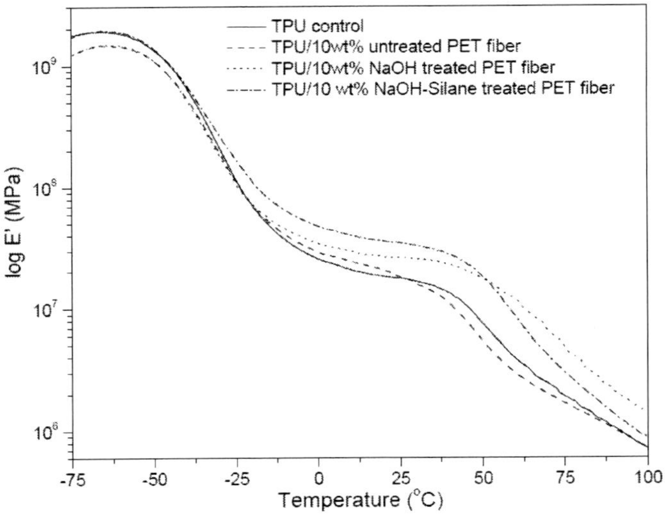

Figure 3. DMA storage modulus (log E') of PET fiber/TPU composites as a function of temperature.

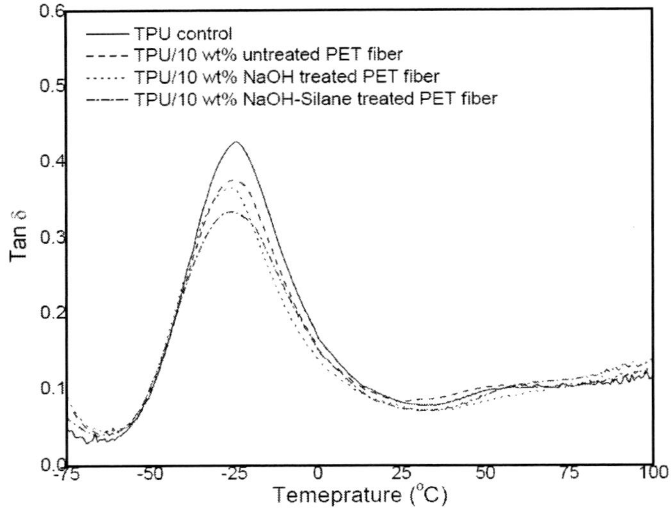

Figure 4. DMA tan δ of PET fiber/TPU composites as a function of temperature.

SEM Photomicrographs of Fracture Surfaces

The interfacial adhesion can also be assessed qualitatively through SEM fractography (figure 5). To understand the fiber-matrix adhesion, the cryogenically fractured surfaces of the extruded TPU composites (containing 10 wt % PET fiber) are studied under SEM. The cryogenically fractured surfaces of the composites with untreated fiber are shown in figure 5(a) at magnifications of ×250 and ×1000. SEM reveals that some fibers are pulled out of the composites and the holes left after fiber pullout are clean. The extracted fiber surfaces are clean with the proximal ends buried in the matrix with poor adhesion between fiber and the matrix; relatively few step structures are observed around the holes or fibers. These images indicate the poor interfacial adhesion between the untreated PET fiber and the TPU matrix.

The cryogenically fractured surfaces of the composites with PET fiber treated with 25 wt% NaOH solution are shown in figures 5(b). In addition to fiber pullout, fiber breakage is also observed. The proximal ends of the fibers are buried in the matrix with good sticking between the fiber and matrix. A number of trenches are also observed on the fracture surfaces of the composites, indicating that large amount of energy is absorbed and dissipated through shearing of the polyurethane matrix during the failure process. This level of adhesion accounts for the higher observed mechanical properties of the composites with NaOH Treated PET fiber. Figure 5(c) shows the cryogenically fractured surface of composites in which PET fiber was treated with 25 wt% NaOH followed by 10 wt% silane solution, at magnifications of ×250 and ×1000. The proximal ends of the fibers were buried in the matrix with better sticking between the fiber and the matrix. The diameter of the extracted fiber from the matrix in figure 5(c) is greater than those of untreated and alkali treated PET fibers in TPU matrix, indicating that the silane coupling agent attacks the TPU matrix and more TPU is adhered on PET fiber surfaces. Due to better bonding with TPU polymer surfaces, fibers protruding from polymer matrix are covered with a polymer layer; this behavior differs from those of the untreated and sodium hydroxide treated fiber composites. These SEM morphologies are consistent with the results obtained by dynamic mechanical analysis and stress-strain properties of the composites. It is well known that the mechanical properties are strongly dependent on the composites composition and the adhesion or interaction at the fiber/matrix interface. The importance of good adhesion between the fiber and matrix has long been recognized. Good adhesion between the fiber and matrix results in efficient stress transfer from the continuous polymer matrix to the dispersed fiber and increase the ability of the material to absorb energy.

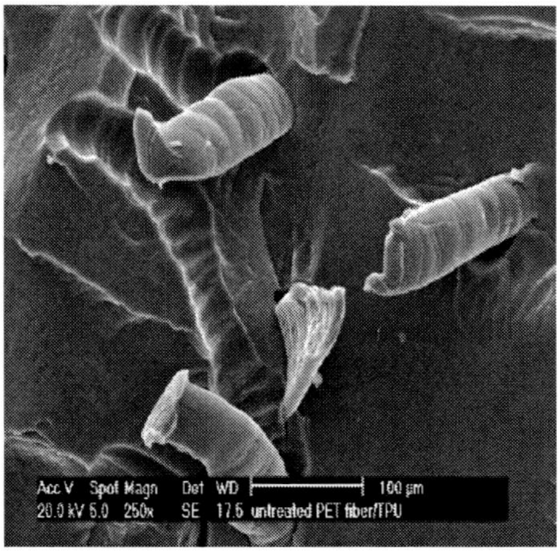

(a) Untreated PET fiber in TPU matrix

Figure 5 (Continues)

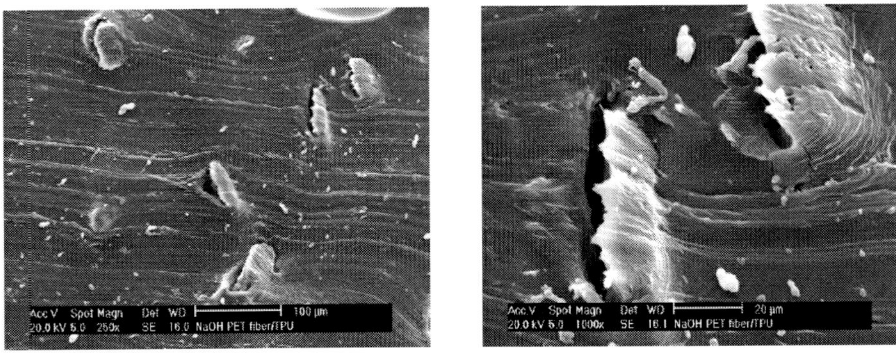

(b) NaOH treated PET fiber in TPU matrix

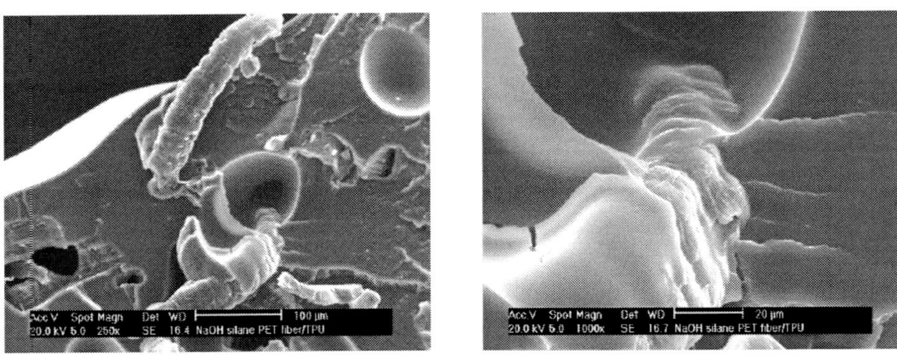

(c) NaOH-Silane treated PET fiber in TPU matrix

Figure 5. SEM photomicrographs of untreated and chemically treated PET fibers in TPU matrix.

CONCLUSIONS

1. Chemically treated PET fabric exhibits stronger interfacial adhesion and better mechanical properties with TPU as compared to their untreated counterparts. The combination of alkali and silane treatment method provides dramatic improvement of interfacial adhesion between PET fabric and TPU matrix. As expected alkali treatment first produces some carboxylate ions on PET fiber surface, facilitating the formation of covalent linkages with hydrolyzed silane coupling agent, which is probable H-bonded with TPU matrix.
2. The dynamic mechanical behavior and stress-strain properties of untreated and treated PET fiber/TPU composites are in good agreement with the adhesion peel strength and scanning electron microscopic studies. The incorporation of the PET fiber gives rise to a considerable increase of the DMA storage modulus and tensile modulus, and a decrease of the DMA tan δ values. These results clearly demonstrate the enhanced interfacial adhesion between chemically treated PET fabric and TPU matrix.

ACKNOWLEDGEMENT

The authors wish to express their sincere gratitude to BioPlastics Co., N. Ridgeville, Ohio for its financial support.

REFERENCES

[1] Plueddemann, E.P. (ed) "Silane coupling agents" second edition, 1991 Plenum Press, New York.
[2] Fung, W (ed), "Coated and laminated textiles" first edition, 2002 CRC Press, England.
[3] http://www.architecturalfabrics.com/home.html
[4] Licari, J.J.; Hughes, L.A. (eds), "Handbook of Polymer Coatings for Electronics, 1990 William Andrew Publishing/Noyes, Chapter 1.
[5] Ghosh, A; Schiraldi, D.A., "Improving interfacial adhesion between thermoplastic polyurethane and copper foil using amino carboxylic acid" *J. Appl. Polym. Sci.* (Submitted).
[6] Ishida, H., Koenig, J.L. *Polym. Eng. Sci.* 1978, **18**, 128.
[7] Sung, N.H.; Kaul, A.; Chin, I.; Sung, C.S.P. *Polym. Eng. Sci.* 1982, 22, 637.
[8] Kaul, A.; Sung, N.H. *Polym. Eng. Sci.* 1985, 25, 1171.
[9] Ishida, H. in *Molecular Characterization of Composite Interfaces*, H. Ishida and G. Kumar, Eds., Plenum Press, New York, 1985, p. 25.
[10] Harding, P.H.; Berg, J.C. J. *Appl. Polym. Sci.* 1998, 67, 1025.
[11] Gentle, T.E.; Schmidt,R.G.; Naasz, B.M.; Gellman, A.J.; Gentle, T.M. *J. Adhesion. Sci. Technol.* 1992, 6, 307.
[12] Chou, C.T.; Penn, L.S.; *J. Adhesion.* 1991, 36, 125.
[13] Mittal, K.L. *Pure Appl. Chem.* 1980, 52, 1295.
[14] Plueddemann, E. P. (ed.), *Interface Phenomena in Polymer Matrix Composite Materials*, Academic Press, New York, 1974, Ch. 6, pp. 173-216.
[15] Plueddemann, E. P., *Am. Chem. Soc.*, Div. ORPL (now *Div. Polym. Mater. Sci. Eng.*), 50 (1984) 130-134.
[16] Arkles, B.C. *Chem. Technol.* 1977, 766.
[17] Rosen M.R. *J. Coat Technol.* 1978, 50, 72.
[18] Walker, P. in A.D. Wilson, J. W. Nicholson and H.J. Prosser (eds.), *Adhesion Promoters-Surface Coatings I*, Elsevier Applied Science, London, 1987, pp. 189-232.
[19] Boxall, J. *Polym. Paint Colour J.* 1985, 175 (4157), 890.
[20] Walker, P. *J. Oil Colour Chem. Assoc.* 1982, 65 (1), 415.
[21] Walker, P. *J. Oil Colour Chem. Assoc.* 1982, 65 (12), 436.
[22] Walker, P. *J. Oil Colour Chem. Assoc.* 1983, 66 (7), 188.
[23] Schrader, M.E.; Cardamone, J.A. *J. Adhes.* 1978, 9, 305.
[24] Kaul, A.; Sung, N.H.; Chin, I.J.; Sung, C.S.P. *Polym. Prepr.* 1985, 26 (2), 113.
[25] Cleveland Society for Paint Technology, *J. Coat. Technol.* 1979, 51 (653), 53.
[26] PPG Industries Ltd., US Patent No. 4 146 585 (1979).
[27] Bikiaris, D.; Matzinos, P.; Larera, A.; Flaris, V.; Panayiotou, C. *J. Appl. Polym. Sci.* 2001, 81, 701.

[28] Kotha, S.P.; Lieberman, M.; Vickers, A.; Schmid, S.R.; Mason, J.J., *J. Biomedical. Materials Research Part A* 2006, 76A, 111.

[29] Dow Corning Corp., US Patent No. 4 228 061; *Off. Gaz.* 999 (2) (1980) 696.

[30] Schering AG, Ger. Patent No. 2 857 425; *Paint Resin Patent,* 17 (11) (1980) Abstr. 1070.

[31] Schering AG, Ger. Patent No. 2 84 923. p. 5; *Ger. Patent Abstr.,* 1980. Vol. C, No. 35, GpG. 3.

[32] Namboori, C. G. G. *Text Chem. Color.* 1969, 1, 24.

[33] Waters, W. *J. Soc. Dyers Color.* 1950, 66, 609.

[34] Pficer, J. *Man-made Textiles*. 1964, 41, 47.

[35] Kallay, N.; Grancaric, A. M.; Tomtic, M. *Text Res. J.* 1990, 60, 663.

[36] Heidemann, G. *Melliand Textilberichte.* 1979, 60, 350.

[37] Hsiao, K. –J.; Jen, Z.-F.; Lu, C.-L. *J Appl. Polym. Sci.* 2002, 86, 3601.

[38] Teh, S.F.; Liu, T.; Wang, L.; He,C. *Composites: Part* A 2005, 36, 1167.

[39] Zeronian, S.H.; Collins, M.J., *Text Prog.* 1989, 20, 1.

[40] Solbrig, C.M.; Obendorf, S.K. *Text Res. J.* 1991, 61, 177.

[41] Fujimoto, K.; Iohara, K.; Ohwaki, S.; Murase, Y. J. *Appl. Polym. Sci.* 1991, 42, 1509.

[42] Van den Heurel, C.J.M.; Heuvel, H.M.; Faassen, W.A.; Veurink, J.; Lucas, L.J. *J. Appl. Polym. Sci.* 1993, 49, 925.

[43] Hutchinson, I.J.; Mard, I.M.; Willis, H.A.; Zichy, V. *Polymer.* 1980, 21, 55.

[44] Ghosh, A.; Gupta, M.; Schiraldi, D.A., unpublished work.

[45] Hull, D. An introduction to composite materials. Cambridge, UK: Cambridge University Press, 1981.

[46] Lopez-Manchado, M.A.; Arroyo, M. *Polymer.* 2000, 41, 7761.

In: Progress in Chemistry and Biochemistry
Editors: E. M. Pearce and G. E. Zaikov
ISBN: 978-1-60692-344-3

Chapter 11

RECYCLING OF SUGARCANE BAGASSE FOR PRODUCING CELLULOSE ACETATE

G. R. Filho,[1,*] D. A. Cerqueira,[1,2] R. M. N. Assunção,[1] C. S. Meireles[1] and A. J. M. Valente[2]
[1] Instituto de Química da Universidade Federal de Uberlândia
P.O. BOX 593, Zip Code 38400-902, Brazil
[2] Department of Chemistry, University of Coimbra,
3004-535. Coimbra, Portugal

ABSTRACT

The prospects for future applications of cellulose acetate and cellulose esters can be assessed by analysis of literature databases; according to the ISI Web of Knowledge, during the period between 2003-2006, 838 papers related with cellulose acetate were published; for a polymer initially reported in literature in 1945, those numbers clearly suggest a long-life term polymer, which can only be justified by the permanent interest and new findings of potential applications. Besides outstanding properties of these polymers, other factors such as sources and price are also important. In this review, we describe physical chemical properties of cellulose acetate and other cellulose derivatives obtained from a waste product: sugarcane bagasse.

Keywords: cellulose, cellulose acetate, cellulose derivatives, sugarcane bagasse.

INTRODUCTION

Due to its high crystallinity (up to 70%), cellulose is insoluble in most of the traditional organic solvents, which significantly reduces its application in its original form. One of the aspects that makes cellulose an important commercial polymer is its capacity to undergo chemical modifications, since most of the cellulosic derivatives can be dissolved in several common solvents. Cellulosic derivatives, such as esters and ethers of cellulose, may be produced due to the reactivity of the hydroxyl groups in the glucosidic units.

Among these derivatives, cellulose acetate presents great commercial importance. The main reasons and advantages for using cellulose acetate (CA) as a polymeric matrix, to be used alone or for incorporation of other compounds to form blends or composite materials are: neutral properties, capacity for transparent film formation, and low cost. In addition, water concentration and water diffusion coefficients in cellulose acetate can be controlled by changing thickness, solvent, etc. [1]. CA can be used in several processes such as in membranes for hemodialysis, nanofiltration and reverse osmosis [2], matrices for drug controlled release [3], sensors and as protection for optical films [4-7], gas separation [8], and preparation of alumina films [9], among others. Another interesting and growing field of CA applications is the use of this polymer as a matrix for other functional polymers or compounds, such as polyaniline [10] and macrocycles [11]. Blends containing PANi with cellulose esters can be used as hole transport layers for electronic devices, such as multilayer polymeric and organic light emitting diodes (LEDs) [12], whilst a number of applications of cellulose acetate-PANi blends have been reported in the literature in the last 10 years, including electronic devices [*e.g.*,13], selective electrodes [14], supports for urease immobilization for biochemical applications [15] and electrochromic devices [16]. Because of the reasons described above, cellulose acetate and, in general, cellulose-derivatives still are a great field of research and development, not only for fundamental but essentially for practical purposes.

The acetylated derivatives of cellulose may be produced by two kinds of acetylation reactions: homogeneous and heterogeneous [17]. Both reactions are usually carried out by using acetic acid as solvent, acetic anhydride as acetylating agent, and sulfuric and perchloric acids as catalysts. However, in the heterogeneous route, toluene or any other non swelling agent is used, what preserves the integrity of the cellulosic fibers and results in a more crystalline material then that obtained by the homogeneous route. Homogeneous route is the most common in the industry and the acetylation reaction usually leads to a product where nearly all hydroxyl groups are acetylated, and thus presents a degree of substitution (DS) next to 3 - the cellulose triacetate (CTA). CTA is then hydrolyzed to reduce its DS down to 2.4, which is called cellulose diacetate (CDA). The CDA is quite often called, in commercial terms, by simply cellulose acetate. The differences in the crystallinity and the degree of substitution (DS) of materials are the main factors related with the applicability of cellulose acetate to both the processability and the potential of biodegradability of the polymer; the latter increases as DS decreases [18].

The development of new cellulose derivative applications in the future depends on many factors, such as nature of the resource and products characteristics. Additional factors may involve societal preferences, legislative actions, market forces, prices, etc. [19]. Cellulose acetate and its derivatives may become the raw material for products resulting from secondary modifications.

In recent literature there is a series of papers about this important derivative, which focus in the synthesis of cellulose acetate from renewable sources [20-23], new routes of synthesis [22, 24-26], preparation of nanocomposites [27] and new materials [28-30].

Brazil is the world's major producer of sugarcane, with an estimate production of 469.8 million metric tons for the 2006/2007 harvest. From this production, 227.1 million tons of sugarcane will be used to produce sugar, while 186.3 million tons will be destined to produce 17.8 billion liters of ethanol (hydrated, anhydride and neutral). The remaining 46.4 million tons will be destined to other uses, such as production of animal feed and seeds. The planted area has increased in 5.4%, from 5.8 million hectares in 2005/2006 to 6.2 million hectares. This growth results from the occupation of areas previously used by cattle rising and grain farms.

As each sugarcane metric ton that is consumed to produce ethanol (biofuel) or sugar generates about 280 kg of residue (sugarcane bagasse), several alternatives have been proposed to reuse bagasse. These focus at the production of energy through burning (energetic recycling), increase in the production of ethanol by acid or enzymatic hydrolysis [31-33], and production of composites and cellulosic derivatives that aggregate value to the agro-industrial residue [34-37]. The Group of Polymer Recycling (GPR) in the Uberlândia University works with the latter, and has been developing and/or adapting methodologies that allow not only demonstrate the viability of recycling sugarcane bagasse for producing this important cellulosic derivative: they also look for demonstrating the performance of the obtained derivatives for some applications such as reverse osmosis, sensors, additives for mortar and concrete, and biodegradable seedling tubes for agro-industry.

Production and Application of Cellulose Acetate from Cellulose Extracted from Sugarcane Bagasse

Cellulose acetate is produced in Brazil by Rhodia S.A. from cellulosic pulps obtained from wood, imported from the United States of America. The use of agro-industrial residues as an alternative source of cellulose has been increased lately due to the necessity of searching for renewable sources which produce a lower environmental impact.

Research on the synthesis of cellulose acetate from sugarcane bagasse started, in GRP, with the extraction of cellulose and production of membranes of regenerated cellulose [38]. Posteriourly, cellulose from sugarcane bagasse was used in the heterogeneous production of cellulose acetate [35]. This material was characterized by Fourier Transformed Infra-Red (FT-IR), Wide-Angle X-Ray Diffraction (WAXD), Differential Scanning Calorimetry (DSC) and water vapor flux. The materials were produced using 24 (CA24) and 48 h (CA48) of acetylation, acetic acid as solvent, acetic anhydride as acetylating agent, perchloric acid as catalyst and toluene as non swelling agent. CA48 FTIR spectrum is shown in figure 1. The weak intensity of hydroxyl group bands in the region 3880 to 3200 cm^{-1} (due to the substitution of the hydroxyl groups by acetyl groups) and a new band at 1750 cm^{-1} assigned to the stretching of carbonyl groups (C=O) are typical of cellulose triacetate [35].

The results showed the viability of using sugarcane bagasse for producing cellulose acetylated derivatives. Besides, as seen in figure 2 [35], a fundamentally amorphous WAXD pattern is seen for CA24, while CA48 is semicrystalline. In the latter, the main characteristic is the presence of the maximum in 2θ approximately 8°, called low van der Waals halo, which corresponds to a certain chain parallelism.

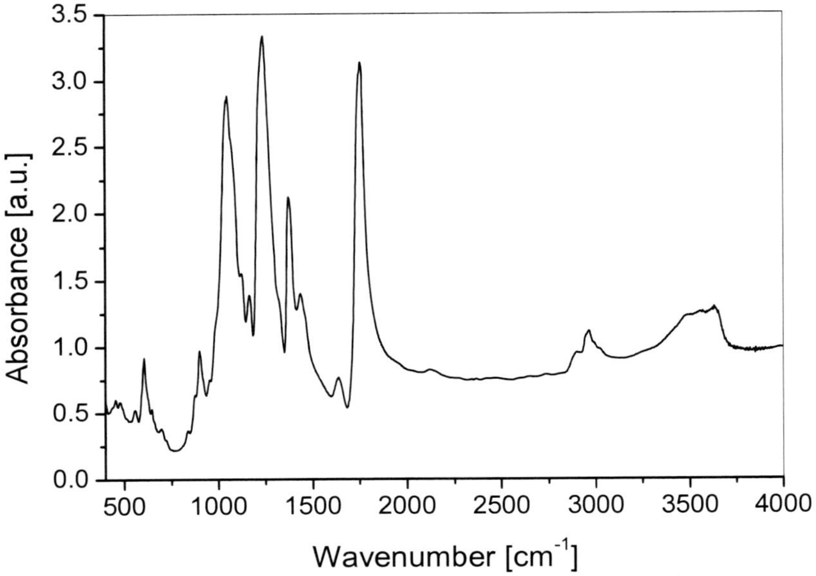

Figure 1. FTIR spectrum of cellulose acetate produced by sugarcane bagasse cellulose [35].

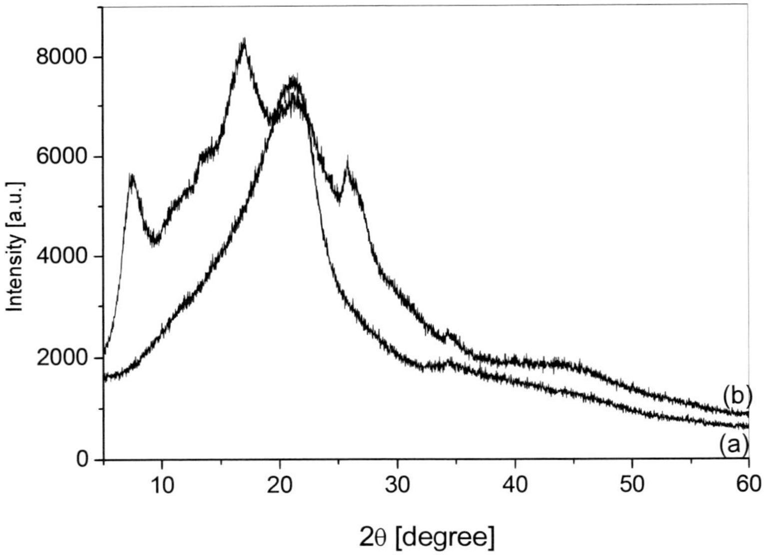

Figure 2. WAXS of CA24 (a) and CA48 (b) [35].

Figure 3 presents the DSC thermograms, where (a) CA24, is typically an amorphous material, as already shown through WAXD, and (b) CA48, is a semicrystalline material. The latter presents an endotherm peak (ΔH_f of 30 J g^{-1}) at 300 °C, due to the fusion transition, corresponding to a typical value for cellulose triacetate.

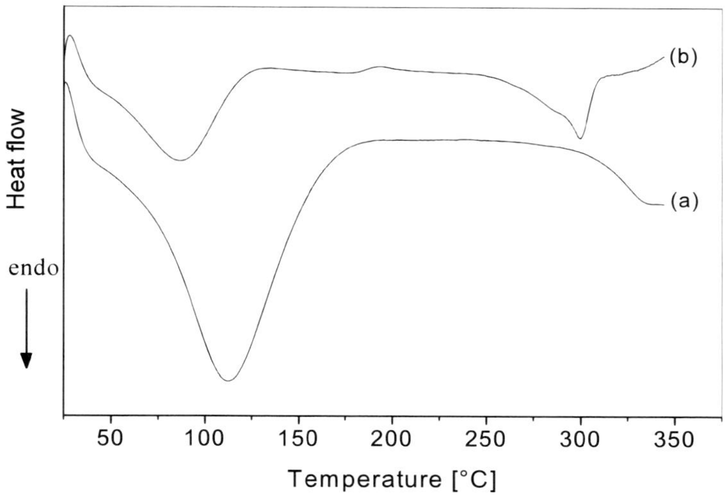

Figure 3. DSC curves for CA24 (a) and CA48 (b) [35].

These results are important not only for the characterization of cellulose acetate produced from an alternative source of cellulose, but also for the discussion about the crystallinity of cellulose acetates it brought about. Up to this point, the enthalpy of fusion (ΔH_f) of cellulose acetates had always presented very low values, especially considering the enthalpy of fusion for a perfect crystal (ΔH_f°), which value has been proposed by Takahashi *et al.*, 34 Jg^{-1} [39]. However, since ΔH_f of CA48 is 30 Jg^{-1}, which is very near to the ΔH_f° value and therefore this sample would be nearly 90 % crystallinity, what is not corroborated by WAXD. The value of 34 Jg^{-1} was thus questioned, resulting in a study that presented a new value for cellulose acetate ΔH_f°, 58 ±0.5 Jg^{-1} [40]. In these studies, the crystallinity of the samples was calculated using mathematic models of deconvolution of the WAXD curves, which made it possible to test hypotheses using the two-phase model, i.e., semicrystalline and amorphous. This crystallinity was then compared to ΔH_f of the samples. Figure 4 [40] shows the final result for the hypothesis that presented the best fit to the correlation between ΔH_f and crystallinity of the samples.

As the production of cellulose acetate from sugarcane bagasse cellulose is possible using this initial methodology, the next step was the producing of a membrane for applying the new material. Thus, membranes of CA48 were produced using dichloromethane as solvent. Even though these membranes presented low resistance, it was possible to study them by water vapor transport measurements.

In order to improve the quality of cellulose acetate and mainly the quality of the produced membranes for reverse osmosis systems, the parameters of synthesis have been altered such as the route to obtain cellulose acetate (heterogeneous or homogeneous) and in another step, the proportion of reactants, catalyst and reaction time was altered. In this sense, homogeneous acetylation was first evaluated. The membranes made with this material, as well as the membranes produced using the heterogeneous route, were usually brittle, what made them unusable for separation processes dealing with high pressures. An alternative proposal for using this material was the use of another polymer in the preparation of the membrane that would reinforce it and increase its workability. In this sense, polystyrene from waste cups was used. The results showed that the use of polystyrene in fact improved the quality of the membranes [41,42].

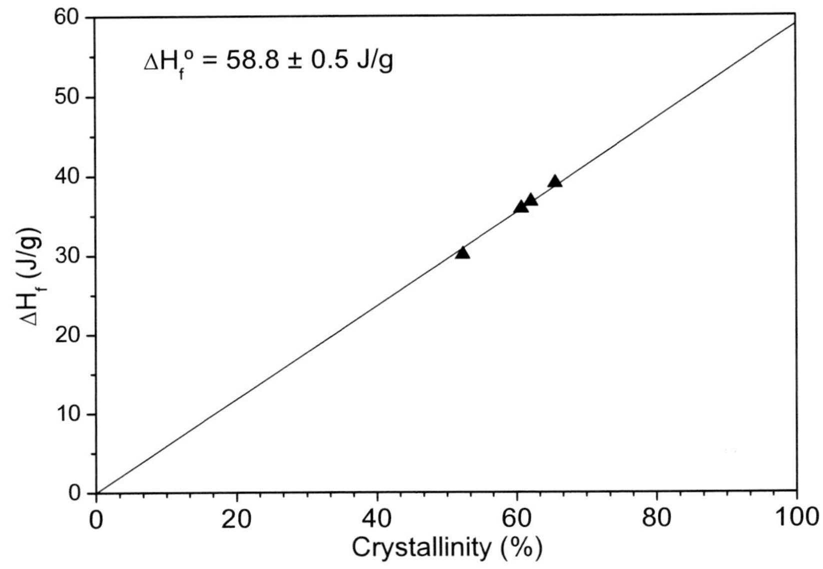

Figure 4. Relationship between ΔH_m and crystallinity of cellulose acetate samples [40].

Even though the processing of the membrane improved, in order to apply this system it was necessary to study the interaction between the two constituents of the blend. This study was carried out using FTIR aiming to understand in which structural level the interaction would occur between these two fundamentally incompatible materials. FTIR spectrum, figure 5, shows alterations of position and intensity of some bands. Particularly the band in 1429 cm^{-1}, which is related to the crystallinity regions of cellulose acetate, suffers drastic modification. This band does not appear, or it is very weak, in the blend and its change is attributed to PS modifying the crystallinity of cellulose acetate, thus indicating miscibility regions where van der Waals forces predominate. These results were also corroborated by DSC and WAXD [41]. Using other techniques such as TGA, MEV and water diffusion through the blends, it was possible to confirm the existence of miscibility regions and that these interactions occur in a molecular level, in a short extension between the polymers [42].

At the same time, an optimization process of the cellulose acetylation reaction was performed in order to obtain a higher molecular weight product [43]. The results have shown that it is possible to improve the synthesis of cellulose acetate in order to obtain membranes with better workability. In spite of this improvement in the production of the membranes using only cellulose acetate, the use of polystyrene in the formulation was an important aspect due to the possibility of modification in the properties of the membrane, mainly those related to the transport properties. Furthermore, it allows the recycling of two materials, waste polystyrene and sugarcane bagasse.

Cellulose acetate membranes as matrices in the process of controlled release of drugs have also been studied. The production of these matrices was investigated with the use of cellulose acetate and polyethylene glycol 600 g mol^{-1} (PEG600). PEG600 had two functions:

pore former agent and plasticizing admixture in order to improve the flexibility and workability of the membranes. One of the fundamental aspects in this work was to evaluate and demonstrate the potential of this system, produced from recycled material, in medical applications since the cytotoxicity assay shows that the system is nontoxic, as observed in figure 6 [44].

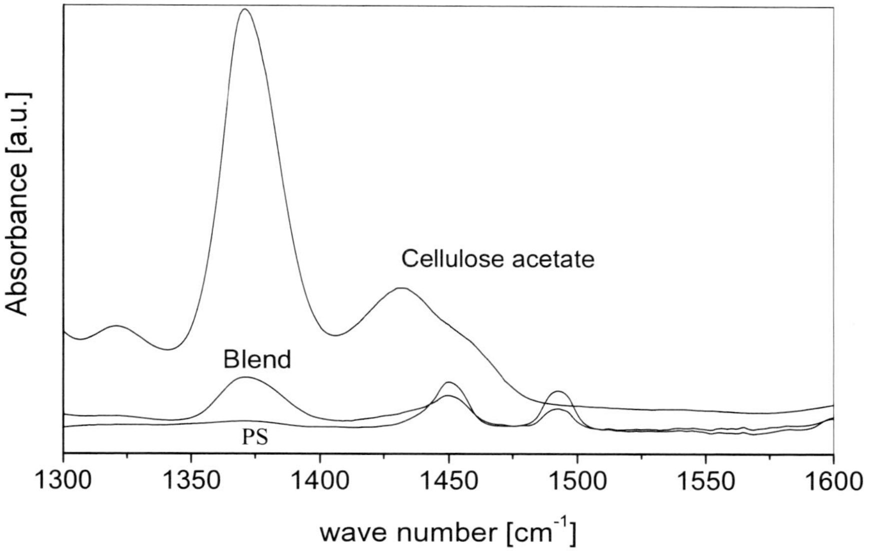

Figure 5. FTIR spectra, in the range from 1300 to 1600 cm^{-1}, for cellulose acetate (CA), polystyrene (PS) and the blend [41].

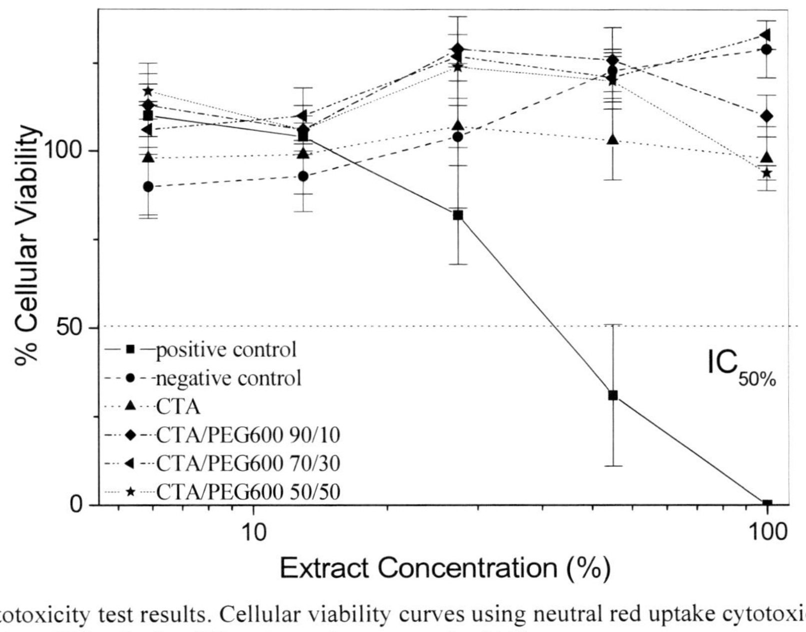

Figure 6. Cytotoxicity test results. Cellular viability curves using neutral red uptake cytotoxicity assay, with NCTC clone 929 cells for different membrane samples [44].

The improvement in workability of the membrane may be attributed to the role of PEG in its producing. In fact, the enthalpy of fusion (ΔH_f) and the temperature of fusion (T_f) increase by increasing PEG content, at the same time as the enthalpy of crystallization (ΔH_c) decreases during the heating scan (Figure 7, [44]). This clearly suggests that PEG can act as a crystallinity inductor. This was firstly reported by us in the ref. [44].

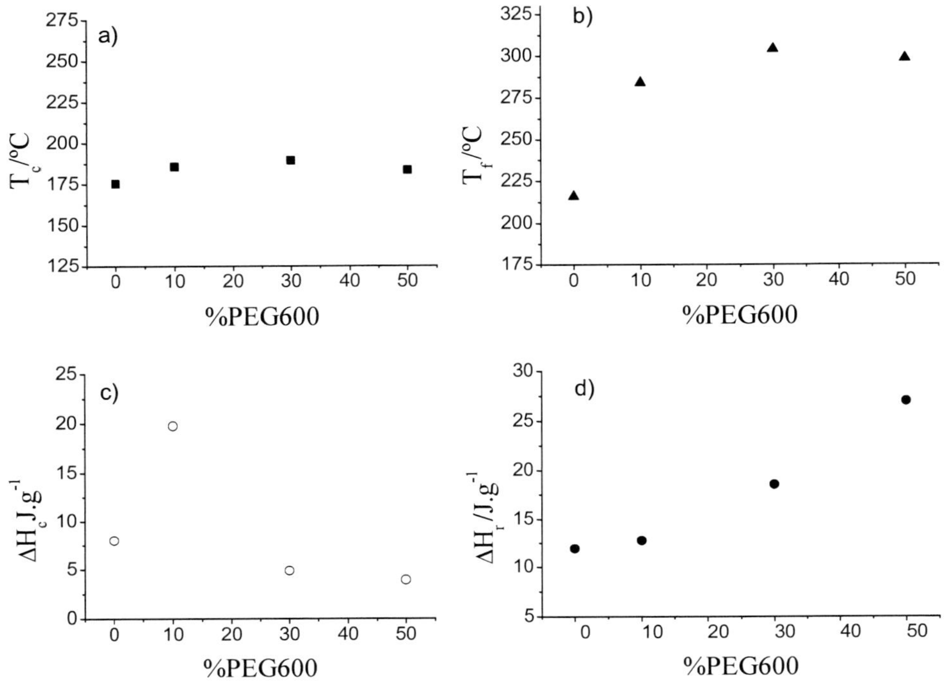

Figure 7. a) Tc versus % PEG 600; b) T_f versus % PEG 600; c) ΔH_c versus % PEG 600; d) ΔH_{fv} versus % PEG 600 [44].

The optimization process was continued using mathematical methods for a more criterious investigation of the reaction parameters, using simplex optimization [43]. This methodology led to a material with higher molecular weight and better workability of the produced membranes, what is allowing the development of scientific research cooperation with the Chemistry Department of the University of Coimbra (Portugal).

The optimization and the study with PEG600 also allowed the investigation of transport properties of the membranes in separation processes using high pressures. Even though membranes of cellulose acetate resist to high pressures for some time, it was noticeable on the tests carried out that, by using PEG600 in the membrane preparation, the flexibility of the matrix was increased and it was possible to use them in high pressures. Membranes containing or not PEG600 resisted to the highest operational pressure of the equipment, but the membrane without PEG600 broke after a few minutes, while the others resisted for many hours. The membrane produced with cellulose acetate before the optimization step would resist only to pressures up to 1 bar [45].

CONCLUSION

The research that has been carried out shows the viability of using cellulose acetate from renewable sources such as sugarcane bagasse for producing cellulose acetate. Therefore the ongoing research work points to the possibility of, e.g., using this recycled material for producing biodegradable seedling tubes, as well as in the substitution of the commercial material, in blends with polyaniline for producing sensors for the detection of surfactants.

ACKNOWLEDGMENTS

Financial support from FINEP/SEBRAE (0535/2007) and CNPq is gratefully acknowledged. D.A. Cerqueira thanks CAPES/Brazil for financial support (BEX 0368/07-5) of his stay at the University of Coimbra.

REFERENCES

[1] A.J.M. Valente, A.Ya. Polishchuk, H.D. Burrows, V.M.M. Lobo. *Eur. Polym. J.* 41 (2005) 275.
[2] H. Strathmann. *Pure Appl. Chem.* 46 (1976) 213.
[3] K.J. Edgar. *Cellulose.* 14 (2007) 49.
[4] Z. Liao, E. Klein, C.K. Poh, Z. Huang, J. Lu, P.A. Hardy, D. Gao. *J. Membrane Sci.* 256 (2005) 176.
[5] F.-J. Wang, Y.-Y. Yang, X.-Z. Zhang, X. Zhu, T.-S. Chung, S. Moochhala. *Mater. Sci.* Eng. C 20 (2002) 93.
[6] K. J. Edgar, C. M. Buchanam, J. S. Debenham, P.A. Rundquist, B.D. Seiler, M.C. Shelton, D. *Tindall. Prog. Pol. Deg.* 26 (2001) 1605.
[7] T.K. Dey, B.M. Misra. *J. Polym. Mater.* 16 (1999) 13.
[8] S. Sridhar, B. Smitha, T. M. Aminabhavi. *Separat. Purif. Rev*. 36 (2007) 113.
[9] Y. Kobayashi, T. Ishizaka, Y. Kurokawa. *J. Mater. Sci*. 40 (2005) 263.
[10] H.D. Burrows, V.M.M. Lobo, J. Pina, M.L. Ramos, J. Seixas de Melo, A.J.M. Valente, M.J. Tapia, S. Pradhan, U. Scherf, S.I. Hintschich, C. Rothe, A.P. Monkman. *Colloids Surf. A: Physicochem. Eng. Aspects.* 270/271 (2005) 61-66.
[11] A.J.M. Valente, A. Jiménez, A.C. Simões, H.D. Burrows, A.Ya. Polishchuk, V.M.M. Lobo. *Eur. Polym. J.* 43 (2007) 2433.
[12] R.W.T. Higgins, N.A. Zaidi, A.P. Monkman. *Adv. Funct. Mater.* 11 (2001) 407.
[13] J. Planes, A. Wolter, Y. Cheguettine, A. Pron, F. Genoud, M. Nechtschein. *Phys. Rev.* B 58 (1998) 7774.
[14] T. Lindfors, P. Sjoberg, J. Bobacka, A. Lewenstam, A. Ivaska. *Anal. Chimica Acta.* 385 (1999) 163.
[15] J. Laska, J. Wlodarczyk, W. Zaborska. J. Molec. *Catalysis B-Enzymatic.* 6 (1999) 549.
[16] M.A. DePaoli, E.R. Duek, M.A. Rodrigues. *Synth. Met*. 41 (1991) 973.
[17] J.F. Sassi,.H. Chanzy. *Cellulose.* 2 (1995) 111.
[18] E. Samios, R. K. Dart , J. V. Dawkins. *Polymer*. 38 (1997) 3045.

[19] W.G. Glasser. *Macromol. Symp.* 208 (2004) 371.
[20] J.M. Song, Y.J. Zhang, A.W. Xu, S. H. Yu. *Langmuir.* 23 (2007) 7321.
[21] Z.T. Liu, X. Fan, J. Wu, L. Zhang, L. Song, Z. Gao, W. Dong, H. Xiong, Y. Peng, S. *Tang. React. Funct. Polym.* 67 (2007) 104.
[22] A. Biswas, B.C. Saha, J.W. Lawton, R.L. Shogren, J.L. Willett. *Carbohydrates.* 64 (2006) 131.
[23] B.A.P. Ass, G.T. Giacco, E. Frollini, *Bioresource Technology.* 97 (2006) 1696.
[24] B.A.P. ASS, E. Frollini, T. Heinze. *Macromol. Biosci.* 4 (2004) 1008.
[25] J. Luo, Y. Sun. *J. Appl. Polym. Sci.* 100 (2006) 3288.
[26] L.F. Yan, W. Li, Z.H. Qi, S.Y. Liu. *J. Polym. Res.* 13 (2006) 375.
[27] M. Yoshioka, K. Takabe, J. Sugiyama, Y. Nishio. *J. Wood Sci.* 52 (2006) 121.
[28] F. Fischer, A. Rigacci, R. Pirard, S. Berthon-Fabry, P. *Achard. Polymer.* 47 (2006) 7637.
[29] K.G. Varshney, M. Drabik, A. Agrawal. Indian J. Chem. Section A – Inorg. Bioinorg. Phys. Theor. Analyt. Chem. 45 (2006) 2045.
[30] H. Kamitakahara, T. Enomoto, C. Hasegawa, F. Nakatsubo. *Cellulose.* 12 (2005) 527
[31] R. Basanta, M.A.G. Delgado, J.E.C. Martinez, *et al.* Ciencia y Tecnologia Alimentaria 5 (2007) 293.
[32] S.P. Singh, R.K. Asthana, A.P. Singh. *Energy Policy.* 35 (2007) 4164.
[33] C. Martín, M. Galbe, C. F. Wahlbom, B. Hahn-Hagerdal, L. J .Jonsson. *Enzyme Microbial Techn.* 31 (2002) 274.
[34] D. Pasquini, E. de M, Teixeira, A. A. da S. Curvelo, M. N. Belgacem and A. Dufresne. *Compos. Sci. Techn.* 68 (2008) 193.
[35] G. Rodrigues Filho, S.F. Cruz, D. Pasquini, D.A. Cerqueira, V.S. Prado, R.M.N. Assunção. *J. Membrane Sci.* 177 (2000) 225.
[36] K. Bilba, M-A. Arsene and A. Ouensanga. *Cement Concrete Comp.* 25 (2003) 91.
[37] G. Rodrigues Filho, R. M.N. de Assunção, J. G. Vieira, C. da S. Meireles, D. A. Cerqueira, H. S. Barud, S. J.L. Ribeiro, Y. Messaddeq. *Polym. Degrad. Stab.* 92 (2007) 205.
[38] G. Rodrigues Filho, R. M. N. Assunção. *J. Membrane Sci.* 82 (1993) 43.
[39] A. Takahashi, T. Kawaharada, T. Kato. *Polymer J.* 11 (1979) 671.
[40] D.A. Cerqueira, G. Rodrigues Filho, R.M.N. Assunção. *Polym. Bull.* 56 (2006) 475.
[41] G. Rodrigues Filho, R.C. da Silva, C.S. Meireles, R.M.N. Assunção, H. Otaguro, *J. Appl. Polym. Sci.* 96 (2005) 516.
[42] C.S. Meireles, G. Rodrigues Filho, R.M.N. Assunção, K. Mello, M. Zeni. *J. Appl. Polym. Sci.* 104 (2007) 909.
[43] D.A. Cerqueira, C.S. Meireles, G. Rodrigues Filho, *Carbohydrate Polym.* 69 (2007) 579.
[44] G. Rodrigues Filho, L.C. Toledo, D.A. Cerqueira, R.M.N. Assunção, C.S. Meireles, H. Otaguro, S.O. Rogero, A.B. Lugao, *Polym. Bull.* 59 (2007) 73.
[45] D.A.Cerqueira, G. Rodrigues Filho, R. M. N. de Assunção, C. S. Meireles, L. C. Toledo, M. Zeni, K. Mello and J. Duarte, *Polym. Bull.* In press, DOI 10.1007/s00289-007-0856-8.

In: Progress in Chemistry and Biochemistry
Editors: E. M. Pearce and G. E. Zaikov

ISBN: 978-1-60692-344-3

Chapter 12

DEVELOPMENT OF A RELIABLE, EFFICIENT AND AUTOMATED METHOD FOR MEASURING NANOFIBER DIAMETER IN ELECTROSPUN WEBS

M. Ziabari, V. Mottaghitalab and A. K. Haghi*
University of Guilan, P.O.BOX 3756, Rasht, Iran

ABSTRACT

This paper describes the new distance transform method used for measuring fiber diameter in electrospun nanofiber webs. The results obtained by our method were significantly better than the distance transform indicating that the new method could successfully be used to measure electrospun fiber diameter.

1. INTRODUCTION

Fibers with a diameter of around 100 *nm* are generally classified as nanofibers. What makes nanofibers of great interest is their extremely small size. Nanofibers compared to conventional fibers, with higher surface area to volume ratios and smaller pore size, offer an opportunity for use in a wide variety of applications. To date, the most successful method of producing nanofibers is through the process of electrospinning. The electrospinning process uses high voltage to create an electric field between a droplet of polymer solution at the tip of a needle and a collector plate. When the electrostatic force overcomes the surface tension of the drop, a charged, continuous jet of polymer solution is ejected. As the solution moves away from the needle and toward the collector, the solvent evaporates and jet rapidly thins and dries. On the surface of the collector, a nonwoven web of randomly oriented solid nanofibers is deposited [1]-[8]. Figure 1 illustrates the electrospinning setup.

* Haghi@Guilan.ac.ir

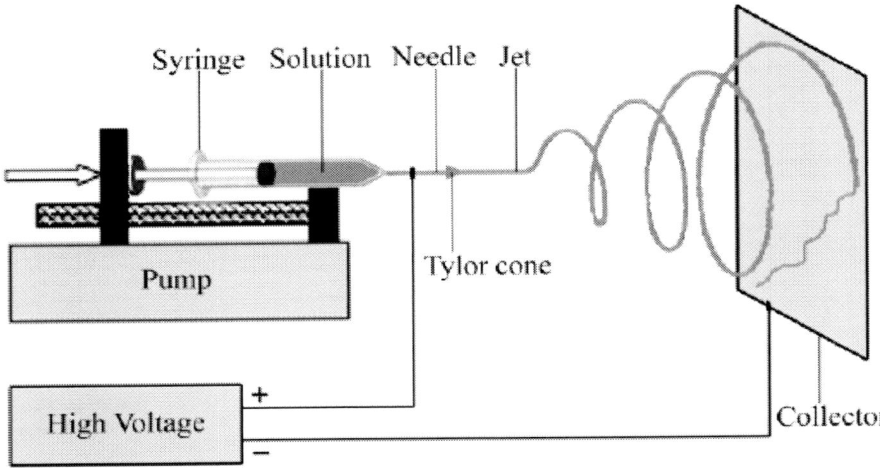

Figure 1. Electrospinning setup.

The properties of electrospun nanofiber webs depend not only on the nature of the component fibers but also on its structural characteristics. In the last few years, image analysis methods have been developed in order to identify fibers and measure nonwoven characteristics such as fiber orientation [9]-[16], fiber diameter [17], [18], pore size [19], [20], uniformity [21] and other structural features [15], [22]. Note however that, since these are new techniques and their accuracy and limitations have not been verified, samples with known characteristics are required to evaluate the accuracy of the methods which can be produced by simulation schemes [9], [23].

Fiber diameter is the most important structural characteristic in electrospun nanofiber webs. Despite the importance, thus far there is no successful method for determining fiber diameter and a few works have been conducted to develop a method for measuring fiber diameter. Furthermore, large scale production of nanofibers requires unique on-line quality control. Hence, developing an accurate and automated fiber diameter measurement technique is useful and crucial. In a method proposed by Pourdyhimi et al. [17] image analysis has been used to measure fiber diameter in nonwoven textiles. Nevertheless, the method has some problems at the intersections of fibers making it inefficient for measuring electrospun nanofiber diameter. In this contribution, an attempt has been made to circumvent the problems associated with this method thereby developing a reliable, efficient and automated method for measuring nanofiber diameter in electrospun webs.

2. Methodology

2.1. Fiber Diameter Measurement

Understanding how fiber diameter and its distribution are affected by the electrospinning variables is essential to produce nanofibers with desired properties. The extremely small fiber size and random production of nanofiber made its diameter measurement very difficult. Most commercially available measurement equipment cannot work with nanofibers [18]. In order to measure fiber diameter, images of the webs are required. These images called micrographs usually are obtained by Scanning Electron Microscopy (SEM), Transmission Electron Microscopy (TEM) or Atomic Force Microscopy (AFM). Dealing with fiber diameter requires high-quality images with appropriate magnifications. The methods for measuring fiber diameter are presented as follows.

2.1.1. Manual Method

Routine measurement of fiber diameter and its distribution are carried out by manual method using micrographs obtained from SEM. First the length of a pixel in the image is determined; in other words the scale is set. Then a fiber is selected and pixels between two edges of the fiber perpendicular to the fiber axis are counted. The number of the pixels is then converted to *nm* using the scale and the resulting diameter is recorded. This procedure is repeated for other selections until any fiber is processed. Typically 100 diameters are measured (Figure 2). Finally the histogram of fiber diameter distribution is plotted.

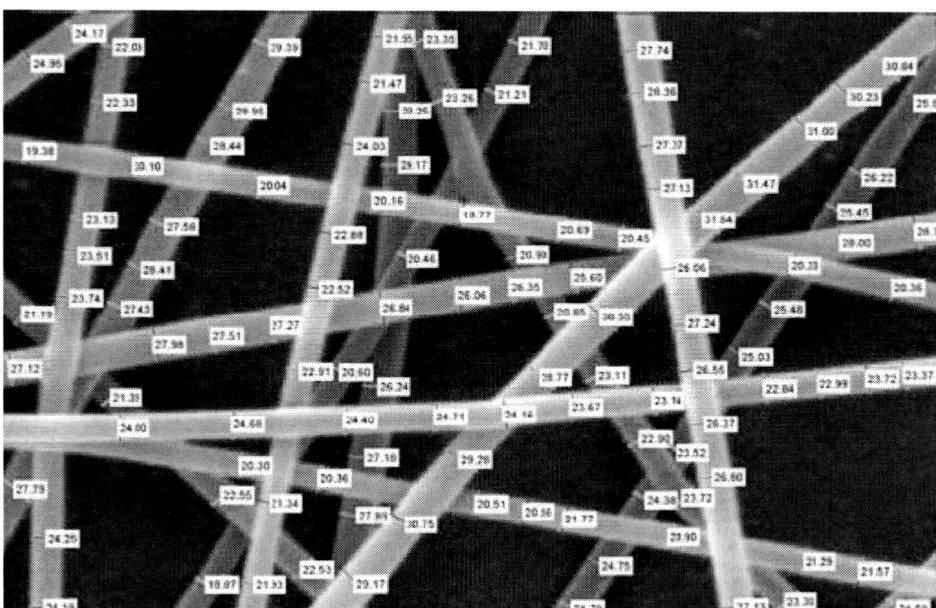

Figure 2. Manual method.

This process is very time-consuming and operator consistency and fatigue can reduce the accuracy. Identifying the edges of the fibers needs attention and the measurements are not exactly made perpendicular to the fiber axis. Furthermore, since it is an operator based method, it cannot be used as on-line method for quality control. Automating the fiber diameter measurement which eliminates the use of an operator is a natural solution to this problem.

2.1.2. Distance Transform Method

The *skeleton* of an object in a binary image, which provides helpful information about the shape of the object, is defined as the corresponding object with one-pixel width. There are two approaches for assessing the object's skeleton: *skeletonization* and *thinning*. In the first, using medial axis transformation (MAT), the center points of the object which are equidistant form two closest points of the object's boundary are attained and set to as skeleton [24]. Whereas, in the second, the pixels on the boundary of the object are removed without allowing it to break apart thereby shrinking a thick object to a centrally located one-pixel width object. In thinning operation, the following conditions must be satisfied:

1. An object must not break into pieces.
2. The end points must not be removed so that the object does not become shorter.
3. An object must not be deleted [25].

Both of these two operations result in line-like structures with one-pixel in thickness preserving the topology of the object. However, the skeleton obtained by skeletonization is often different with that of thinning and has more branches. Figure 3a shows a binary image to which skeletonization and thinning is applied and the resultant skeletons are depicted in Figure 3b and Figure 3c respectively. Note that these operations often produce short *spurs* (also called *parasitic components*) which may further be cleaned up by a postprocessing called *pruning* procedure (Figure 3d). Identifying and removing the spurs iteratively, this procedure is an essential complement to skeletonization and thinning [26].

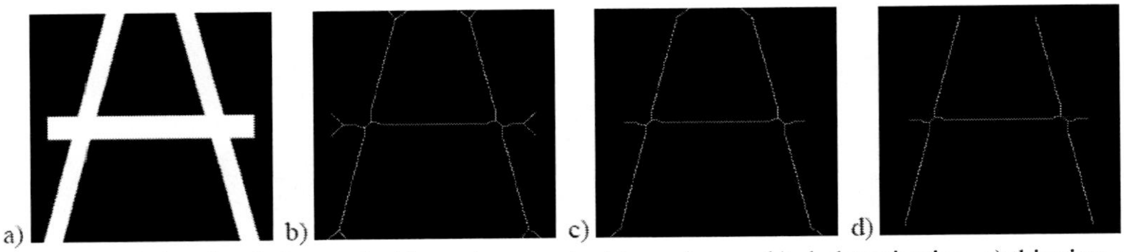

Figure 3. Obtaining the skeleton of a binary image: a) a binary image, b) skeletonization, c) thinning, d) resulting skeleton after pruning.

The distance transform is an operation which is applied to a binary image consisting of 1s and 0s corresponding to objects and background respectively, and results in a grayscale image often called *distance map* (known also as distance transformed image). For each pixel in the binary image, the corresponding pixel in the distance map has the value equal to the minimum distance between that pixel and the closest object pixel, that's to say, the distance from that pixel to the nearest non-zero valued pixel [25], [26]. There are several different sorts of distance transform according to which distance metric is being used in order to measure the distance between the pixels. Three common distance metric used in this approach are: city block, chessboard and Euclidean. The city block distance gives the length of a path between the pixels according to a 4-connected neighborhood (moving only in horizontal and vertical directions). The city block distance between (x_1,y_1) and (x_2,y_2) is given by:

$$D_{Cityblock} = | x_1 - x_2 | + | y_1 - y_2 | \tag{1}$$

In contrast, the chessboard distance metric measures the path between the pixels based on an 8-connected neighborhood (diagonal move is also allowed) as if a King moves in chess. This metric is given by:

$$D_{Chessboard} = Max(|x_1 - x_2|, |y_1 - y_2|) \tag{2}$$

With the city block metric, distances in the direction of diagonals are longer resulting in diamond-shaped structures. In the case of chessboard metric is used. Square-shaped structures are obtained [25]. Even though they could be used in certain applications, Euclidean metric is more practical and relevant, since it is the only one that preserves the isotropy of the continuous space (Figure 4) [27]-[29]. The Euclidean distance which is the straight line distance between two pixels is defined as:

$$D_{Euclidean} = \sqrt{(x_1 - x_2)^2 + (y_1 - y_2)^2} \tag{3}$$

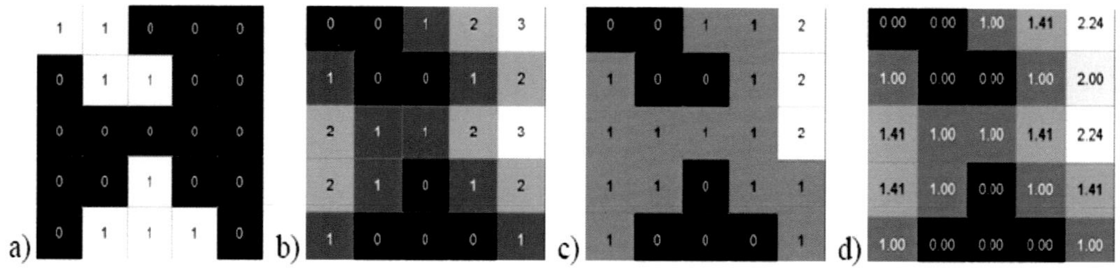

Figure 4. Distance map of a binary image: a) a small binary image and its distance map obtained by b) city block, c) chessboard, d) Euclidean metric.

The center of an object in distance transformed image has the highest value which coincides with the axis of the object. Interestingly enough, an object's skeleton will lie exactly over the maximum of the distance transform for that object [17]. This fact is clearly demonstrated in Figure 5. Note that the z-position of the skeleton in Figure 5d is quite arbitrary, just for showing the coincidence. Serving as the basic component of the methods, this remarkable feature will later be utilized for determining nanofiber diameter distribution in electrospun webs.

The algorithm for determining fiber diameter uses the skeleton and distance map of binary input image. Since our images consist of light fibers on dark background, they first need to be complemented. In the complement of a binary image, zeros become ones and ones become zeros; black and white are reversed [26]. Thus fibers become black and background white. The complemented image is used to create a distance transformed image. Then the skeleton of the objects is created from the input binary image by the process of skeletonization or thinning. Fiber diameter is then determined using the distance transformed image and the skeleton. The skeleton acts as a guide for tracking the distance map and distances at all points along the skeleton (which coincide with the center of the objects in the distance transformed image) are recorded to compute fiber diameters. Finally, the recorded results are doubled and fiber diameters are obtained. The values (in pixels) may further be converted to *nm* and the histogram of fiber diameter distribution is plotted. This method was proposed by Pourdeyhimi et al. [17]. Figure 6 shows a simple simulated image together with its skeleton and distance map including the histogram of fiber diameter obtained by this method.

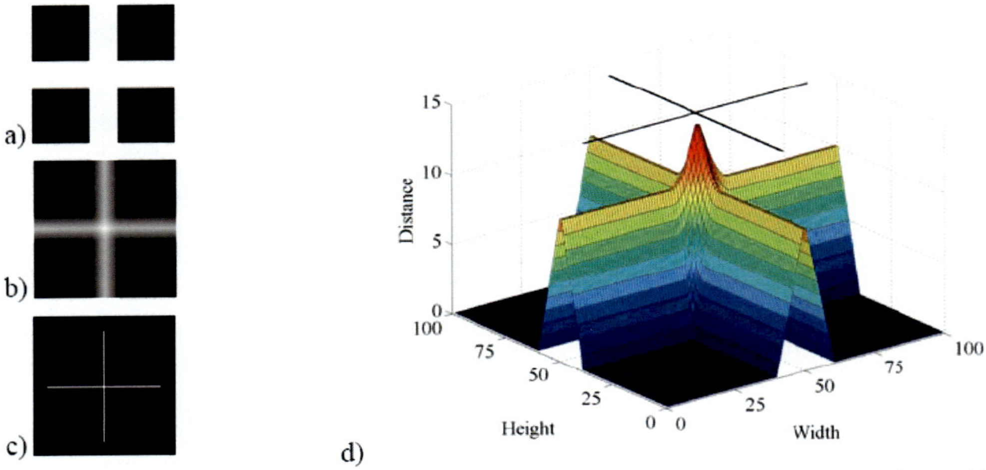

Figure 5. Skeleton lies exactly over the center of distance map: a) a simple 100×100 binary image, b) Euclidean distance map, c) skeleton obtained by thinning after pruning, d) 3-D plot showing the coincidence of skeleton and center of the object in distance map.

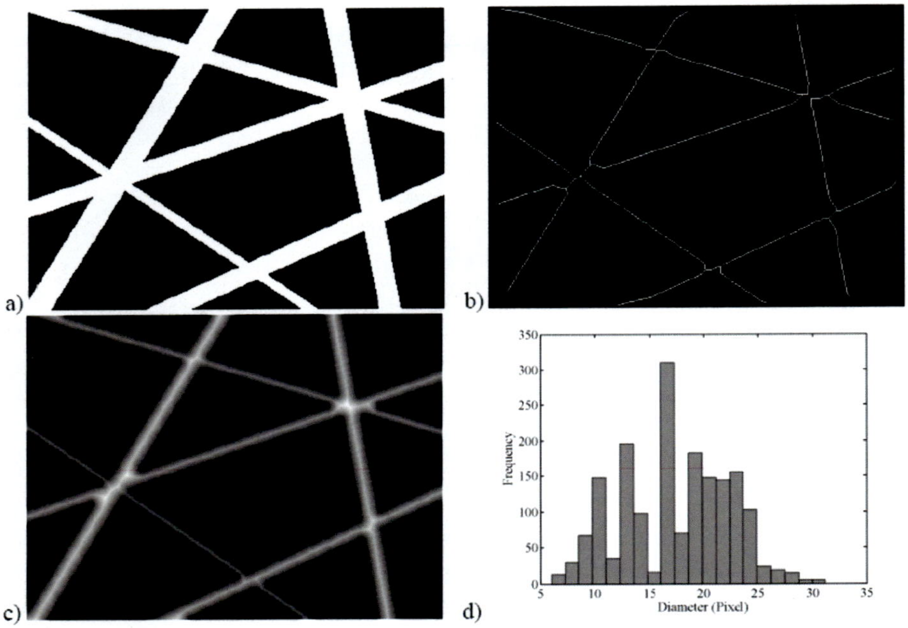

Figure 6. Distance transform method: a) a simple simulated image, b) skeleton of (a), c) distance map of (a) after pruning, d) histogram of fiber diameter distribution.

2.1.3. New Distance Transform Method

The problem of the distance transform method is that skeletons are often broken at intersection points. Furthermore, since two or more fibers cross each other at the intersections, the value of the center of the object in the distance transformed image doesn't coincide with the fiber diameter because it isn't corresponding to a single fiber. As it is depicted in Figure 7a, the intersections in the distance map are brighter than where a single fibers present. This demonstrates that higher values than expected were returned at these points. Figure 7b shows the broken skeleton at intersections. This problem becomes more pronounced as fibers get thicker and for points where more fibers cross each other. Hence, the distance transform method fails in measuring fiber diameter at intersections.

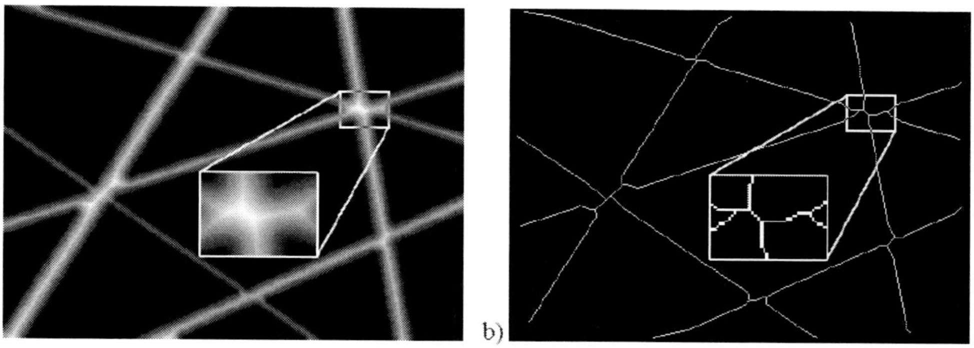

Figure 7. Distance transform method failure at intersection points: a) distance map of the image shown in Figure 6a, b) Broken skeleton obtained from thinning of Figure 6a (area around an intersection has been magnified for the sake of clarity).

We modified the distance transform method so that the problems associated with the intersections are solved. Furthermore, in the method proposed by Pourdeyhimi et al. [17], city block distance transform was used which as mentioned earlier, is not a realistic metric since it does not preserve the isotropy. In order to provide more rational results, in this approach we used Euclidean distance metric. The method uses a binary image as an input. Then, the distance transformed image and its skeleton are created. In order to solve the problem of the intersections, these points are identified and deleted from the skeleton.

First, in order to find the intersection points, a *sliding neighborhood operation* is employed. A sliding neighborhood operation is an operation which is applied to a pixel at a time; the value of that pixel in the output image is determined by the implementation of a given function to the values of the corresponding input pixel's neighborhood (Figure 8). A neighborhood about a pixel, which is usually called the center point, is a square or rectangular region centered at that pixel. The operation consists of five steps:

1. Defining a center point and a neighborhood block.
2. Starting from the first (normally top left) pixel in the image.
3. Performing an operation (a function given) that involves only the pixels in the defined block.
4. Finding the pixel in the output corresponding to the center pixel in the block and setting the result of the operation as the response at that pixel.
5. Repeating steps 3 to 4 for each pixel in the input image [26].

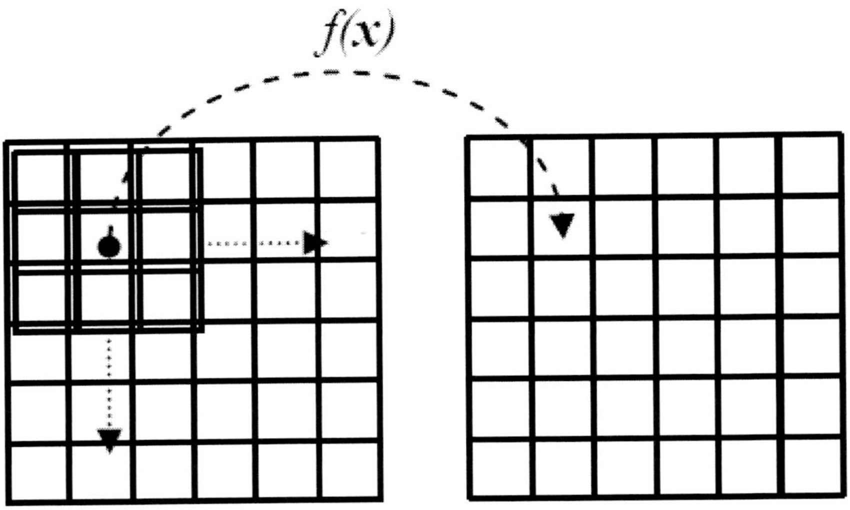

Figure 8. Sliding neighborhood operation with a 3-by-3 neighborhood block.

Since at an intersection point, two or more fibers meet each other, it could be defined as a location where a white pixel in the skeleton has more than two neighboring pixels each leading a branch. Hence, performing a sliding neighborhood operation on the skeleton with a 3-by-3 sliding block and summation as the function (which is applied over all pixels in the block), the intersections could be identified as the points having values more than 3. This is demonstrated in Figure 9 (the intersections are shown with arrows).

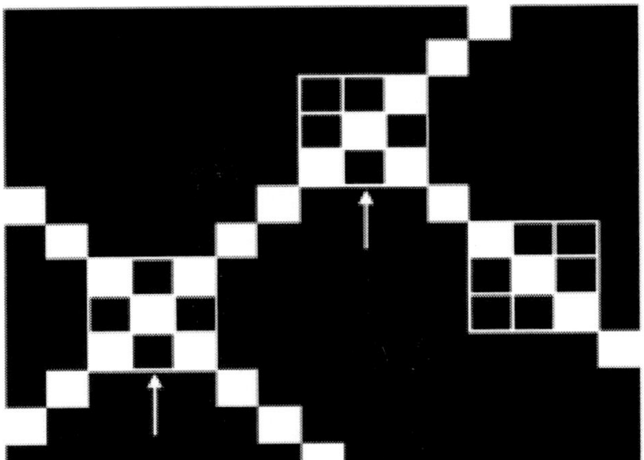

Figure 9. Identifying intersection points using a sliding neighborhood operation with a 3-by-3 neighborhood block.

After the intersection points are located, the next step is to find the width of each one. This is carried out using the distance map of the binary input image via finding the pixel corresponding to that intersection point. The value of the distance map at the pixel is then considered as the width of that intersection. After that, the pixels in the skeleton which lie inside a square with that width around the intersection point are cleaned. This procedure is replicated until each intersection is identified and cleaned. Figure 10 a exhibits the skeleton of the simple simulated image shown in Figure 6a after deleting the intersection points followed by a pruning procedure.

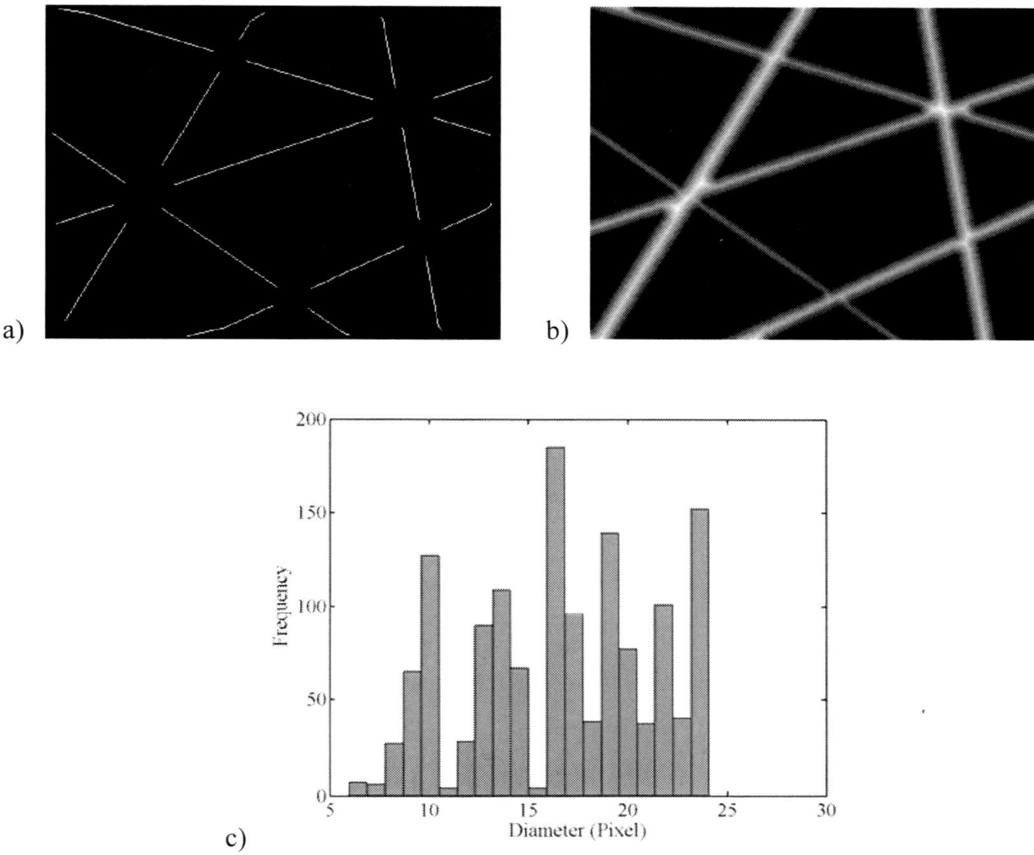

Figure 10. New distance transform method: a) the skeleton of the simple simulated image shown in Figure 6a after deleting the intersection points, b) the distance map, c) histogram of fiber diameter distribution.

Finally, the resultant skeleton (of which the intersections are deleted) is used as a guide for tracking the distance transformed image and fiber diameters are obtained by recording the intensities to at all points along the skeleton (white pixels in in Figure 10a show the skeleton) and doubling the results. The distance map of image in Figure 6a is also shown in 10b for better understanding of the procedure. Setting the length of a pixel in the image, the values may then be converted to *nm* and the histogram of fiber diameter distribution is plotted. Figure 10c demonstrates the histogram of fiber diameter (in term of pixel) obtained by this method. The procedure for determining fiber diameter via this approach is summarized in Figure 11. The method is efficient, reliable, accurate and so fast and has the capability of being used as an on-line method for quality control.

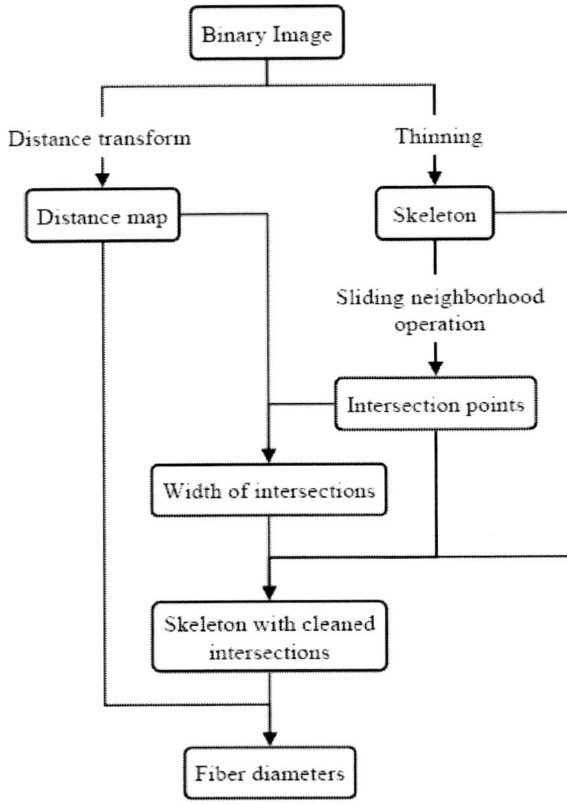

Figure 11. Flowchart of the new distance transform method.

2.2. Validation of the Methods

In order to validate the methods for determining fiber diameter, test samples with known characteristics are required. Since it is almost impossible to obtain real electrospun webs with specific characteristics through the experiment and there is not a method which measures fiber diameter precisely to compare the results with, the method will not be well evaluated using real webs merely. To that end, a simulation algorithm has been employed for generating test samples (which are binary images resemble electrospun webs) with known characteristics. A geometric model has been considered here to simulate electrospun fiberwebs. There are three widely used methods for generating random network of lines. These are called S-randomness, μ-randomness (suitable for generating a web of continuous filaments) and I-randomness (suitable for generating a web of staple fibers). These methods have been described in elaborations by Abdel-Ghani and Davis [23] and Pourdeyhimi et al. [9]. Since the physical characteristics of simulations are known exactly, one can employ them to test the usefulness of algorithm used in characterizing diameter and other structural features [9]. In this study, *μ-randomness* procedure has been used for generating simulated images with known characteristics. Under this scheme, a line with a specified thickness is defined by the perpendicular distance d from a fixed reference point O located in the center of the image and the angular position of the perpendicular α. Distance d is limited to the diagonal of the image.

Figure 12 demonstrates this procedure.

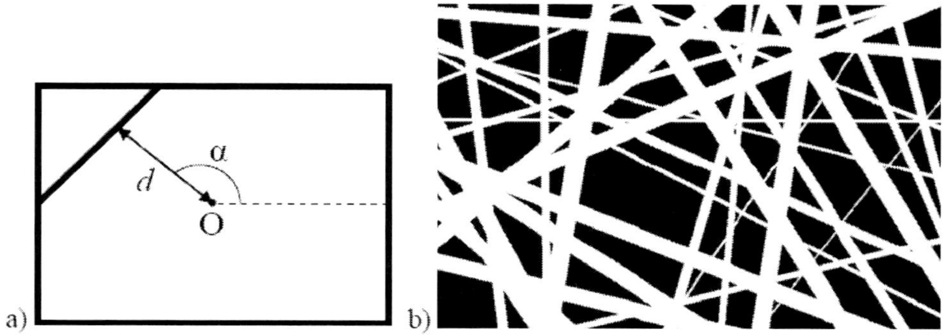

Figure 12. μ-randomness: a) schematic view of the procedure, b) a typical simulated image generated using this approach.

One of the most important features of simulation is that it allows several structural characteristics to be taken into consideration with the simulation parameters. These parameters are: web density (controlled as line density), angular density (sampled from a normal or random distribution), distance from the reference point (sampled from a random distribution), line thickness (sampled from a normal distribution) and image size.

2.3. Thresholding

Fiber diameter determination by the use of image analysis requires the initial segmentation of the micrographs in order to produce binary images. This is a critical step because the segmentation affects the results dramatically. The typical way of producing a binary image from a grayscale image is by *global thresholding* where a single constant threshold is applied to segment the image. All pixels up to and equal to the threshold belong to object and the remaining belong to the background. One simple way to choose the threshold is picking different thresholds until one is found that produces a good result as judged by the observer. Global thresholding, however, is very sensitive to any inhomogeneities in the gray-level distributions of object and background pixels [24]-[26].

Figure 13a illustrates a typical micrograph obtained from electron microscopy. As it is shown in

Figure 13b, global thresholding resulted in some broken fiber segments. In order to eliminate the effect of inhomogeneities, *local thresholding* scheme could be used. In this approach, the image is divided into subimages where the inhomogeneities are negligible. Then, optimal thresholds are found for each subimage [24]-[26].

A common practice in this case is to use morphological *opening* to compensate for nonuniform background illumination. The morphological opening is a sequential application of an *erosion* operation followed by a *dilation* operation (i.e., opening = erosion + dilation) using the same *structuring element*. Dilation is an operation that grows or thickens objects in a binary image by adding pixels to the boundaries of objects. Erosion shrinks or thins objects in a binary image by removing pixels on object boundaries. The specific manner and extent of the thickening or thinning is controlled by the size and shape of the structuring element which is a matrix consisting of 0's and 1's having any arbitrary shape and size. Opening the image produces an estimate of the background provided large enough structuring element is used so that it does not fit entirely within the objects (

Figure 13c). Subtracting the opened image from the original image, which is called *top-hat* transformation, results in an image with a reasonably even background (

Figure 13d) [24]-[26]. Now that the background is homogeneous and the edges of the objects are clearer, a global thresholding could be applied to provide the binary image. It could be shown that this process is equivalent to segment the image with locally varying thresholds [26].

In order to automatically select the appropriate threshold, *Otsu*'s method [30] is employed. This method is a simple but efficient technique in which the optimal threshold is chosen automatically by the discriminant criterion thereby maximizing the interclass variance and minimizing the intraclass variance of the black and white pixels.

Figure 13e depicts the binary image obtained using this approach. As it is apparent, the problem associated with the global thresholding has totally been solved. Note that, since the process is extremely sensitive to noise contained in the image, before the segmentation, a procedure to clean the noise and enhance the contrast of the image is necessary.

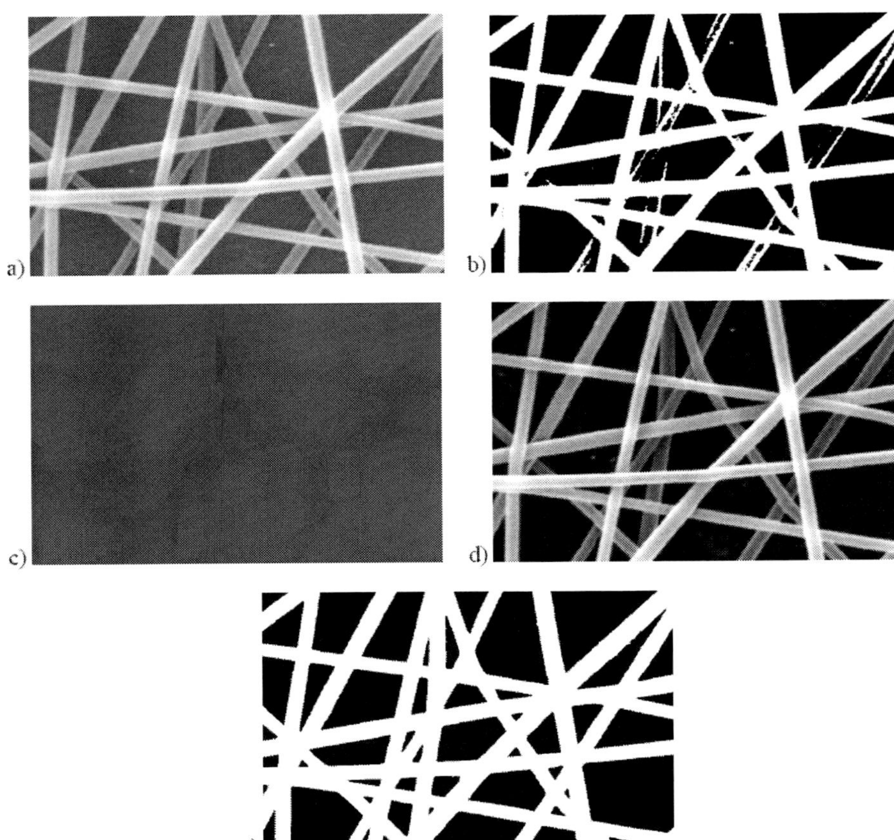

Figure 13. Thresholding: a) a typical micrograph of electrospun web, b) Global thresholding, c) Opened image, d) Top-hat transformation, e) Local thresholding.

3. EXPERIMENTAL

Electrospun nanofiber webs used as real webs for image analysis were obtained from electrospinning of PVA with average molecular weight of 72000g/mol, purchased from MERCK Company, at different processing parameters. The micrographs of the webs were obtained using Philips (XL-30) environmental Scanning Electron Microscope (SEM) under magnification of 10000X after being gold coated. Figure 14 shows the micrographs of the electrospun webs used as real webs in this study.

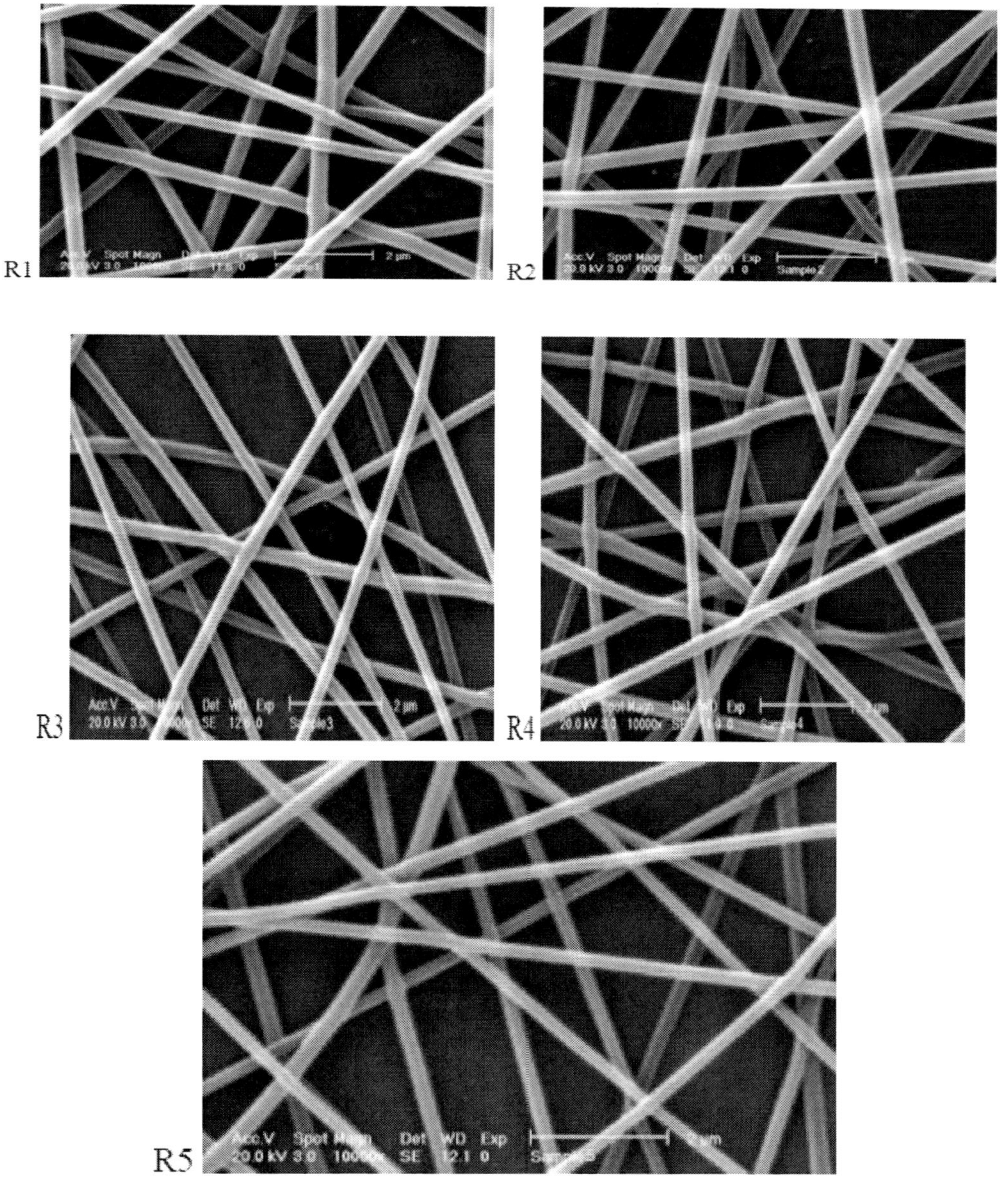

Figure 14. Micrographs of the electrospun webs.

4. Results and Discussion

We have evaluated the accuracy of the fiber diameter determined by the two methods using two sets of test samples each composed of 5 simulated images. The first set images had constant diameters which increased from 5 to 25 pixels for different samples. The second set images had varying diameter sampled from normal distributions with a mean of 15 pixels and standard deviation of 2 to 10 pixels. For both cases the line density was set to 30 and the angular density sampled from a random distribution in the range of 0-360°. The simulation parameters for the two sets are presented in Table 1 and Table 2. Figure 15 and Figure 16 show the simulated images in the two sets.

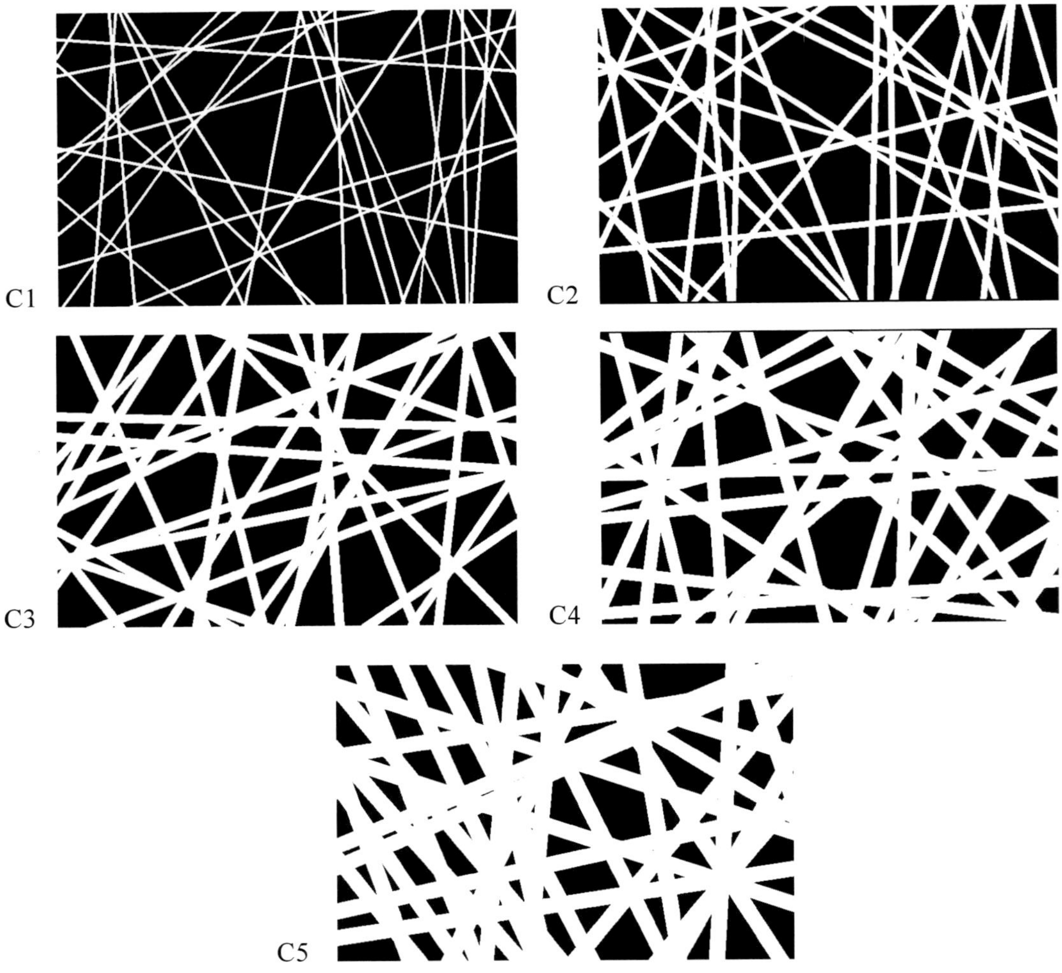

Figure 15. Simulated images with constant diameter.

Figure 16. Simulated images with varying diameter.

Table 1. Structural characteristics of first set images

Image No.	Angular Range	Line Density	Line Thickness
C1	0-360	30	5
C2	0-360	30	10
C3	0-360	30	15
C4	0-360	30	20
C5	0-360	30	25

Table 2. Structural characteristics of second set images

Image No.	Angular Range	Line Density	Line Thickness	
V1	0-360	30	15	2
V2	0-360	30	15	4
V3	0-360	30	15	6
V4	0-360	30	15	8
V5	0-360	30	15	10

The results for the simulated images are given in Table 3 and Table 4. Figure 17 and Figure 18 show the distribution of fiber diameter for the two sets of simulated images obtained by the methods. A normal distribution was also fitted on the histogram in each case.

Table 3. Mean and standard deviation for series 1

		C1	C2	C3	C4	C5
Simulation	mean	5	10	15	20	25
	std	0	0	0	0	0
Distance Transform	mean	5.486	10.450	16.573	23.016	30.063
	std	1.089	2.300	5.137	6.913	10.205
New Distance Transform	mean	5.366	9.917	15.106	20.013	24.645
	std	0.747	1.053	1.707	1.765	2.837

Table 4. Mean and standard deviation for series 2

		V1	V2	V3	V4	V5
Simulation	mean	15.247	15.350	15.243	15.367	16.628
	std	1.998	4.466	5.766	8.129	9.799
Distance Transform	mean	16.517	16.593	17.135	17.865	19.394
	std	5.350	6.165	7.597	9.553	11.961
New Distance Transform	mean	14.876	15.020	14.812	14.651	15.546
	std	2.403	4.797	6.047	7.851	9.942

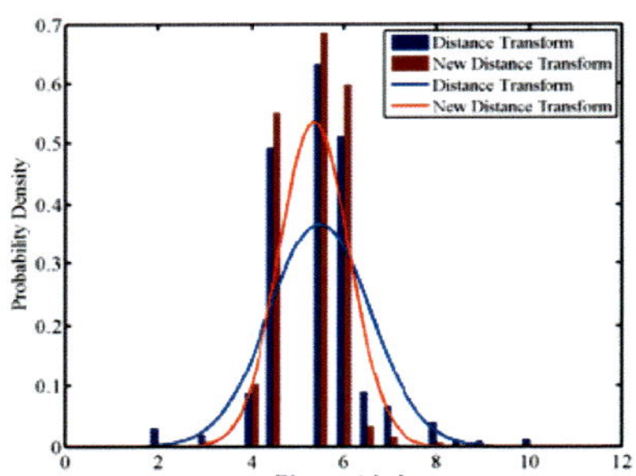

C1

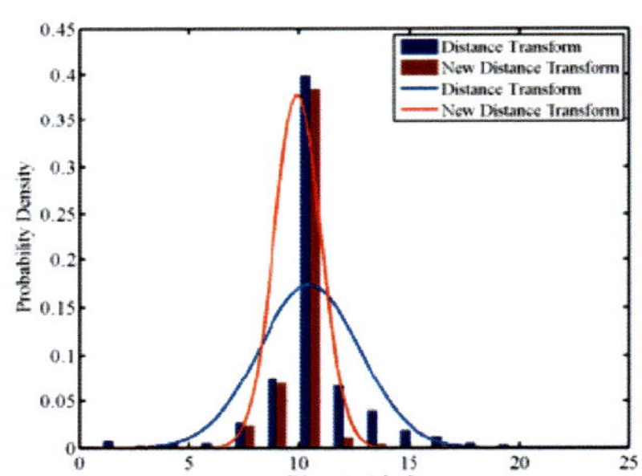

C2

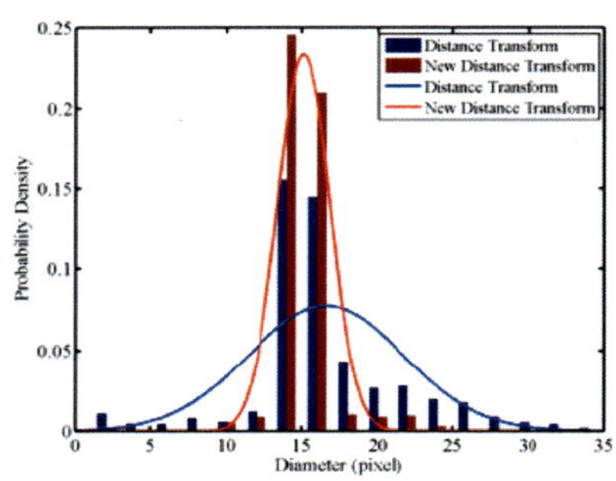

C3

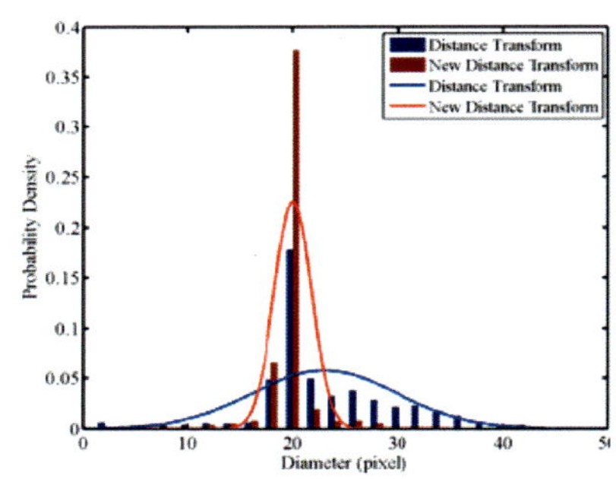

C4

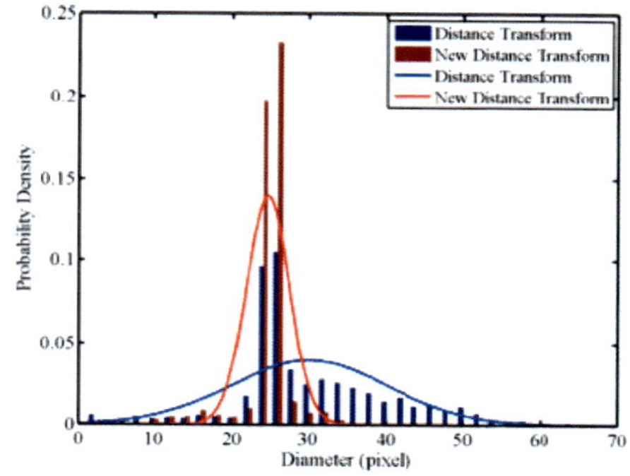

C5

Figure 17. Histograms for simulated images with constant diameter.

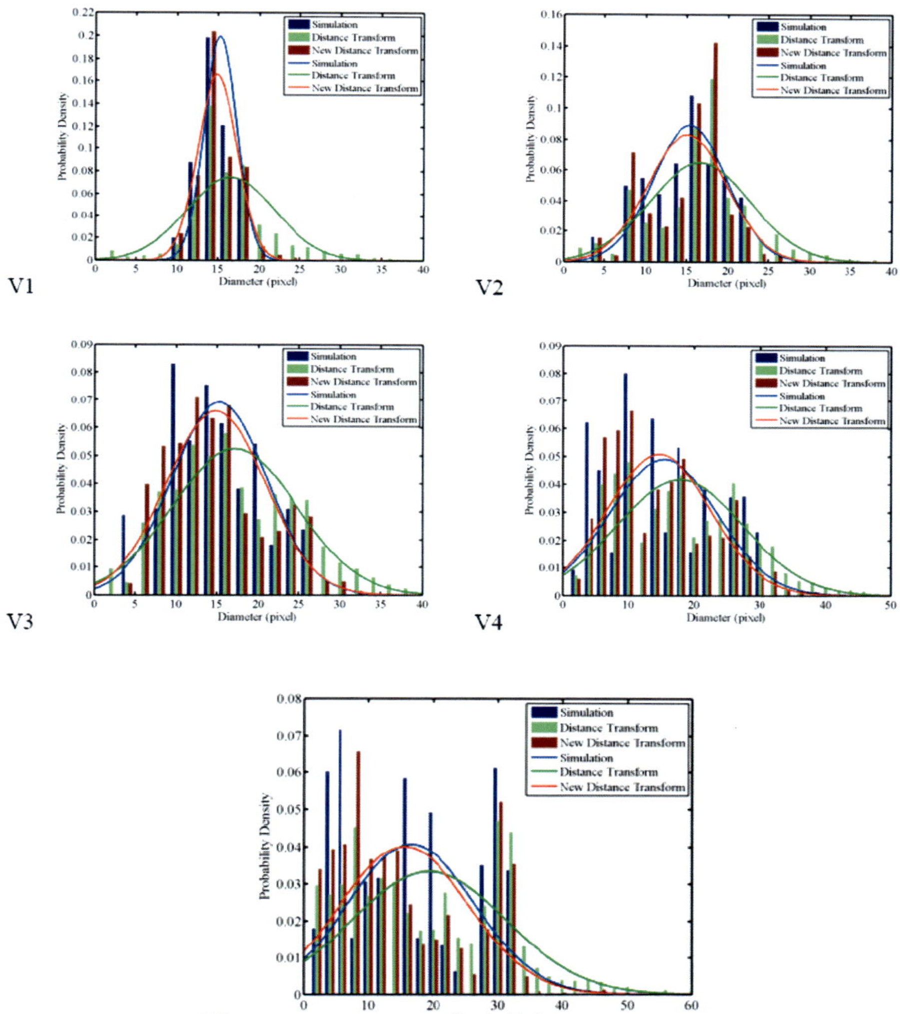

Figure 18. Histograms for simulated images with varying diameter.

From Table 3 and Table 4 it is apparent that for all the simulated samples, the results obtained by new method were significantly better than the old one. For both sets of simulated images, Mean and standard deviation of fiber diameter obtained by new distance transform were so close to those of simulation. It is noteworthy that the true mean and standard deviation of diameter in samples with varying line thickness slightly differ from those of used as simulation parameters (see Table 2).

As described earlier, distance transform method fails in measuring the fiber diameter at intersection points. The intersections cause to overestimate fiber diameter. Since in our developed method, these points were deleted, the effects of intersections which cause in imprecise measurement of fiber diameter were eliminated. Therefore, the fiber diameter was determined more accurately.

Both methods do not distinguish multiple fibers being joined together. This can happen in simulation by laying one line over the other. In real webs, fiber bundling sometimes happens (often in high density webs which consist of lots of fibers). Since there is not any black pixel (associated with the background) between joined fibers, they are segmented as a single fiber in the step of thresholding.

There may also be up to half a pixel error in both directions which turns out up to 1 pixel error in measuring fiber diameter. The error may be more significant when the fibers are thicker.

The slight difference between the diameters obtained by our method and simulation can be attributed to the 1-pixel measurement error, bundling of fibers, remaining some parts of branches after pruning and slight variations of skeleton adjacent to intersections which weren't deleted. The last two problems could easily be solved by further pruning for the former and increasing the area to be deleted for each intersection for the later. But due to the fact that it causes a decrease in the number of diameter measurements, authors preferred to leave it in this manner. It can be shown that the errors due to these problems do not play an important role in variation of the diameter since they are in the range of measurement error. Furthermore, some parts of the image due to the presence of intersections are deleted and not counted in diameter measurement; this can be another reason for the variation observed. In most cases except where thick fibers are present, the difference between our method and simulation was within the 1-pixel measurement error. That is to say the effects of other errors are negligible.

In addition there were 5 real webs in this study for testing the applicability of the methods for real samples. The results for the real webs obtained by two methods together with manual method are presented in Table 5 . Figure 19 shows the diameter distribution of these samples in term of *nm*. The curved line over each histogram corresponds to its fitted normal distribution.

The results for the real webs are in complete agreement with the trends observed by the simulation. Mean and standard deviation of fiber diameter obtained by our method are so close to those obtained by manual method. In addition to above mentioned reasons, the differences here can also be attributed to different number of measurements. For each sample, our method measured over 2000 fiber diameter whereas the operator could only measure 100 fiber diameters (in manual method) because of the time-consuming nature of this work. Despite all of these facts, the differences here are also within the 1-pixel measurement error which suggests that other errors are less significant.

Table 5. Mean and standard deviation for real webs

			R1	R2	R3	R4	R5
Manual	mean	pixel	24.358	24.633	18.583	18.827	17.437
		nm	318.67	322.27	243.11	246.31	228.12
	std	pixel	3.193	3.179	2.163	1.984	2.230
		nm	41.77	41.59	28.30	25.96	29.18
Distance Transform	mean	pixel	27.250	27.870	20.028	23.079	20.345
		nm	356.49	364.61	262.01	301.94	266.17
	std	pixel	8.125	7.462	4.906	7.005	6.207
		nm	106.30	97.62	64.18	91.64	81.21
New Distance Transform	mean	pixel	24.741	25.512	18.621	20.100	18.299
		nm	323.681	333.767	243.610	262.954	239.395
	std	pixel	3.854	3.961	2.826	2.903	2.795
		nm	50.417	51.821	36.976	37.980	36.571

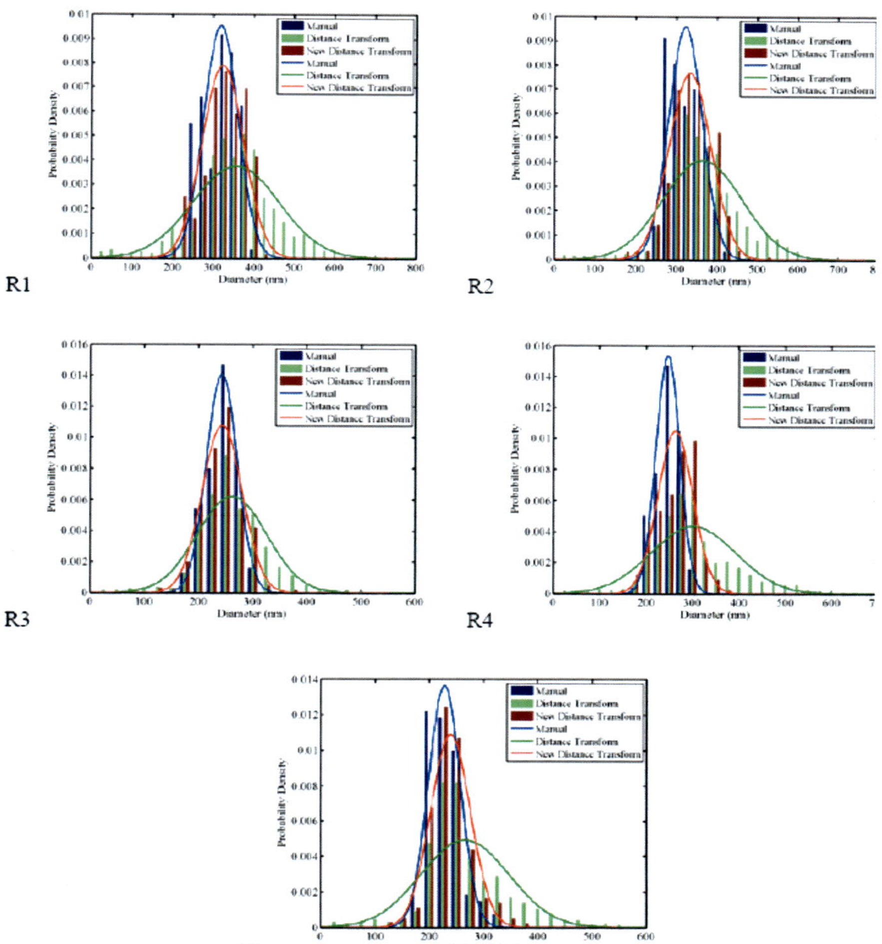

Figure 19. Histograms for real webs.

5. Conclusion

The results show the effectiveness of our method for diameter measurement. For the entire simulated images, a new algorithm resulted in mean and standard deviation of fiber diameter closer to the simulation. In addition, 5 electrospun webs of PVA were used in order to test the general applicability of the methods for real webs. Due to the need of binary input image, local thresholding was applied to the micrographs of the webs obtained from SEM. Otsu's method was used in order to automatically determine the appropriate threshold. The results for the real webs show that mean and standard deviation of fiber diameter obtained by new algorithm were in perfect agreement with manual method. That is to say, our attempt to develop a method for measuring fiber diameter has been quite successful.

References

[1] A. K. Haghi and M. Akbari, *Phys. Stat. Sol. (a)*, 204, 1830 (2007).
[2] J. Doshi and D. H. Reneker, *J. Electrostatics*, 35, 151 (1995).
[3] H. Fong and D. H. Reneker, Electrospinning and the Formation of Nanofibers, In: D. R. Salem, Structure Formation in polymeric Fibers, Hanser, Cincinnati (2001).
[4] D. Li and Y. Xia, *Advanced Materials*, 16, 1151 (2004).
[5] Z. -M. Huang, Y. -Z. Zhang, M. Kotaki and S. Ramakrishna, *Compos. Sci. Technol.*, 63, 2223 (2003).
[6] H. S. Park and Y. O. Park, *Korean J. Chem. Eng.*, 22, 165 (2005).
[7] G. T. Kim, Y. J. Hwang, Y. C. Ahn, H. S. Shin, J. K. Lee and C. M. Sung, *Korean J. Chem. Eng.*, 22, 147 (2005).
[8] G. T. Kim, J. S. Lee, J. H. Shin, Y. C. Ahn, Y. J. Hwang, H. S. Shin, J. K. Lee and C. M. Sung, *Korean J. Chem. Eng.*, 22, 783 (2005).
[9] B. Pourdeyhimi, R. Ramanathan and R. Dent, *Text. Res. J.*, 66, 713 (1996).
[10] B. Pourdeyhimi, R. Ramanathan and R. Dent, *Text. Res. J.*, 66, 747 (1996).
[11] B. Pourdeyhimi, R. Dent and H. Davis, *Text. Res. J.*, 67, 143 (1997).
[12] B. Pourdeyhimi and R. Dent, *Text. Res. J.*, 67, 181 (1997).
[13] B. Pourdeyhimi, R. Dent, A. Jerbi, S. Tanaka and A. Deshpande, *Text. Res. J.*, 69, 185 (1999).
[14] B. Pourdeyhimi and H. S. Kim, *Text. Res. J.*, 72, 803 (2002).
[15] B. Xu and Y. L. Ting, *Text. Res. J.*, 65, 41 (1995).
[16] I. Krucinska and S. Krucinski, *Text. Res. J.*, 69, 363 (1999).
[17] B. Pourdeyhimi and R. Dent, *Text. Res. J.*, 69, 233 (1999).
[18] D. M. Luzhansky, Quality Control in Manufacturing Electrospun Nanofiber Composites, *International Nonwovens Technical Conference*, Baltimore, Maryland (September 15-18, 2003).
[19] H. S. Kim and B. Pourdeyhimi, *Intern. Nonwoven J.*, 15-19 (Winter 2000).
[20] A. H. Aydilek, S. H. Oguz and T. B. Edil, *J. Comput. Civil Eng.*, 280-290 (2002).
[21] R. Chhabra, *Intern. Nonwoven J.*, 43-50 (Spring 2003).
[22] E. Ghassemieh, H. K. Versteeg and M. Acar, *Intern. Nonwoven J.*, 26-31 (Summer 2001).
[23] M. S. Abdel-Ghani and G. A. Davis, *Chem. Eng. Sci.*, 40, 117 (1985).

[24] W. K. Pratt, *Digital Image Processing*, 3rd Ed., John Wiley and Sons, New York (2001).

[25] B. Jähne, *Digital Image Processing*, 5th Ed., Springer, Germany (2002).

[26] R. C. Gonzalez and R. E. Woods, *Digital Image Processing*, 2nd Ed., Prentice Hall, New Jersey (2001).

[27] H. Breu, J. Gil, D. Kirkpatrick and M. Werman, *IEEE T. Pattern Anal.*, 17, 529 (1995).

[28] N. Sudha, S. Nandi, P. K. Bora and K. Sridharan, Efficient Computation of Euclidean Distance Transform for Applications in Image Processing, *Proceedings of IEEE Region 10 International Conference on Global Connectivity in Energy, Computer, Communication and Control*, 49-52 (1998).

[29] Q. -Z, Ye, The Signed Euclidean Distance Transform and Its Applications, *Proceedings of the 9th International Conference on Pattern Recognition*, Rome, Italy, 495-499 (1988).

[30] N. Otsu, *IEEE T. Syst. Man. Cy.*, 9, 62 (1979).

In: Progress in Chemistry and Biochemistry
Editors: E. M. Pearce and G. E. Zaikov

ISBN: 978-1-60692-344-3

Chapter 13

TRENDS AND ACHIEVEMENTS IN NEW GENERATION OF INTELLIGENT GARMENTS

M. Rashidi, M. Saberi,
V. Mottaghitalab and A. K. Haghi*
University of Guilan, P. O. Box 3756, Rasht, Iran

ABSTRACT

In this chapter a detailed review on new developments in smart textiles is presented. Different applications of currently available wearable computer clothes are clearly discussed. As textile materials are very flexible and easily deformable in all directions, so the sensors used should be able to support, often all at the same time, tensile, shear, bending and even compression deformations.

1. INTRODUCTION AND OVERVIEW

Electronic textiles (e-textiles) are fabrics that have electronics and interconnections woven into them. Components and interconnections are a part of the fabric and thus are much less visible and, more importantly, not susceptible to becoming tangled together or snagged by the surroundings. Consequently, e textiles can be worn in everyday situations where currently available wearable computers would hinder the user. The fabrication of electronic systems onto substrates which are not only flexible, but also conformable to the human body represents a break-through in many areas of application such as teleoperation, ergonomics and rehabilitation engineering [1] as well as in several health-focused disciplines, such as biomonitoring, telemedicine, teleassistance and sports medicine.

The possibility to realize sensing textiles by coating traditional fabrics with smart materials (piezoresistive, piezoelectric and piezocapacitive polymer) has opened the way to implement a new type of man-machine interface technology. The sensors are "smart" because

* Corresponding author e-mail: Haghi@Guilan.ac.ir

of their capacity to adapt to the specific mechanical properties of textile structures that are lightweight, highly flexible, stretchable, elastic, etc. Because of these properties, textile structures are continuously in movement and easily deformed, even under very low stresses.

Wearable sensing of human physiological signals is becoming an essential part of current monitoring systems, These revolutionary systems should allow clinicians, physiologists, psychologists, social institutions and people themselves to have access to a lot more information on the human body function than ever before.

Familiarity with clothing allows sensors and computational elements to be naturally integrated into garments such that wearability and usability is preserved. The dynamics of the human body and the wide variety of sensor and processing choices render the typical prototype-based design methodology prohibitively difficult and expensive. Simulation of e-textile systems not only reduces these problems but allows for thorough exploration of the design space, faster design cycles, and more robust applications. Gait analysis, the measurement of various body motion parameters during walking for medical purposes, and context awareness, the recognition of user motions, are two immediate applications that e-textiles can impact and emphasize the feasibility of e-textiles as a medium for sensor deployment on the human body One of the most compelling needs for wearable technology is in the continuous monitoring of the human body, be that for medical monitoring or to inform the operation of a context-aware computerized application. Sensing in the wearable environment is crucial for many applications, but often existing sensor technologies pose significant wearability problems when integrated into the user's peri-personal space. However, because of their ubiquitous, constant-wear nature, body sensors must prioritize the effects of the technology on the user's physical and social comfort.

Traditional sensing technologies applied to the wearable environment are rarely designed for continuous, on-body use: those that require skin contact are generally designed to be used in a hospital or doctor's office, and those that do not are generally designed for use in stationary devices. Consequently, the achievement of certain design goals for existing sensors (such as durability) is ultimately detrimental to the user's comfort when applied to the wearable environment. For example, durability often equates to stiffness, which results in a solid device that can cause discomfort by localizing pressure [2]. Figure 1 shows a generic platform for monitoring patients; full mobility is offered to the patient. The system comprises sensors, processing and communication functions worn by the user, interconnected with a portable base station, which is the gateway to the Internet where the collected data are made available to the authorized professionals, i.e., doctors, nurses, etc. The portable system includes local pre-processing close to the physiological signal collection nodes and global processing which takes benefit of the availability of various types of physiological (electrocardiogram (ECG), partial oxygen bloodsaturation (SpO2), skin temperature, etc.) and physical activity signals (acceleration, position, etc.) to perform efficient denoising and classification of the biosignals.

The architecture of the system may be described as a number of discrete layers. Distribution of functions in the layers and communication between the layers must be done in such a way that the relevant information is always properly extracted and transferred at the appropriate rate.

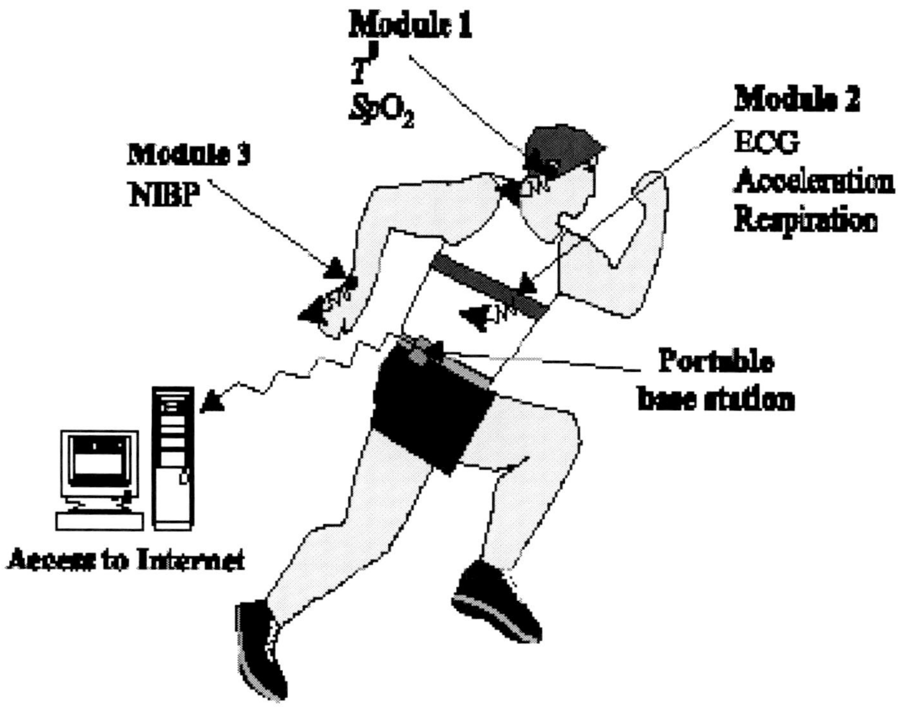

Figure 1. Overview of a possible architecture comprising, sensors, processors and communication units[3].

Target users of the intelligent biomedical clothes include patients during rehabilitation and early release from hospitals, professional personnel at risk and people during sport activities (professional or leisure). Sensing biomedical wearable can for instance:

1. Provide an integrated view of normal and abnormal patterns of activity, which would be otherwise difficult to detect and in situations that are usually uncontrollable by physicians.
2. Improve the quality of care for patients by monitoring health status during rehabilitation activities, allowing them to perform their everyday activities.
3. Support citizen and athletes by providing monitoring and processing of physiological parameters while they are performing sport activities.

In the last years, several major sport and leisure electronics companies have developed alliances with clothes industries to combine electronic functions and fashion into the clothes as shown in figure 2, where respiration and shoulder movements are captured by sensors in the fabric [3].

Figure 2. Prototype of shirt with electrodes for movement measurement at shoulders and respiration bands [3].

2. Smart Textiles

Textiles with added functionality over and above the conventional protective purpose may be regarded as smart. In practice, smart textiles may be classified depending upon their functionality and adaptation to the changing situation:

I) Passive Smart: Providing additional features in a passive mode irrespective of the change in the environment a highly insulating coat which remains insulating to the same degree irrespective of the outside temperature [4].
II) Active Smart: Tuning functionality to specific agent or environment a breathable fabric would allow perspiration to pass but would not allow rain droplets to enter the

fabric, variable insulation (Dawson et al., 1999). In this category one should include fabrics which incorporate "fibrous actuation elements" base on EAPs to provide external muscle-type functions to complement biological muscles.

3. Sensing Fabric

Sensors enable electronic devices to interact with our physical world. Therefore, non-electrical signals are converted into the electrical domain utilizing transducer effects. Many coupling mechanisms are known and used to transform signals from the physical, biological and chemical domain into electrical variables such as voltage, current, resistance or capacity.

A new generation of smart sensors has been designed for such flexible materials, because conventional sensors and gauges were not suitable as their general mechanical characteristics were incompatible with those of textile structures. Indeed, textile materials are very flexible and easily deformable in all directions, and the sensors used should be able to support, often all at the same time, tensile, shear, bending and even compression deformations. The sensors should thus be intimately integrated in the textile structure to be able to follow all these mechanical deformations, without affecting the original textile characteristics such as softness, feel, etc[5]. Based on the described experimental reference measurements, the following requirements for a strain sensor can be defined:

I) The sensor must be able to measure at least 30% elongation plus a certain pre-stretch of the textile.
II) The sensor must have textile stress properties. This means that the force needed to stretch the sensor must be in the same range as the force of the fabric. To have a convenient wearing comfort, force in the textile should not exceed 2N measured at a 5cm band.
III) It is important that the sensor property is not influenced by bending the sensor down to a bending radius of 0.5m.
IV) The sensor should be washable.[6]

4. Smart Materials

Optical fibers [7-9], piezoelectric polymer fibers [9], conductive polymers used in coating or spinning [10-13], and Conductive Polymer Composites (CPC) [14] are some of the developments which seem to respond to these specifications. A new class of sensors that may be integrated into textile structures in order to measure different physical values is based on *Conductive Polymers.* Two sub-classes can be identified: *intrinsically conductive polymers (ICP) and conductive polymer composites (CPC).* ICP are composed of polymer chains containing long conjugated double bonds which give rise to highly conducting properties, and they have been considered as promising materials [15,16]. They can be simply prepared using electrochemical or chemical oxidative methods but most ICP are infusible or insoluble in common organic solvents. However, advanced solution processing of ICP has been developed significantly over the last decade to improve their low solubility. For example, chemical modification of monomers with dopants has enhanced the solubility in the case of polythiophene and polyaniline [17,18]. For their part, CPC are obtained by blending - generally by melt mixing - an insulating polymer matrix (thermoplastic or thermosetting plastic) with conductive fillers like carbon black (CB), carbon fibres or nanotubes, metallic particles or conductive polymers. However, the presence of filler particles in the matrix may have a negative impact on the mechanical properties of the final composite [19,20], and this should be taken into account in the design of the sensor system. An example of application of CPC can be found in electrically conductive elastomer sensors with piezoresistive properties which are integrated into fabrics to obtain wearable kinesthetic garments: used as strain sensors, they are able to detect the posture and movement of the user [21-23]. The advantage of conductive polymer-based sensors is that they have potentially the flexibility and elasticity which are compatible with textile structures. Moreover, it was found that the variation of electrical conductivity of these materials under stress, temperature variation, UV radiation, moisture or other vapour [24,25] is observable and can therefore be used as a measuring parameter. However, this sensor also offers a great potential for use in many other applications where there is a need to measure force or strain levels in light and flexible materials. Possible applications are numerous and varied and can be found in: sportswear, airbags or other inflatable structures, safety belts, car seats, light-planes, wind sailing, medical applications, etc.[5]

5. General Structure

Figure 3 shows the schematic representation of the sensor structure. These geometry and dimensions have been chosen for two reasons: the first one is to obtain a good longitudinal sensitivity of the sensor, and the second, to have a good compatibility of the sensor to the common textile structures.

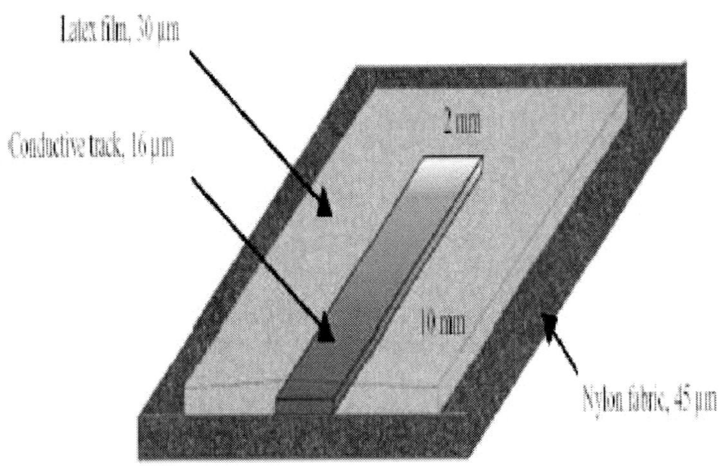

Figure 3. Schematic representation of the structure and dimension of the sensor integrated on the fabric[5].

Moreover, the sensor system thickness is relatively small in order to preserve the flexibility of the textile structure, and hence to have the best "conformance" of the sensor to the substrate, or in other words, to obtain the most "invisible" possible sensor which will not affect the host structure's mechanical characteristics. Figure 4 illustrates the flexibility of the Nylon fabric with the integrated CPC sensor (black track), with the setting up of the electrical connections.[5]

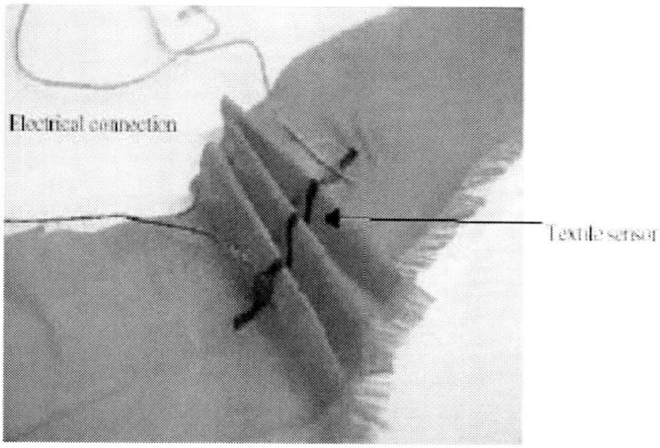

Figure 4. Nylon fabric and sensor, with electrical connections[5].

6. Conducting Electroactive Polymer(CEP)

Many textile-based sensors are actually sensing materials used to coat a textile [26] or conductive/sensing materials formed into fibers and woven or knitted into a textile structure [27-30]. The properties required by textile-based sensors can include flexibility, surface area, washability, stretch, and hand (texture of textile).

However, they must also include the properties required for the electronic device, including durability, power consumption, and ease of connection into a circuit. Metallic components, designed to function in rigid environments, often do not satisfy these needs. For instance, a metallic element in a high-flex environment (such as a garment) will soon break. However the recently discovered conducting electroactive polymers (CEP), offer a potential solution to this problem. CEPs such as polypyrrole (PPy), polyaniline and polythiophene constitute a class of polymeric materials which are inherently able to conduct charge through their polymeric structure. They can be reversibly switched from the doped conducting state to the undoped insulating state upon chemical or electrochemical treatment. The result is that the substrate is covered with a thin layer of conducting polymers rendering the whole object conducting without compromising the mechanical properties of the substrate. PPy coated textiles are often used in wearable sensing applications [27-29]. These applications generally use knitted textiles, resulting in a sensor with the ability to respond to stretch. [2]

CEP show a drastic change in electrical conductivity and in the dimensions associated with changes in ionic doping inside the polymer. For these characteristics, conducting polymers are potentially useful in several fields of applications [31]. In particular, it has been shown that conjugated electroactive polymers can exert high forces, much greater than those of natural muscle, and that they undergo volume changes with noticeable variations of elastic moduli when ionic species are forced to penetrate inside their network by electrodiffusion.[32]

Electrically conducting polymers have been developed for chemical and biological sensors as well as the perhaps for the more straightforward task of strain or stress feedback.

Electrical feedback from stress or strain may active as in piezoelectric systems or passive as in piezoresistive systems. In all cases polymers offer a tunable gauge factor, large recoverable strain ie > 1% typical of non-polymeric strain sensors and compatibility with other polymeric components such as actuators. High gauge factors (fractional change in resistance due to an applied strain) minimize or eliminate the need for costly and sensitive electronic amplification of signals. It fact the development of plastic transistors could lead to an integrated low cost all polymer device. The low cost and ease of deployment are vital factors in the development of distributed sensing system which essential for complete biomimetic impressions. Conducting polymers such as polypyrrole exhibit a linear strain dependence of the conductivity over a strain of more than 1% and the gauge factors lie in the range 0.2 to 5. However, as these are semi-conductors the overall resistance is very temperature sensitive. A wide range of piezoresistive systems have been studied. These may be as apparently straightforward as metal filled and/or coated deformable polymers. Recent studies have shown that wire loops are particular useful conducting fillers. Other approaches involve a simple coating using a metal or a conducting polymer.

7. Typical Methodology for Sensor Fabrication

The development of textile yarns with electronic functions is usually obtained using conjugated polymers (essentially organic semi-conductors), such as polyaniline, polypyrrole and polythiophene. The conductivity in these polymers is due to the ease of electronic jumps between chains, owed to the presence of dopant agents, able to modify the electrons amount in the bands.The main advantages of using conductive polymers are the low actuation voltages, and their ease of integration into textile fabrics, provided the material used has mechanical properties not differing too much from those of textile substrates . After a suitable material selection, the goal of obtaining a smart appears to be achievable in two ways:

I) Electronic/electrical functionalized fabrics are obtained by depositing the selected materials on it, according to a predetermined geometry
II) Fabric are obtained starting from functionalized fibers such as those made of actuating Polyaniline fibers

The typical approach to both the above strategies is to achieve the optimization of conducting polymer technology; this represents a possible route to create an actuator having muscle-like performance. Such an optimization can be reached by modifying the electrolyte, the electrode geometry or the material microstructure. An alternative approach consists of designing the artificial muscle from the molecular level up, where force and displacement generation comes from chemically engineered molecule building blocks changing shape upon application of an electrical stimulus .

8. Carbon Loaded Rubber Based

A new generation of high-performance strain sensors has been obtained by coating yarns and fabrics with carbon loaded elastomers or rubber (CLR), typically consisting of a silicone matrix filled with carbon black powder. Sensors are fabricated on a Lycra®/cotton textile by masked smearing of the conducting mixture. The same polymer/conductor composite is also used as material for the tracks of connection between sensors and an acquisition electronic unit, avoiding the stiffness of conventional metal wires. Such devices have been used to demonstrate prototype sensorised garments, including gloves, leotards, knee bands and sleeves.[figure 5]

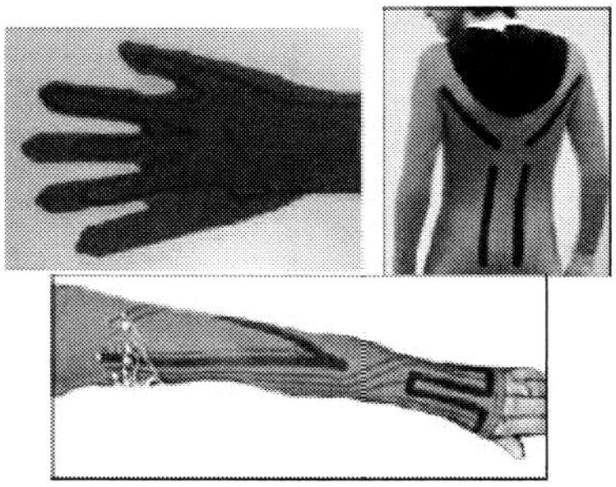

Figure 5. Sensorised glove, leotard and sleeve[33].

All these systems are truly wearable fabrics, incorporating compliant polymer sensors, capable of recording body posture and gesture, with no discomfort for the subject wearing the garment and with negligible motion artifacts.

9. Material Characterization

1. CLR Coated Fabric

Several studies have been dedicated to the characterization of presented materials [30-34]. Figure 6 shows an example of response of the CLR-based sensor under stepwise stretching.[32]

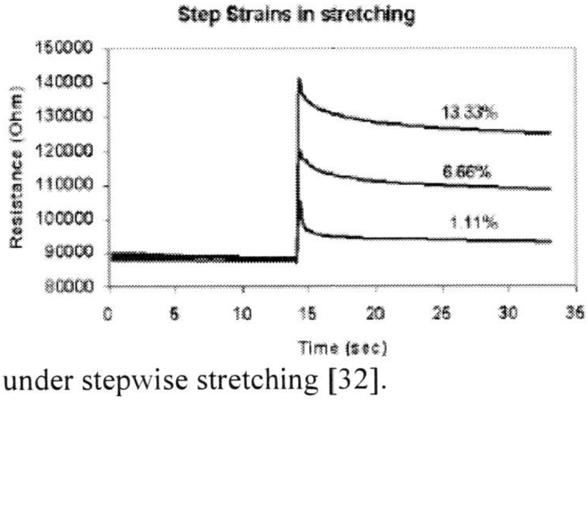

Figure 6. CLR-based sensor under stepwise stretching [32].

2. CEP-Coated Fabrics

The characterization of CEP coated fabrics has shown a gauge factor (GF) of about 13 (negative and similar to nickel) and a temperature coefficient of resistance (TCR) of about $0.018C^{-1}$.The numerical value of GF was calculated from a linear interpolation of data (before saturation) reported in figure 7.

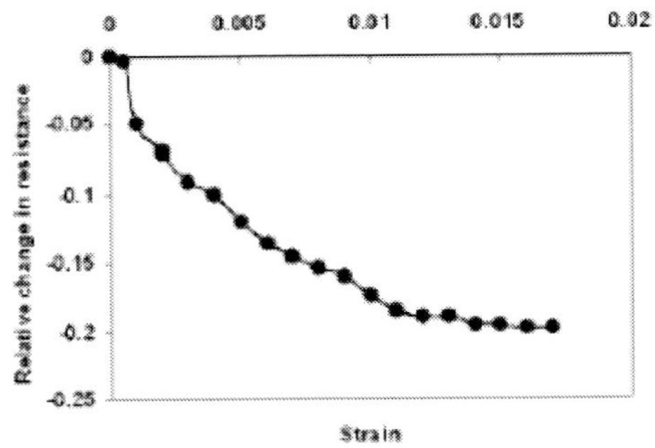

Figure 7. Quasistatic response in terms of percent change in electrical resistance versus uniaxial strain for a CEP based sensor[39].

In figure 8, a quasi-static response in terms of percent change in electrical resistance versus uniaxial strain, both in stretching and shortening configuration, for a CEP-based sensor is reported. It can be noted that a hysteresis loop is present, i.e., the fabric sensor response when it is stretched lies on a different path against shortening solicitation. The sensor response is also affected by noise artifacts. The noise disturbance has been estimated as an uncertainty band on the signal coming from the fabric sensors. This implies, obviously, that the intrinsic noise of the sensor is not decoupled from interferences due to electronic devices used for the acquisition. In the light of these considerations, the estimated error signal is about 3%. Despite the fact that a high GF value is suitable for strain gauge implementation, two serious problems affect CEP-coated fabric sensors. The first problem resides in the drift with time of the sensor resistance. This process is called aging, and is due to the slow oxidation of the polymer deposited on the textile. The second problem is the slow response time of the sensors; in fact, after sudden application of a mechanical stimulus the resistance reaches a steady state only after several minutes (see figure 8); this makes these fabrics unusable in most applications.

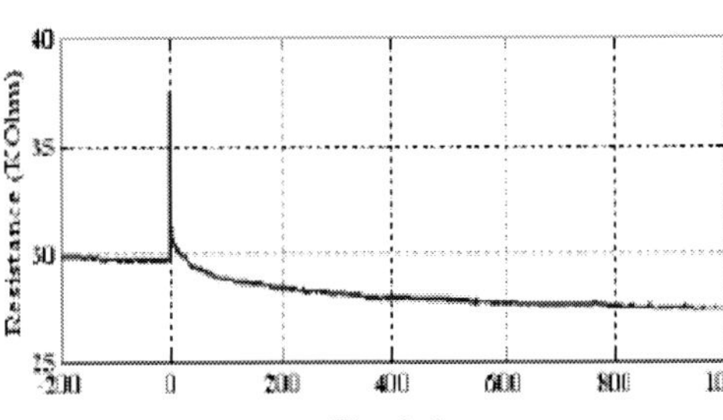

Figure 8. Response in terms of change in electrical resistance under stepwise stretching for a CEP-based sensor[39].

Both limitations have been partially overcome by analyzing the resistance response in the first second after the imposition of a step-wise deformation; it is possible to derive the applied strain in an aging invariant way. it can be considered a right-angled triangle (figure 9) where the cathetus height is equal to the excursion of the response peak and the slope of the hypotenuse is equal to the time derivative of the resistance calculated at the mid-point between the peak and the final value of such range. [40]

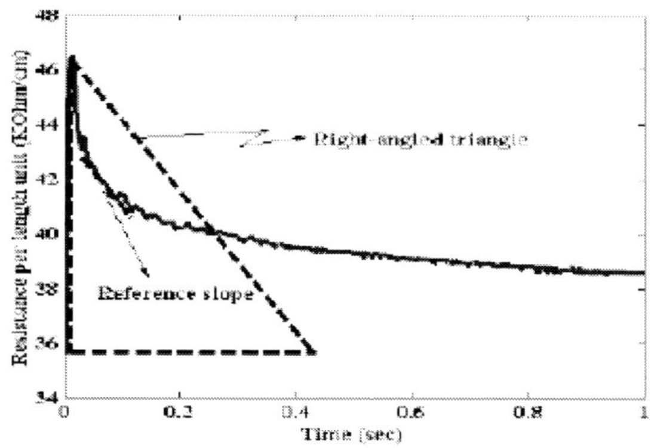

Figure 9. CEP sensor resistance per unit of length vs. time after the application of the mechanical strain [39].

10. Sensing Body Movement and Joint Angle

Sensing of large-scale body movements is useful in many applications: for sports, medicine, device interface, and context awareness, to name a few. Motion capture and joint angle can be recorded using many methods, including accelerometers [41-43] gyroscopes [42], strain gauges [41-44], piezoelectric materials [43], and fiber-optics [45], and pressure sensors [46]. Of these, textile -integration has been achieved in the cases of piezoresistive elements, strain gauges, and pressure sensors. However, most prior research into joint movements has relied on sensors either strapped to the body [41, 42] adhered to the skin, or integrated into a skin-tight garment [43,44].

11. Smart Fabrics for Motion Capture

Wearable instrumented garments, capable of recording body kinematic maps with no discomfort to the subject and showing negligible motion artifacts caused by sensor-body mechanical mismatch, are crucial in several fields of application.

In most applications, a bottleneck is caused by devices too cumbersome and invasive for the subject, hence a well-fitting functionalized garment would provide strong advantages. In this context a sensing leotard and glove sensors were constructed. They permit movements of the arms and upper trunk to be recorded (figure 10), as well as movements of the fingers. Figure 10, shows that during a trunk flexion the resistance of the right sensor (RS) increases, while that of the left one (LS) decreases. For an opposite movement, the outputs of the two sensors would be exchanged.

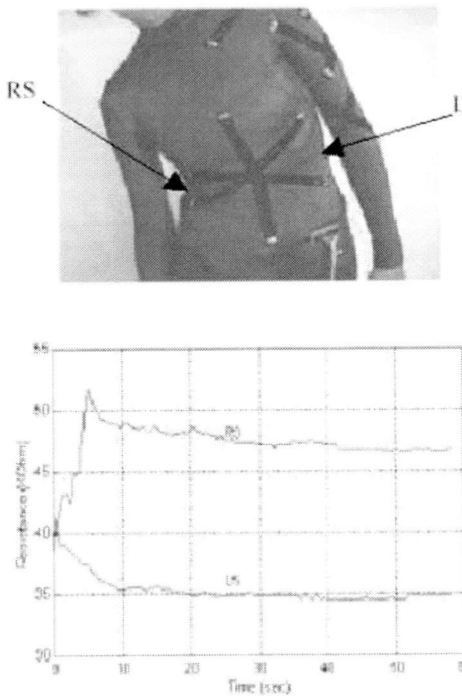

Figure 10. Flexion towards right from rest position (top) and output signals from the sensorised leotard (down).

Rehabilitation, sports medicine, multimedia and virtual reality are examples of potential fields of application. An elastic glove was sensorised by an array of CLR piezo-resistive sensors (figure 11 top), and it was used with a pseudo-3D graphic interface (figure 11 down).

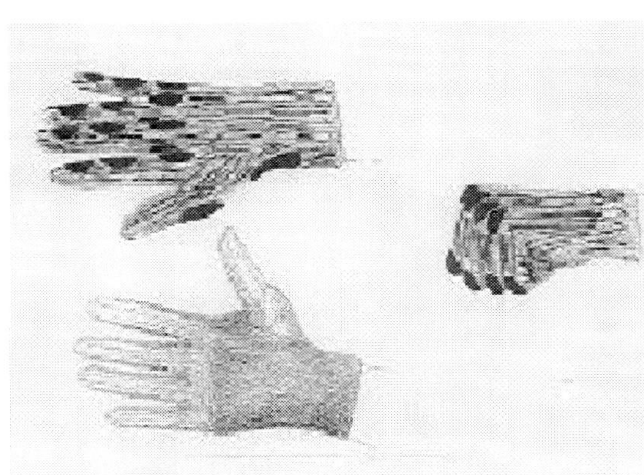

Figure 11. Sensor position on the sensing glove (top) and use of the glove with a pseudo-3D graphic interface (down)[32].

The data path from the sensor to the posture classification can be split into the three parts: sensors, data acquisition, and classification.

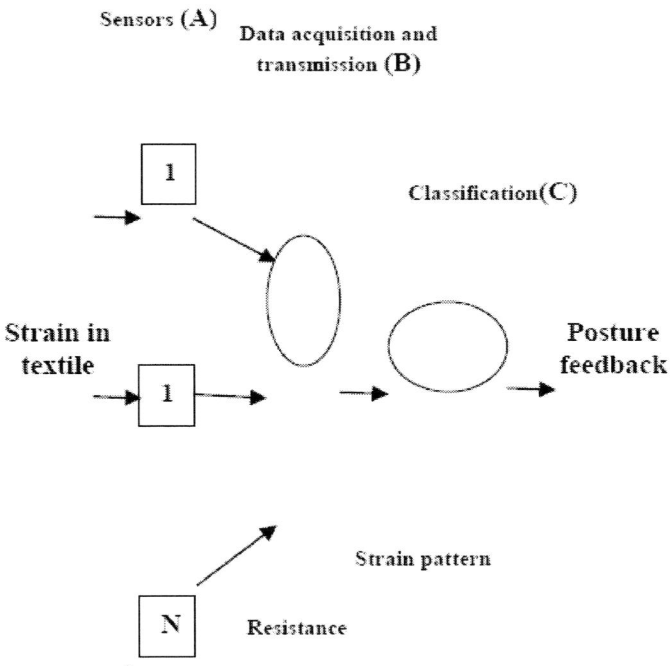

Figure 12. Data path of the textile classification system[47].

Local deformations in the textile caused by a change in posture are measured by textile strain sensors. We propose to integrate the sensors in a tight-fitting undershirt so that no additional subjacent layers influence the measurement.

Strain sensitive material normally reacts with a changing electrical resistivity to strain. This signal is measured, transformed to a strain value and transmitted to a central processing unit where a classification is done .To minimize parasitic coupling, resistance measurement directly at the sensor would be ideal. However, due to comfort issues, no rigid components can be placed on the back.

12. Concluding Remarks

Intelligent textiles, variously known as smart fabrics, electronic textiles, or e-textiles, have attracted considerable attentions worldwide due to their potential to bring revolutionary impacts on human life. Despite many promising progresses in this exciting newly emerged research field, there still exist a number of important obstacles. One of the most challenging issues is the conflict between the flexibility of the textiles and the need to incorporate sensing and computation modules. To address this critical issue, an innovative intelligent textile technology is proposed. The central hypothesis is that practical intelligent textiles can be realized by integrating fabrics with flexible transducers/electronics that are made using a unique, ¡®flexible-skin¡¯ technology.

The unique features of the silicon-based flexible skins are extremely desirable for intelligent textiles. A novel approach of making intelligent textiles by integrating the silicon-based flexible skins with textiles is proposed. The most important advantage of this novel technology is its compatibility with current MEMS and IC technologies, since MEMS devices (micromachined transducers) and ICs can be fabricated on the silicon wafer before the formation of the skin. This not only saves significant R&D efforts by avoiding re-invention, but also enables the integration of abundant sensing and computational capabilities offered by the silicon-based technology.

A wearable physiological monitoring system introduces many benefits for health care and fitness applications. For these systems, various technologies are available to implement sensors, interconnects, electronic circuits, and software for analyzing physiological signals.

This paper focuses on the development of a wearable Physiological Sensor System (PSS). The PSS was required to be non-invasive, lightweight, and low-power. In this work a description of novel sensors developed to be integrated in sensorised garments for monitoring vital signs and body gesture/posture has been presented. The main advantage ensured by these systems is the possibility of wearing them for a long period of time without discomfort. This work was intended to demonstrate the feasibility of creating smart garments capable of monitoring vital signs and human posture. Further efforts are needed to reach a full integration of actuating and sensing elements in a woven device. Wearable sensing is far from being mature and needs lots of improvements concerning the power distribution and the ergonomics of the systems. User functionality performances and clinical relevance of the extracted information will impinge on the final market potentials, rather than the individual sensor technical performances. Communication between wearable units is largely governed by the special needs of signal processing of sensory information. It can be found that fabrics coated with conducting polymer have piezoresistive properties. In particular, these fabrics are easily integrated into truly wearable, instrumented garments, capable of recording kinaesthetic maps of human function with no discomfort for the subject.

Nevertheless, the union of electronics and textiles to form electronic textiles (e-textiles) provides a promising substrate upon which motion analysis applications can be developed and implemented. Strain gauges have been used as sensors and have been deposited directly onto textile substrate, demonstrating one way to realize a wearable sensor system. The implementation of truly wearable instrumented garments capable of recording biomechanical variable is crucial to several fields of application from physical rehabilitation to sporting. Long term monitoring of body kinematics that can be analyzed posture and gesture together is a field of increasing interest in bioengineering. This sensory function of the garments is achieved by fabric strain sensors, through change of resistance based on threads coated with conductive elements that give them intelligent properties, enabling the detection of local strain on the fabric. Acquired signal through this simple process can be transmitted to an offshore system for further processing and producing of highest level of interpretation of the extracted information and possible biofeedback to the user.

REFERENCES

[1] Lumelsky V. J. 2001. "Sensitive skin", *IEEE Sensors Journal,* vol.1, pp. 41-51.
[2] L. Dunne, R. Tynan, Gregory M. P. O'Hare, B. Smyth, S. Brady, D. Diamond "Coarse Sensing of Upper Arm Position Using Body- Garment Interactions "Department of Computer Science University College Dublin Belfield, Dublin 4, Ireland (2005)
[3] P. Celka, R. Vetter, P. Renevey, C. Verjus, V. Neuman, J. Luprano, J. Decotignie, and C. Piguet " Wearable biosensing: signal processing and communication architectures issues "*journal of telecommunications and information technology,* (2005)
[4] R.H.C. Bonser, Centre for Biomimetics, W.S. Harwin, Department of Cybernetics W. Hayes, School of Chemistry G. Jeronimidis, Centre for Biomimetics G.R. Mitchell, Department of Physics C. Santulli, Centre for Biomimetics "EAP-BASED ARTIFICIAL MUSCLES AS AN ALTERNATIVE TO SPACE MECHANISMS "ESA/ESTEC (2004)
[5] C. Cochrane, V. Koncar ,M. Lewandowski and C. Dufour "Design and Development of a Flexible Strain Sensor for Textile Structures Based on a Conductive Polymer Composite "ISSN 1424-8220 ©(2007) by MDPI
[6] C. Mattmann, G. Tröster, "Design Concept of Clothing Recognizing Back Postures "Manuscript received June 16,(2006).
[7] El-Sherif, M. In Wearable electronics and photonics; Tao, X., Ed.; Woodhead Publishing Limited: Cambridge, 2005, Chapter 6, p 105.
[8] El-Sherif, M.; Fidanboylu, K.; El-Sherif, D.; Gafsi, R.; Yuan, J.; Richards, K.; Lee, C. A novel fiber optic system for measuring the dynamic structural behaviour of parachute. *J. Intell. Mater. Syst. Struc.* 2000, 11, 351-359.
[9] King, M.; Sun, H.; Gruverman, A. The piezoelectric properties of bicomponent PVDF fibres for use as artificial muscle actuators. 5th World Textile Conference AUTEX June 27-29, 2005, Portorož, Slovenia
[10] Kim B.; Koncar V.; Devaux E.; Dufour C.; Viallier P. Electrical and Morphological properties of PP and PET conductive polymer fibers. *Synth Met.* 2004, 146, 167-174.
[11] Kim B.; Koncar V.; Dufour C. Polyaniline (PANI) coated PET conductive yarns: Study of Electrical, Mechanical and Electro-Mechanical properties. *J. Applied Polym. Sci.* 2006, 101, 1252- 1256.

[12] Kim B.; Koncar V.; and Bruniaux P. Electro mechanical and morphological properties of PET conductive polymer fibers – modelling and simulation. *J. of Adv. Mat.* 2006, 38, 12-20.
[13] Kim B.; Koncar V.; Dufour C. In Intelligent Textiles and Clothing; Mattila, H.R., Ed.; Woodhead Publishing Limited: Cambridge, 2006, Chapter 16, p 308.
[14] Flandin, L.; Brechet, Y.; Cavaille, J.Y. Electrically conductive polymer nanocomposites as deformation sensors. *Composites Sci. and Tech.* 2001, 61, 895-901.
[15] Heeger, A.J. Semiconducting and metallic polymers: the fourth generation of polymeric materials. *Synth. Met.* 2002, 125, 23-42.
[16] Kumar, D.; Sharma, R.C. Advances in conductive polymers. *Eur. Polym. J.* 1998, 34, 1053-1060.
[17] Haba, Y.; Segal, E.; Narkis, M.; Titelman, G.I.; Siegmann, A. Polymerization of aniline in the presence of DBSA in an aqueous dispersion. *Synth. Met.* 1999, 106, 59-66.
[18] Gettinger, C.L.; Heeger, A.J.; Pine, D.J.; Cao, Y. Solution characterization of surfactant solubilized polyaniline. *Synth. Met.* 1995, 74, 81-88.
[19] Krupa, I.; Chodak, I. Physical properties of thermoplastic/graphite composites. *Europ. Polym. J.* 2001, 37, 2159-2168.
[20] Novak, I.; Krupa, I.; Chodak, I. Investigation of the correlation between electrical conductivity and elongation at break in polyurethane-based adhesives. *Synth. Met.* 2002, 131, 93–98.
[21] Lorussi, F.; Rocchia, W.; Scilingo, E.P., Tognetti, A.; De Rossi, D. Wearable, redundant fabricbased sensor arrays for reconstruction of body segment posture. *IEEE Sensors J.* 2004, 4, 807-818.
[22] Tognetti, A.; Lorussi, F.; Bartalesi, R.; Quaglini, S.; Tesconi, M.; Zupone, G.; De Rossi, D. Wearable kinesthetic system for capturing and classifying upper limb gesture in post-stroke rehabilitation. *J. Neuro Engineering and Rehabilitation.* 2005, 2, 1-16.
[23] Lorussi, F.; Scilingo, E.P.; Tesconi, M.; Tognetti, A.; De Rossi, D. Strain Sensing Fabric for Hand Posture and Gesture Monitoring. *IEEE Transactions on Information Technology in Biomedicine.* 2005, 9, 372-381.
[24] Krupa, I.; Novak, I.; Chodak, I. Electrically and thermally conductive polyethylene/graphite composites and their mechanical properties. *Synth. Met.* 2004, 145, 245-252.
[25] Omastová, M.; Kosina, S.; Pionteck, J.; Janke, A., Pavlinec, J. Electrical properties and stability of polypyrrole containing conducting polymer composites. *Synth. Met.* 1996, 81, 49-57.
[26] De Rossi D, Carpi F, Lorussi F, Mazzoldi A, Paradiso R, Scilingo EP, Tognetti A: Electroactive Fabrics and Wearable Biomonitoring Devices. *AUTEX Research Journal.* 3:4.(2003)
[27] Hertleer C, Grabowska M, Van Langenhove L, Catrysse M, Hermans B, Puers R, Kalmar A, van Egmond H, Matthys D: Towards a Smart Suit. In Proceedings of Wearable Electronic and Smart Textiles (2004)
[28] Malinauskas, A.: Chemical depositiong of conducting polymers. *Poly.* 42 (2001) 3957-3972
[29] azzoldi, A., De Rossi, D., Lorussi, F., Scilingo, E. P., Paradiso, R.: Smart Textiles for Wearable Motion Capture Systems. *AUTEX Research Journal.* 2:4. (2003)

[30] Farringdon, J., Moore, A., Tilbury, N., Church, J., Biemond, P.: Wearable Sensor Badge & Sensor Jacket for *Context Awareness. Proc.* of ISWC99 (1999)

[31] Baughman R., H., and Sacklette L., W., Science Applications of Conducting Polymers, Salaneck W., R., et al., Ed., Adam Higler, New York, 47-61, (1991).

[32] D. Rossi, F. Carpi, F. Lorussi, A. Mazzoldi, R.ita Paradiso,E. Scilingo, A. Tognetti "ELECTROACTIVE FABRICS AND WEARABLE BIOMONITORING DEVICES "AUTEX Research Journal, Vol. 3, No4, December(2003) © AUTEX

[33] T. Martin, M. Jones, J. Edmison, R. Shenoy B. Dept. "Towards a design framework for wearable electronic textiles"Proceedings of the Seventh IEEE International Symposium on Wearable Computers (ISWC'03) 1530-0811/03 $ 17.00 © (2003)

[34] D. Della Santa, A. Mazzoldi, D. De Rossi, *Mat. Sci. and Eng.* C 7, 31-35 (1999)

[35] F. Carpi, F. Lorussi, A. Mazzoldi, G. Pioggia, E.P. Scilingo, D. De Rossi, "Electroactive polymer based skin and muscles for man machine interfaces, MRS Fall Meeting (Boston, 2001)

[36] D. De Rossi, F. Carpi, F. Lorussi, A. Mazzoldi, E.P. Scilingo, A. Tognetti, "Electroactive fabric for distributed, conformable and interactive systems", IEEE SENSORS 2002 Conference (Hyatt Orlando, Kissimmee, Florida, 11-14 June 2002), pag. 5.1-6

[37] A. Mazzoldi, D. De Rossi, F. Lorussi, E. P. Scilingo, R. Paradiso, *Autex Research Journal,* 2 n° 4. 199-203, (2002)

[38] G. G. Wallace, G. M. Spinks, L. A. P. K. Maguire, P. R. Teasdale, Conductive Electroactive Polymers: Intelligent Materials Systems, 2 nd Ed., CRC Press, Boca Raton, FL 2003.

[39] A. Mazzoldi, D. De Rossi, F. Lorussi, E. P. Scilingo, R. Paradiso "SMART TEXTILES FOR WEARABLE MOTION CAPTURE SYSTEMS "*AUTEX Research Journal,* Vol. 2, No4, December (2002) © AUTEX

[40] De Rossi D., Della Santa A., Mazzoldi A. 1999. "Dressware: Wearable hardware", *Material Science & Eng.* C, 7:31-35.

[41] Farringdon, J., Moore, A., Tilbury, N., Church, J., Biemond, P.: Wearable Sensor Badge & Sensor Jacket for Context Awareness. Proc. of ISWC99 (1999)

[42] Lee, S.-W., Mase, K.: Activity and location recognition using wearable sensors. *IEEE Pervasive Computing,* 1:3 (2002) 24 – 32

[43] Martin, T., Jones, M., Edmison, J., Sheikh, T., Nakad, Z.: Modeling and simulating electronic textile applications. *Proc. of LCTES* 2004 (2004)

[44] Mazzoldi, A., De Rossi, D., Lorussi, F., Scilingo, E. P., Paradiso, R.: Smart Textiles for Wearable Motion Capture Systems. *AUTEX Research Journal.* 2:4. (2003)

[45] Akin, D., Braden, J.: Neutral Buoyancy Technologies for Extended Performance Evaluation of Space Suits. Proc. of 33rd International Conference on Environmental Systems. (2003)

[46] Toney, A.: A novel method for joint sensing on a wearable computer. Proc. of ISWC00 (2000)

In: Progress in Chemistry and Biochemistry
Editors: E. M. Pearce and G. E. Zaikov

ISBN: 978-1-60692-344-3

Chapter 14

ACHIEVEMENTS IN APPLICATION OF RECYCLED POLYAMIDE FIBRES AND EXPANDED POLYSTYRENE BEADS IN CONCRETE

A. K. Haghi
University of Guilan, P. O. Box 3756, Rasht, Iran
Haghi@Guilan.ac.ir

ABSTRACT

This paper shows the contribution of the polyamide 66 fiber wastes to the improvement of drying shrinkage of EPS (Expanded Polystyrene) concrete. The results are indicative of strong potentials for application of waste fibers for the reinforcement of lightweight EPS concrete products.

INTRODUCTION

Concrete is the most widely used construction material in the world. Its low cost, ease of application and compressive strength are the principle reasons for its universal acceptance. However it has a few shortcomings most of which are attributable to the Portland cement binder. Shortcomings include poor tensile strength, high porosity freeze thaw deterioration, and destruction by corrosive chemicals, etc. [1].

The development of new composite materials possessing increased strength and durability when compared with conventional types is a major requirement of applications in repairs and in the improvement of infrastructure materials used in the civil construction industry. Polymer concrete (PC) is an example of a relatively new material with such high performance[2].

The demand for lightweight concrete in many applications of modern construction is increasing, owing to the advantage that lower density results in a significant benefit in terms of load-bearing elements of smaller cross sections and a corresponding reduction in the size of the foundation [3].

Lightweight aggregate concrete, popular through the ages, was reported to have a comparable or some times better durability even in severe exposure conditions.

Lightweight aggregates are broadly classified in to two types—natural (pumice, diatomite, volcanic cinders, etc.) and artificial (perlite, expanded shale, clay, slate, sintered PFA, etc.). Expanded polystyrene (EPS) beads are a type of artificial ultra-lightweight, nonabsorbent aggregate [4,5]. It can be used to produce low-density concretes required for building applications like cladding panels, curtain walls, composite flooring systems, and load-bearing concrete blocks [6,7].

Expanded polystyrene (EPS) concrete is a lightweight concrete with good energy-absorbing characteristics, consisting a discrete air voids in a polymer matrix. However, polystyrene beads are extremely light, with a density of only 12 to $20\,kg/m^3$, which can easily cause segregation in mixing. Hence some chemical treatment of surface on this hydrophobic material is needed. Other investigators also reported that EPS tends to float and can result in a poor mix distribution and segregation, necessitating the use of admixtures [8,9].

Our aim is to contribute to the understanding of the behavior of EPS concrete reinforced with polyamide 66 fibers.

BACKGROUND

Concrete is durable, inexpensive, readily molded into complicated shapes, and has adequate compressive strength and stiffness, and it is the most heavily used man-made material in the world. However, concrete has a low tensile strength, low ductility, and low energy absorption. Fiber reinforced concrete (FRC) is often made by adding a small fraction (usually below 2% by volume) of short fibers to the concrete mix during mixing. Recycled synthetic fibers such as those from textile industry waste offer low cost alternatives for concrete reinforcement. This work has focused on lightweight EPS concrete reinforced with recycled fibers from hard carpet waste. It suggests that using recycled carpet fiber in construction could be very cost-effective way to improve the durability and performance of concrete structures and at the same time to reduce the needs for the landfill spaces.

Million tons of post consumer and carpet waste is landfilled each year. Many private companies and research institutions have been working on various approaches aimed at alleviating the waste disposal problem and at utilizing this vast amount of resource for new business opportunities. The basic polymeric components in carpets are nylon 6, nylon 66 (for face yarns). A carpet typically consists of two layers of backing (usually fabrics from polypropylene tape yarn), joined by $CaCO_3$ filled styrene-butadiene latex rubber (SBR), and face fibers (majority being nylon 6 and nylon 66 textured yarns). Such nylon fibers can be used for concrete reinforcement.

Nylon fibers, which are man-made, are known to be hydrophilic (water absorptive). Typically, Nylon 6 absorb 4-5% moisture. Ambient moisture from the free water in concrete mix may enter the fiber during the production process. Nylon fibers provide an opportunity to maximize the hydration of cement particles, specifically the cement particles that surround the fibers. Since they absorb moisture, they can be viewed as a reservoir for water within the concrete matrix. This water reservoir is released in the mortar matrix surrounding the fiber during the initial drying stages or when the moisture level in the fiber exceeds that in the mortar. This "new" moisture in the mix can then potentially react with cement particles that have only partially reacted or have yet to react. It must be noted neither the absorption nor the dissipation of moisture alters the dimensions of the fiber. The migration of moisture strengthens the mortar in the contact zone of the fibers. This theorized strength gain at the interface of Nylon fibers and mortar could then increase the bond between the two materials. Thus, it can be expected that nylon fibers produces a greater reduction in shrinkage crack area than polypropylene fibers .

Scope

This study was undertaken to determine issues related to PA reinforced EPS concrete. Consequently, using a waste material as a substitute for natural aggregates results in the beneficial use of waste material. Waste PA 66 fibers were used as a fine concrete aggregate in this project. Adding fibers to the mixer, after which the fibers were found to be uniformely dispersed in the EPS concrete without balling or clumping. Mixing, pouring, and finishing followed standard procedures used conventional equipment. Such concrete containing waste fibers can be used for floor slabs and walls of the buildings. The laboratory study has indicated that the carpet waste fiber was very effective in improving the toughness and shrinkage properties of EPS concrete. Moreover, research needs to be continued to determine other characteristics of PA reinforced EPS concrete. The ultimate objective should be to establish ASTM specification for laboratory testing of PA reinforced EPS concrete and to develop design procedures based on those results.

Experimental Work

Materials

Two types of commercially available spherical EPS beads were used (table 1). As indicated in table 2, the grading of EPS shows that Type A has mostly 6.0-mm-size beads and Type B has mostly 3.0-mm-size beads.

Table 1. EPS beads characteristics

Type	Size mm	Bulk density kg/m^3	Specific gravity
A	6	16	0.014
B	3	20	0.029

The physical properties of PA 66 used in this study are summarized in table 2.

Table 2. physical properties of PA 66

Specific gravity	1.16
Tensile strength	965 MPa
Elastic modulus	5.17 GPa
Ultimate elongation	20%

ADMIXTURE

A naphthalene based superplasticizer (that complies with ASTM C 494) was used to produce the mixes a flowable or highly workable nature, to suit the hand compaction adopted. This admixture is well known in the market and has a good history in international concrete works worldwide.

SPECIMEN PREPARATION

A number of standard test specimens of different sizes were selected for investigating the various parameters. For studying the compressive strengths at 3, 7, and 28 days and also for investigating the absorption tests cubes of 100 mm size were used. Meanwhile, the split tensile strength test was conducted on 100 mm diameter × 200 mm cylinders at 28 days.

Lightweight EPS concrete were made with PA and without PA reinforcement. Initially, the EPS beads were wetted with 25% of the mixing water and superplasticizer before adding the remaining materials. It is expected that by addition of EPS in place of normal aggregate the weight of concrete can be reduced significantly and the interaction of Polyamide 66 fibers with EPS concrete can improve the strength. In this study, the estimated component additions were measured by volume in order to simplify mixing process. The non-absorbent, hydrophobic and closed cellular aggregates (expanded polystyrene beads) were mixed in a planetary mixer in compliance with the recommendation of ASTM C 305. To obtain a very uniform and flowing mixture, the mixing operation was continued on a regular biases. The molds are then filled with fresh EPS concrete and then with PA reinforced EPS concrete. Afterward, the mixture is firmly compacted by hand.

Based on ASTM C 143-98 we measured the slump values of the fresh concrete. Afterward, the compressive strength test was carried out in a testing machine with capacity of 2000 KN. The loading rate used during this test was 2.5 KN/s. According to ASTM C 496-89 we conducted the split tensile strength test on cylinders at 28 days. The absorption test was then carried out as per ASTM C 642-82.

EXPERIMENTAL RESULTS AND DISCUSSION

The results of this investigation are explained in the following subsections.

Fresh Concrete

It appeared that the workability of the concrete in terms of the slump measurements were about 45–80 mm in non-reinforced cases and around 40-75 for polyamide EPS concretes. In addition, it should be noted that the mixes having the higher percentage silica fume can present higher slump values. During the experimental study, all the concretes specimens shown a interesting flexibility and provided a very easy condition to work with. Meanwhile, the specimens could be compacted using just hand compaction and they could be easily finished. Hence,the problems associated with the segregation and finishability of the mixes containing coarse beads as reported earlier [5] were totally solved by ensuring that a minimum amount of cementitious material used. In order to be able to solve the problems associated with the hydrophobic nature of the EPS beads, the silica fume and the superplasticizer were added. This will help to improve the cohesiveness of the mix significantly.

Development of Compressive Strength with the Age

Figure 1 displays the development of compressive strength with the age for reinforced EPS concrete. This illustration clearly states that the compressive strength of reinforced EPS concrete in almost all mixes show a continuous increase with age. It is seen that the rate of strength development was greater initially and decreased as the age increased. Comparison of strengths at 7 days revealed that non-reinforced concretes developed almost 40–50% of its 28-day strength. While it is seen that reinforced concretes developed almost 20–40% of the corresponding 28-day strength.

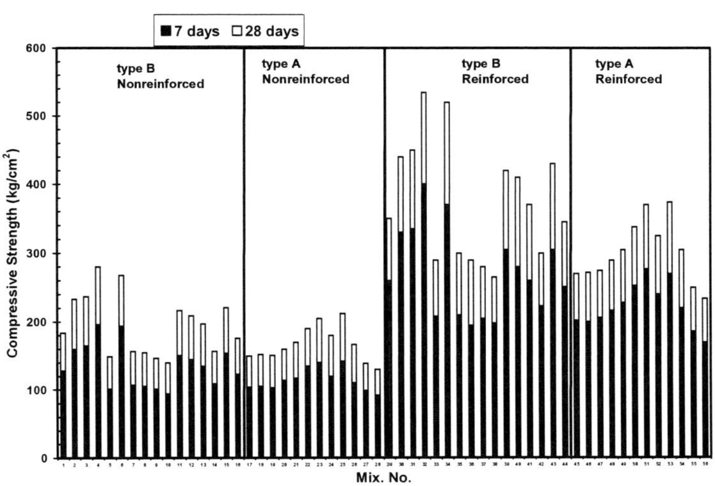

Figure 1. Effects of age on compressive strength.

Effect of Density and EPS Volume

Figures 2 and 3 illustrates the compressive strengths of EPS concretes with different plastic densities for reinforced and non-reinforced concrete and the volume content of EPS .

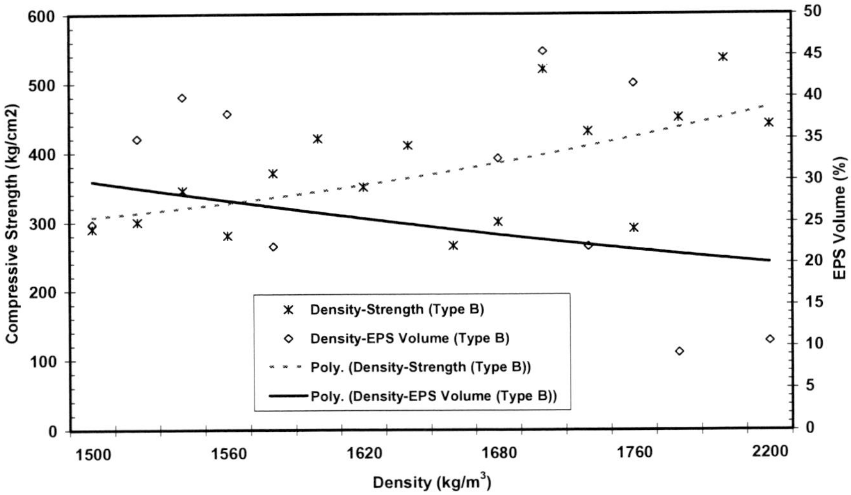

Figure 2. Variation of composite density with EPS volume and Compressive strength (type B).

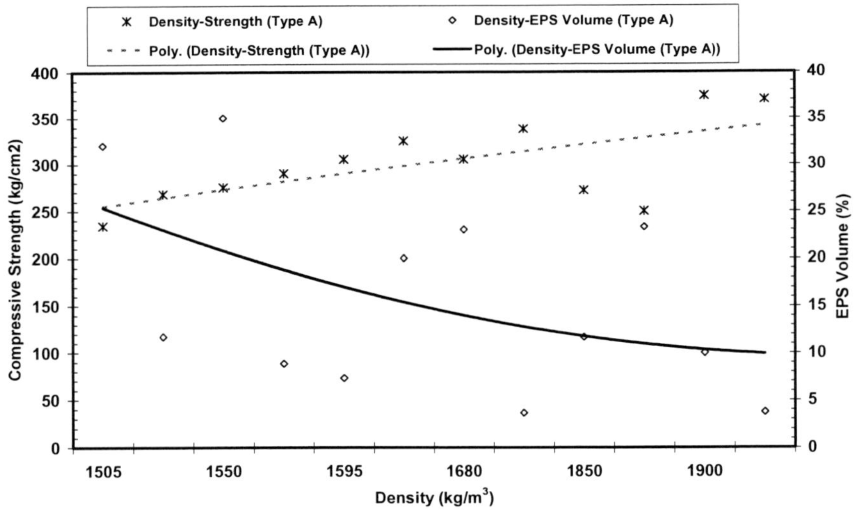

Figure 3. Variation of composite density with EPS volume and Compressive strength (type A).

It is appeared that the strength of non-reinforced and reinforced EPS concrete to increase linearly with a decrease in the EPS volume. Furthermore, it is noted that the density of EPS concretes decreased significantly with an increase in the EPS volume. It is clearly observed that the polyamide EPS concrete have high strength in compression, comparing to that non-reinforced concretes.

In some earlier studies reported on EPS concrete, the densities varied from 1500 to 2000 kg/m^3[13]. At 28 days, it is reported a compressive strength of 100 to 210 kg/cm^2 for these concretes. In our latest research for the same ranges of the densities (*i.e.,* 1500 to 2000 kg/m^3) we obtained the strengths of (130 to 205) and (140 to 280) kg/cm^2, for EPS type A and B, respectively [14]. In the present study, we obtained the strength of (235 to 375) and (265 to 540) with use of polyamide 66 yarns.

Effect of EPS Bead Size

The plot represented in figure 4 show an inverse relationship between the concrete strength and the bead size. That means the strength in EPS concrete increased with a decrease in the EPS bead size for the same mix proportions.

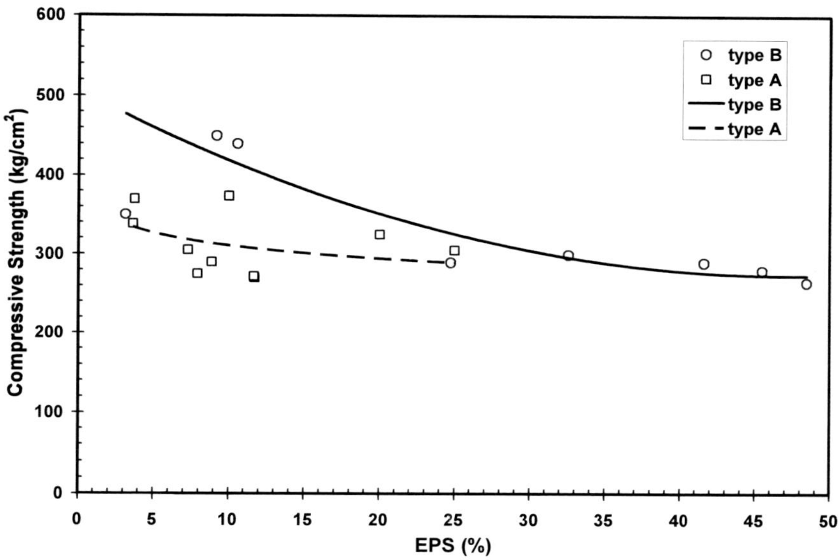

Figure 4. Effects of EPS bead size on compressive strength.

Effect of Silica Fume and Polyamide 66 Fibers

From figure 5 it can be clearly appreciated that in the mixes content fine silica fume the compressive strength can increase significantly (i.e.: mixes no. 4, 6,...). This represents the effects of fine silica fume and polyamide 66 on the compressive strength of EPS concrete. This is clearly visible that fine silica fume can improve the dispersion of EPS in the cement paste and interfacial bonding between EPS and cement paste. Moreover, the compressive strength of polyamide EPS concrete is almost 1.5 to 2 times non-reinforced EPS concretes. As reported previously [12, 15], the strength of concretes containing bigger beads (type A) can be lower than the strength of concretes containing smaller size beads.

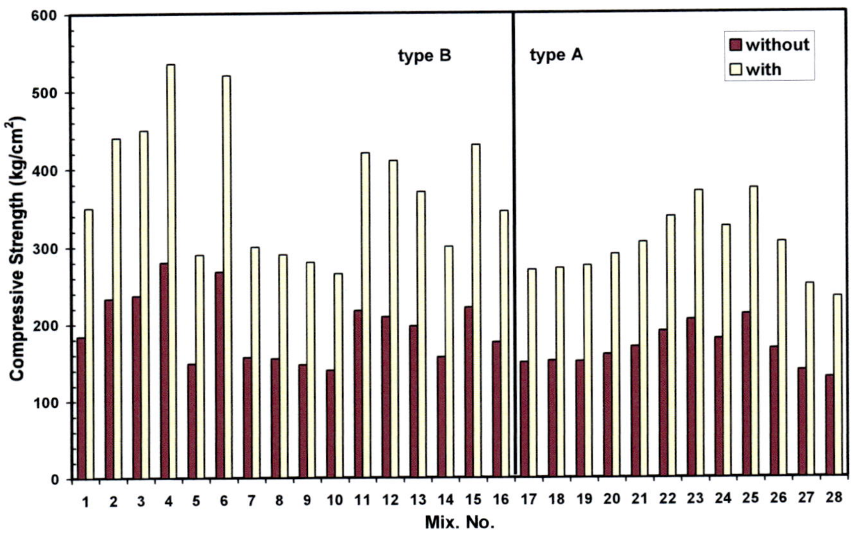

Figure 5. Effects of silica fume and polyamide on compressive strength.

Failure Mode

The failure modes of different specimens are presented in figure 6. Figure (6a) shows the EPS concrete specimen before loading. The failure mode observed in figure (6b) is rather gradual (more compressible), and the specimen is capable of retaining the load after failure, without full disintegration. This clearly indicates the high-energy absorption capacity of EPS concrete which is reported earlier [6].

Figure (6c) illustrates failure mode in polymer concretes containing EPS with polyamide yarns. This is attributed to the fact that the failure mode of this specimen is near to a typical brittle failure.

a)

b)

c)

Figure 6. The EPS concrete samples with and without polyamide 66; (a) Compressive samples before failure, (b) Observed failure mode in polymer concretes containing EPS without polyamide yarns and c) Observed failure mode in polymer concretes containing EPS with polyamide yarns.

Effect of EPS Volume

Figure 7 illustrates the effects of volume content of EPS on the tensile strength of polyamide EPS concretes. This attribute to the fact that there is an inverse relationship between the concrete tensile strength and the EPS volume. This means that the tensile strength of EPS concrete appeared to increase with a decrease in the EPS volume.

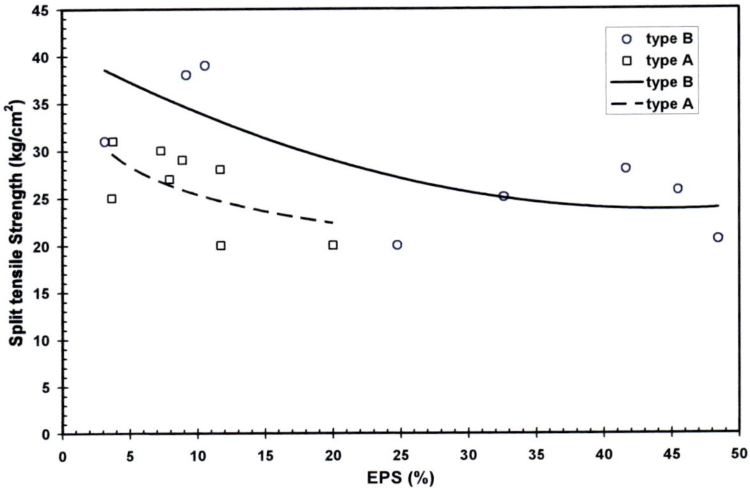

Figure 7. Variation of tensile strength with EPS volume.

Effect of EPS Bead Size

From figure 8, the tensile strength of polyamide EPS concrete (type B) is 40 kg/cm^2 whereas, the maximum tensile strength for polyamide EPS concrete (type A) is only 31 kg/cm^2. But in general, we can expect an increase of 70% in the tensile strength for the lower bead sizes.

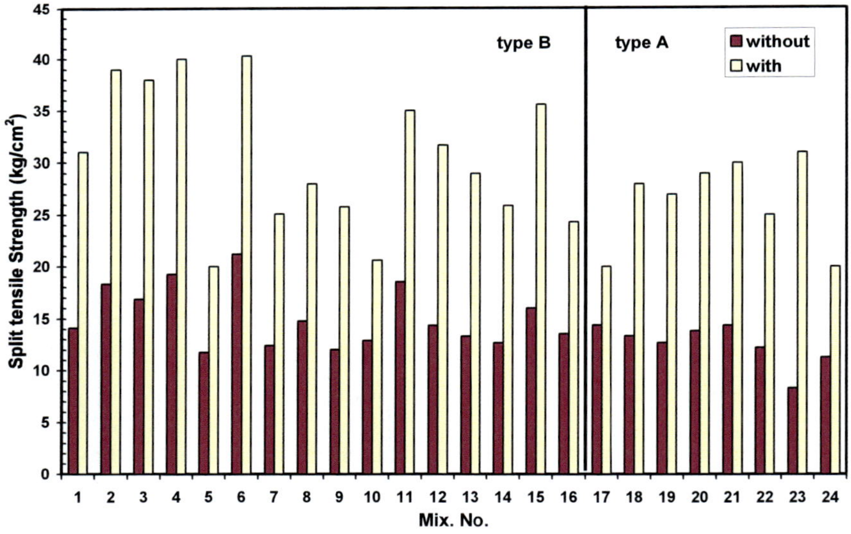

Figure 8. Effects of EPS bead size on tensile strength.

Effect of Polyamide 66 Fibers

The effect of polyamide 66 yarns on the tensile strength of EPS concrete is presented in figure 8. The tensile strength of Polyamide-EPS concrete demonstrated a higher value with respect to EPS concrete. This incremental ratio of strengths is about 60 to 100%.

Variation of Tensile Strength Versus Compressive Strength

Efforts were made to study the variation of tensile strength with the compressive strength (figure 9). It appears that the tensile strength increased with an increase in compressive strength. The rate of strength increasing in polyamide-EPS concrete containing small beads (type B) is higher than other.

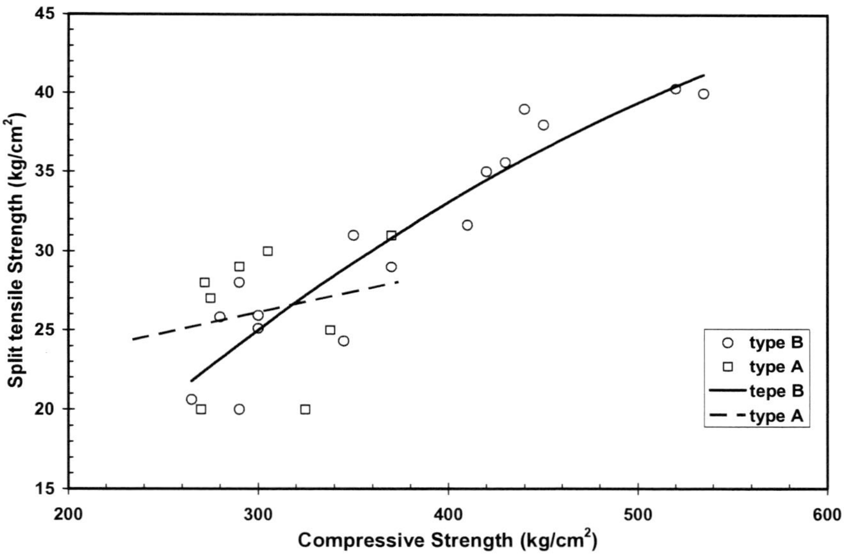

Figure 9. Variation of tensile strength with compressive strength.

Failure Mode

The splitting failure mode of the concrete specimens containing non-reinforced EPS aggregates also did not exhibit the typical brittle failure principally expected in conventional concrete as in compressive strength. As a matter of fact, the failure observed was more gradual (compressible) and the specimens did not separate in two, as it was earlier reported for plastic shredded aggregate concretes [8]. Consequently, the tensile failure mode observed in the polymer concrete containing polyamide 66 yarns was rather a typical brittle failure which is normally resembles to conventional concrete (figure 10).

Figure 10. The failure mode in split tensile test.

Shrinkage

The most significant factor affecting the shrinkage of concrete is the degree of restraint by the aggregate (i.e. its elastic properties and the volumetric proportion of the paste in the mix). It should be noted that the EPS beads principally offer little hindrance to the shrinkage of the paste. Therefore it is expected that as the volumetric proportion of the EPS is increased, the shrinkage would increase as well. This is clearly represented in figure 11. In this study, the drying shrinkage of normal concrete at 90 days was 610 microstrain. Inversely, when the volume content of EPS in the concrete specimen was 20%, the drying shrinkage at 90 days raised to 1000 microstrain, which represents a disadvantage for the application of EPS.

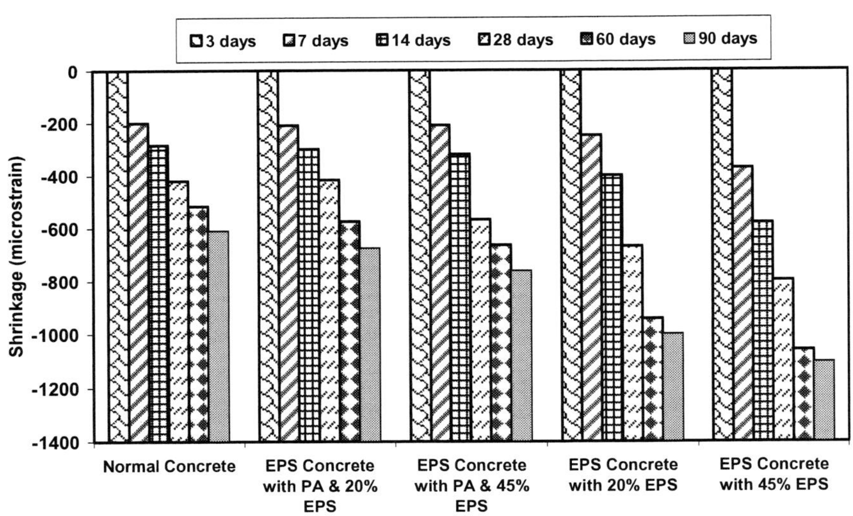

Figure 11. Relationship between drying shrinkage strain of EPS concrete (with polyamide and without polyamide) and reference concrete with age.

In a view of the above, it can be seen that polyamide yarns improved the drying shrinkage of EPS concrete significantly. Even for the case of EPS concrete which contained 20% of its volume by EPS, the drying shrinkage at 90 days was 670 microstrain, which seems to be rather close to normal concrete range.

Absorption

Figure 12 represent the final absorption for polyamide reinforced concretes with different EPS volume. The initial and final absorption for these concretes using type A and B EPS is illustrated in figure 13. From these illustrations, it can be observed that the initial surface absorption of all the EPS concretes are as low as 2.5%, which is specified as "good" concrete by CEB [9]. Nevertheless, the final absorption for these concretes also followed a similar trend. Moreover, the total absorption values observed ranged from 2% to 4.5% for specimens. In general, from the results it can be seen that, lower strength specimens were showing higher absorption compared to the higher strength specimens. From figures 3 and 10, it is revealed that the mixes of 6, 7, have higher strength while their absorption is lower than others.

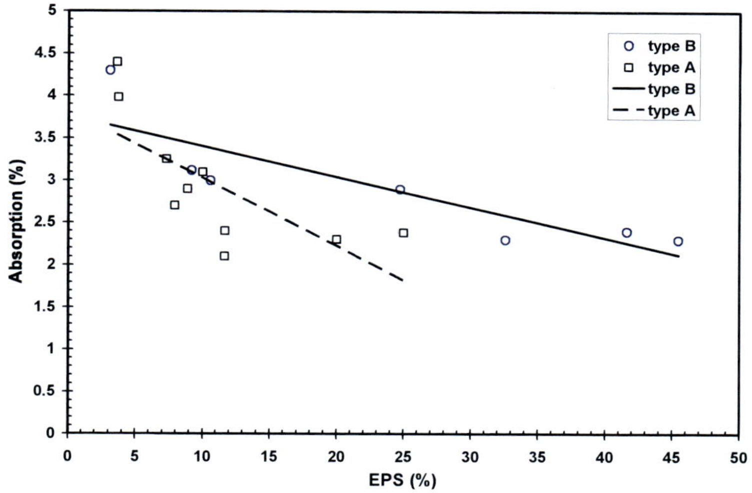

Figure 12. Absorption characteristics of EPS concretes.

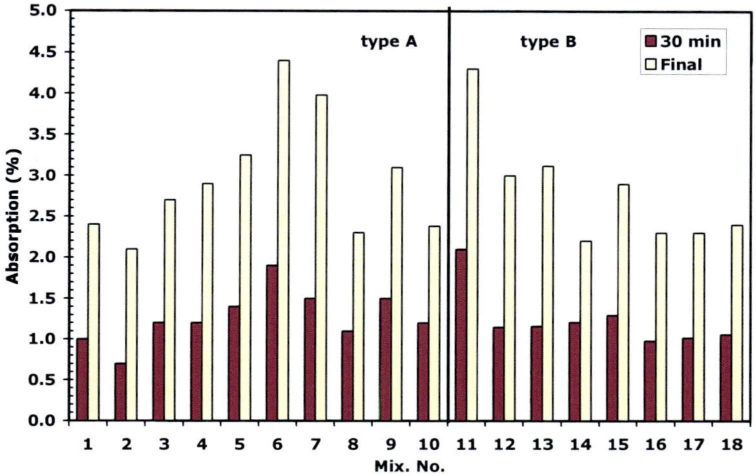

Figure 13. Comparison of initial and final absorption.

Kluge et al. [13] reported the absorption values in the range of 8–25% by weight for concretes made with artificial aggregates of even much higher crushing strength (processed lightweight aggregates like sintered fly ash, pumice, expanded shale, expanded slate, and expanded clay). Also Babu and Babu reported the absorption values in range of 3 to 6% in the concrete containing EPS and silica fume [7]. In this research the maximum range of absorption value obtained was 4.5%. This could be attributed to the fact that the application of polyamide 66 yarns had a significant influence on the absorption value.

Cracking and Adhesion

In our previous study, we shown that there is an inverse relationship between the concrete strength and the EPS content[. This is clearly represented in figure 14.

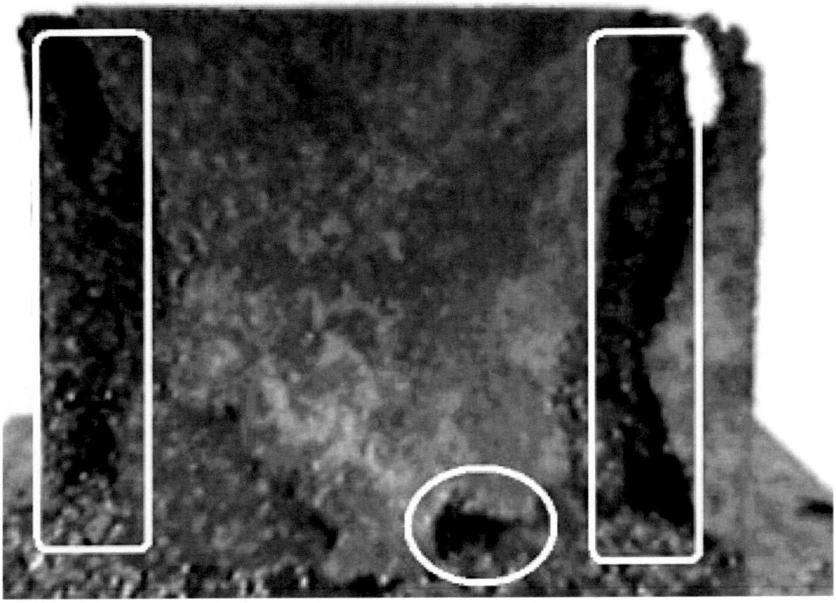

Figure 14. Increasing content of EPS, reduced the sample strength.

The structure of the interacting polystyrene beads from the surface of the specimen was observed after fracture using electron scanning microscope (figure 15). The microstructure of the contact zone of EPS beads are free from deep cavities. This reveals that the adhesion strength of fine beads (Type B) is high.

The application of polyamide 66 yarns in addition of strength increasing, had a significant influence in crack reduction. This is clearly shown in figure 16. The crack propagated much faster in the sample not containing polyamide 66 yarns. Moreover, the sample deformation containing polyamide 66 yarns was noticeably less than others.

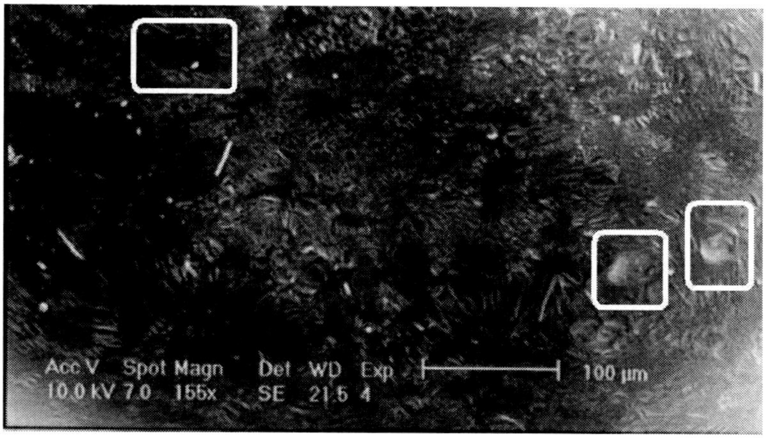

Figure 15. Microstructure of the contact zone of fine EPS (type B).

Figure 16. The adhesion strength is increased when polyamide 66 yarns used.

CONCLUSION

The use of recycled materials in concrete has been on the increase throughout the world due to resource conservation. The results of this study have shown that use of this locally available low-cost PA waste may provide an economical and effective alternative to concrete aggregates. The chief innovation of this work is the significant ability of polyamide 66 fibers to improve the behavior of lightweight EPS concrete. It is concluded that:

1. The interaction of Polyamide 66 fibers with EPS concrete did significantly influence the strength. For a 28 days lightweight Polyamide-EPS concrete, we demonstrated a strength of around 500 kg/cm^2 while the densities varied from1500 to 2000 kg/m^3.
2. The tensile strength of Polyamide-EPS concrete demonstrated a value of about 40 kg/cm^2. This incremental ratio of strengths represents an improvement of about 60 to 100%.
3. Final absorption of Polyamide-EPS concrete demonstrated a value of 4.5%. This value for EPS concrete was about 8.2%.

4. The drying shrinkage can be considered as the weakness of EPS concretes. This can be resolved by addition of Polyamide 66 fibers in EPS concrete. In this study, Polyamide 66 fibers improved the drying shrinkage of EPS concrete significantly. Addition of Polyamide 66 fibers to those EPS concretes containing 20% of EPS demonstrated a drying shrinkage equal to normal concrete. But the presence of polyamide 66 yarns in EPS concretes containing 45% of EPS reduced the drying shrinkage by 30%.
5. The failure mode of Polyamide-EPS concrete demonstrated a plastic behavior., while this characteristics was brittle in EPS concrete.

REFERENCES

[1] M. Muthukumar and D. Mohsen, "Studies on polymer concrete based on optimized aggregate mix proportion", *European Polymer Journal,* Volume 40, 2004, pp. 2167-2177.

[2] J. P. Gorninski, D. C. Dal Mobin and C. S. Kazmierczak, Study of the modulus of elasticity of polymer concrete compounds and comparative assessment of polymer concrete and Portland cement concrete, *Cement and Concrete Research,* Volume 34, 2004, pp. 2091-2095.

[3] ACI Committee 213 R-87. Guide for Structural Lightweight Aggregate Concrete, ACI Manual of Concrete Practice, Part 1, American Concrete Institute, Farmington Hills (1987).

[4] A. Short and W. Kinniburgh, Lightweight Concrete. (3rd ed.),, Applied Science Publishers, London (1978).

[5] V. Sussman, Lightweight plastic aggregate concrete. ACI J. (1975 (July)), pp. 321–323.

[6] C. Bagon and S. Frondistou-Yannas, Marine floating concrete made with polystyrene expanded beads. *Mag. Concr. Res.* 28 (1976), pp. 225–229

[7] K. G. Babu and D. S. Babu, Behaviour of lightweight expanded polystyrene concrete containing silica fume, *Cement and Concrete Research,* Volume 33, Issue 5 , May 2003, pp. 755-762.

[8] A.A. Al-Manaseer and T.R. Dalal, Concrete containing plastic aggregates. *Concr. Int.* (1997 (August)), pp. 47–52.

[9] CEB-FIP. Diagnosis and assessment of concrete structures—State of the art report, CEB Bulletin 1989.

In: Progress in Chemistry and Biochemistry
Editors: E. M. Pearce and G. E. Zaikov
ISBN: 978-1-60692-344-3

Chapter 15

GEOSYNTHETICS AND LIGHTWEIGHT PLASTIC AGGREGATE IN CONCRETE

A.K. Haghi*
University of Guilan, P. O. Box 3756, Rasht, Iran
Haghi@Guilan.ac.ir

ABSTRACT

In the first part of this chapter, different applications of geosynthetics are introduced in detail. In the second part, a case study on application of geofiber in concrete containing expanded polystyrene beads (EPS) as plastic aggregates is presented. The results of this study have shown that use of this locally available low-cost geofiber may provide an economical and effective alternative to concrete aggregates. The chief innovation of this work is the significant ability of polyamide 66 fibers (PA 66) to improve the behavior of lightweight EPS concrete.

1. INTRODUCTION

Geotextile is a synthetic permeable textile material applied with soil, rock, or any other geotechnical engineering related material. Textiles were first applied to roadways in the days of the Pharaohs. Even they struggled with unstable soils which rutted or washed away. They found that natural fibers, fabrics or vegetation improved road quality when mixed with soils, particularly unstable soils. Only recently, however, have textiles been used and evaluated for modern road construction. This chapter clarifies the confusion over terms and definitions of geotextiles, and discusses their common roadway and erosion control applications. In the 1920's the state of South Carolina used a cotton textile to reinforce the underlying materials on a road with poor quality soils. Evaluation several years later found the textile in good workable condition. They continued their work in the area of reinforcement and subsequently concluded that combining cotton and asphalt materials during construction reduced cracking, raveling, and failure of the pavement and the base course. When synthetic fibers became more available in the 1960's, textiles were considered more seriously for roadway construction and maintenance. As these new synthetic fabrics evolved, there was confusion over terms and definitions. Textiles and membranes now have reasonably well accepted definitions in the construction industry. It is produced by synthetic fibers made in a woven or loose nonwoven form. Geotextiles, also named as geosynthetics, are normally applicable to high-standard all-season roads, can also be used to low-standard logging roads. American Society for Testing and Materials (ASTM), describes geotextile as any permeable textile material applied with foundation, soil rock, earth or any other geotechnical engineering related material as an integral part of a: man-made project, structure or system. Geotextiles are mostly used for: Reinforcement of Unpaved Roadways, Paved Roadways, Separation applications in Unpaved Roadways, Paved Roadways, Sediment Control, etc. as part of geo-composites.

1.1. Geotextiles

Geotextiles hace existed for more than fifty years. Though, the development of market sand research work were put in to practice in the early 1960s. Geotextiles are performing progressively in civil engineering construction and are still growing as an alternative, economically viable material. In recent years, the utilization of geotextiles in the world markets has grown at an extraordinary rate. In India, geotextiles have been specially used in road and airport flexible pavements and in overlays. The growth of geotextiles in between 2000 and 2005 was grown at the rate of 4.6% annually, and during the next five years (i.e. up to 2012) it is predicted to 5.3 percent. The geotextile market is increasing in its growth rate, though these are now lower than previously forecasted and compared to other applicators it has relatively little growth for end-users of textiles. In the quantity, geotextiles reported a little growth, more than 250,000 tons in 2000, merely 1.5 percent of the overall technical textile market. Furthermore, this sector with low unit values in small numbers gives a large margin. Geotextiles are mainly made from polyolefin, are light in weight and strong but cheap. These permeable woven geotextiles are generally used for filtration and impermeable membranes to hold out mud when pumping. Certain fabrics provide high puncture resistance and offer a significant recognition in road and rail construction projects, or where the reliability of the sheet is required, as in landfill sites. It is noted that geotextiles have to be made in large quantities and cost-effectively; fibers for geotextiles are normally produced by melt-spinning. High Density Polyethylene (HDPE) is applied to provide reinforcement. Even staple fibers, monofilaments, multifilament yarns and slit films are also applied. The polymers as they are actually made and applied for geotextile production are not available in their chemically pure form. For example, raw polyethylene in its colorless translucent form is rather subjected to light degradation; therefore, it is not applied as geotextiles applicants, but normally includes carbon black as ultraviolet (UV) light stabilizer. It is possibly the most light-resistant polymer in black form. Based on manufacturing process, geotextiles can be categorized as woven, nonwoven, or knitted. Woven fabrics are made by the traditional weaving method, giving a screen-like or mesh material with a variety of sizes of mesh openings and according to the tightness of weave. A woven fabric gives high tensile strength, high modulus, and low strains, but gives poor abrasion resistance and dimensional stability. While nonwoven fabrics have high permeability and high strain characteristics. They are produced in a number of geometric and polymeric compositions to satisfy a various applications. Many geotextiles are prepared by polypropylene. Fabric produced concrete revetment mats; silt filter fences, erosion control blankets, and fabric envelopes for pipe or mat under drains are the illustrations of common geotextile applications. A geotextiles long-term representation is due to the durability and creep characteristics of the polymer structure. The effect of ground, weather, sunlight, and aging conditions must be measured when applying a geotextile for a permanent base. Non-woven geotextiles are available in the form of polypropylene fibers and are needle punched. Nonwoven fabrics possess distinctive ability to lengthen locally to resist damage, superior permeability and frictional resistance, though their tensile strength is lower than that of woven fabrics. Knitted textiles exposed its fewer applications as geotextiles. Though, warp knitted fabrics are important for developed into reinforced soil applied for granular soil and are named as Directionally Structured Fibers (DSF). DSF directed to considerable economies in the application of polymer within the construction, in the form of its evidence for the absorption of tensile stresses. While examine, significantly less stress is to be found on weft element, then there is little grounds. Similarly Directionally Oriented Structures (DOS) are warp knitted fabrics with comparable sets of yarns put into the structure included by loop structures so that load is exactly put on the yarns to use their full potential.

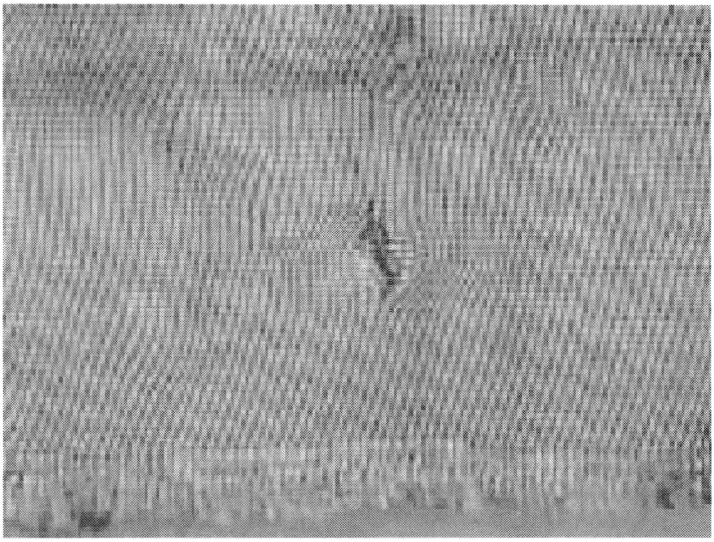

Figure 1. Geotextile.

Geosynthetics have been utilized in numerous civil and environmental engineering applications worldwide for more than 20 years to prevent seepage of liquids. Such seepage control applications as water containment and conveyance, structure waterproofing, and environmental protection make extensive use of geomembranes and geosynthetic clay liners (GCLs) along with other geosynthetics. The materials used for seepage control may be exposed, as with pond linings, or buried, as with landfill linings. They may be subject to significant stresses, as with a pond cover, or exposed to very aggressive environments, as with a chemical tank lining. The wide variety of potential exposure conditions is why there is such a wide variety of geosynthetic barrier materials.

1.2. Geofibers

Geofibers are usually polypropylene fibers blended into soils to create an ideal reinforcement system for the repair of slope failures, reinforcement of pavement subgrades, foundation stabilization, and improvement of retaining wall backfill. By synergistically meshing with the soil already on site, geofibers help create a soil reinforcement system with dramatically improved engineering properties.

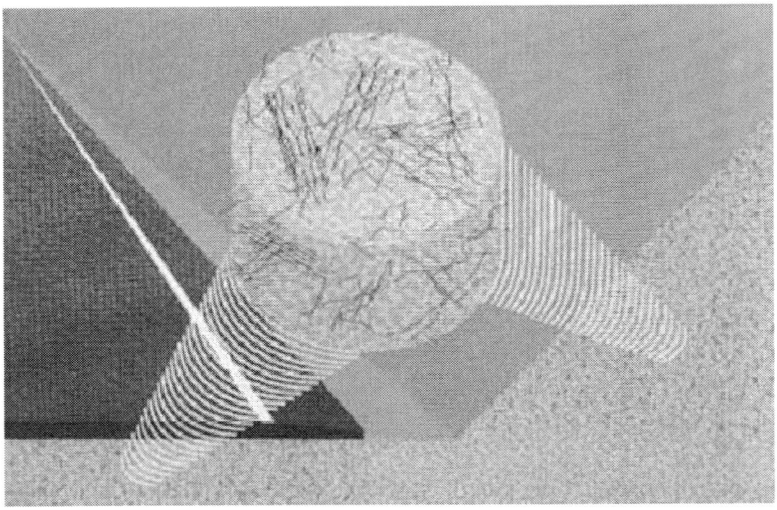

Figure 2. Geofiber.

1.3. Types of Geotextiles

1.3.1. Geomembranes

Geomembranes stand for another form of geosynthetics and are applied mainly for linings and covers of liquid or solid-storage facilities. These are basically a resistant material, in the shape of manufactured sheet, which may be synthetic or bituminous. Therefore, the main task is as a liquid or vapor barrier and is also applicable for various applications. Applied for decorative water feature application and land design and recognized as flexible geomembranes as liners. This is because of the reality that the flexible geomembrane is cheap and flexible for many design ideas, besides having water containment capabilities. Geomembranes are commonly used as barriers in waste containment facilities and landfills due to various benefits associated with their use and because of regulatory requirements. Geomembrane are also increasingly being used in reservoirs, ponds, lined canals and other geotechnical projects. Geotechnical engineers often characterize the shearing resistance along interface between geomembranes and soils using results from interface direct shear tests. The results of these tests are used in an analysis of stability against sliding along the given interface. Interface shear testing between soil and geosynthetics has now become an essential part of the design process in geotechnical and geo-environmental engineering.

Geomembranes are "impervious" thin sheets of rubber or polymeric material used primarily for linings and covers of liquid or solid waste containment facilities. Geomembranes represent the largest category (by cost), of geosynthetics products used in civil engineering applications. The growth in the use of geomembranes can be attributed to the various benefits associated with their application, their relative economy and increasingly stringent environmental regulations. The mechanism of diffusion in geomembrane is on molecular scale which is different from other porous media. Water molecules diffuse through narrow spaces between polymer molecular chains. Geomembranes cannot be regarded as totally impermeable as some amount of diffusion permeation is observed in geomembranes. A typical thermoplastic geomembrane will have diffusion permeability of the order of 10-11 to 10-13 cm/s. Because of their extremely low permeability, their primary function is as a liquid or vapour barriers.

Calendered Geomembranes

Calendered geomembranes are formed by working and flattening a molten viscous formulation between counterrotating rollers. Polyvinyl chloride (PVC), chlorosulfonated polyethylene (CSPE), chlorinated polyethylene (CPE), and, more recently, polypropylene (PP) are the most common calendered geomembranes. Specialty ethylene interpolymer alloy (EIA) geomembranes are used for unique applications. In most cases these engineered films are supported by a textile that provides tensile strength and enhances tear and puncture resistance.

Extruded Geomembranes

Extruded geomembranes are manufactured by melting polymer resin, or chips, and forcing the molten polymer through a die using a screw extruder. The sheet is formed either by a flat horizontal die or through a vertically oriented circular die to form either a flat wide sheet advanced on a conveyor belt, or cylindrical tube of "blown film", filled with air which is collapsed and pulled by nip rollers mounted high above the die. Blown film geomembranes must be slit prior to wind-up. Common extruded geomembranes include high-density polyethylene (HDPE) and various lower density, or very flexible, polyethylenes (VFPE). Polypropylene (PP) is a relatively new type of extruded (as well as calendered) geomembrane. Variations in the manufacturing of geomembranes include texturing to enhance the interface friction between the geomembrane and adjacent soils or other geosynthetics; coextruding different polymers into a single sheet to provide enhanced durability; and the availability of multiple thicknesses and sheet sizes.

Geomembranes are thin, two-dimensional sheets of material with very low permeability. This makes them ideal for forming waterproof or gas proof barriers between adjacent bodies of soil or soil and fluid. Some of their potential applications include sealing against fluid percolation along the coasts, river banks, reservoirs and in water storage. They are also used as buffers against pollutants. The manufacturing of geomembranes begins with the production of raw materials. These are polymer resin, plasticizer accelerators or retarders, filters, and processing aids. The raw materials are blended together and compounded before being extruded in sheet or cylindrical form. Extruders both melt the above materials and homogenize them into a consistent fluid mass in a partial vacuum. The vacuum eliminates air bubbles in the final product.

1.3.2. Geogrids

Geogrids are polymeric structures rather than being a woven, nonwoven or knit textile fabric, in their unidirectional or bidirectional format. They are made in the form of manufactured sheet, including a regular network of integrally associated parts, which may be linked by extrusion, bonding or interlacing, whose openings are larger than the constituents, made into a extremely exposed, network like arrangement, i.e. they have large apertures. They work as reinforcement materials. Coated polyester geogrids have been broadly applied in soil stabilization and geotechnical reinforcement uses. Geogrids are single or multi-layer materials usually made from extruding and stretching high-density polyethylene or polypropylene or by weaving or knitting and coating high tenacity polyester yarns. The resulting grid structure possesses large openings (called apertures) that enhance interaction with the soil or aggregate.

Their physical structure can be categorized in to the following:

- Unidirectional geogrid: Having a great deal of tensile strength in one direction (longitudinal or transversal) than in the other direction.
- Bidirectional geogrid; Having identical strength in both longitudinal and transversal direction.
- Extruded geogrid: Created through stretching uniaxial or biaxial, an extruded integral structure.
- Bonded geogrid: Created through bonding, at right angles, two or more sets of strands.
- Woven geogrid: Created through interlacing, usually at right angles, two or more yarns, filaments or other elements.
-

1.3.3. Geonets

Geonets are normally made by uninterrupted extrusion of corresponding sets of polymeric ribs at acute angles to one another. When the ribs are opened, relatively large apertures are shaped into a netlike pattern. Their pattern work is mostly applicable in the drainage area. Geonets are made of stacked, criss-crossing polymer strands that provide in-plane drainage. Nearly all geonets are made of polyethylene. The molten polymer is extruded through slits in counter-rotating dies, forming a matrix, or "net" of closely spaced "stacked" strands. Two layers of strands are called "bi-planar". Three layers are called "tri-planar".

Geonets are the most recently introduced members of the geosynthetic family. They are grid-like materials, however, distinct, from geogrids by virtue of their function. They do have considerable strength but are used mostly for drainage purposes. All geonets are made of polyethylene. The specific gravity of most geonets is in the range of 0.935 to 0.942. The only other materials in geonets are carbon black and a processing package. In manufacturing, the ingredients are mixed and passed through an extruder, which injects the melt into a die with slotted counter-rotating segments. Over these, the melt flows at angles forming discrete ribs in two planes. As pressure forces the semi-solid mass forward, it is pushed over an increasing diameter core, which forces the ribs apart and opens the net. Diamond-shaped apertures are formed that are typically 12mm long by 8mm wide. Geonets are typically 5.0 to 7.2mm thick. Thickness is a key factor in determining drainage capability. Adding a foam agent to the ingredients can increase thickness of geonets.

1.3.4. Geocomposites

A geocomposite comprises with a mixture of geotextile and geogrid; geogrid and geomembrane; geotextile; or any of these three materials with [another material (e.g. deformed plastic sheets, steel cables, or steel anchors). Geocomposites are accumulated materials, in the appearance of manufactured sheet or strip, compromising of at least one geosynthetic among the components.

1.3.5. Geomat

Geomat is available in the polymeric form, in the shape of a manufactured sheet, compromising of an irregular network of fibers, yarns, filaments, tapes or other elements (thermally or mechanically connected), whose openings are normally greater than the application of the constituents.

1.3.6. Geocell

Geocell is available in a polymeric cellular form including a regular open network of connected strips, linked by extrusion, adhesion or by other methods.

1.3.7. Biomat and Bionet

They are permeable, natural, and accepted as biodegradable polymeric materials, in the shape of a manufactured sheet. Normally biomat comprises with fibers (jute, coir, sisal, straw, or others) set aside collectively by one or two layers of synthetic or natural meshes and bionet comprises with a regular network of knotted or interlaced yarns, whose openings are normally greater than the constituents.

1.4. Natural Forms of Geotextiles

1.4.1. Jute

Jute is natural multi filament fiber, durable and simple to both produce and dispose. Biodegradable woven jute is accessible in a number of weave densities, initially anticipated as a geotextile to avoid land sliding and consequent to deforestation. Jute is available in India in large quantities at a cheaper rate. Jute geotextiles can perform a vital function in the control of soil erosion by revegetation and it has many uses, which are cheaper well as easy to accomplish. It has many benefits as geotextiles, because of its high water absorption capability, flexibility and drapability. It also mixed with other materials, such as in the construction of jute-sand-mat structures. The growth of jute based geotextiles is huge due to its various applications in infrastructure development. There were eight Indian jute mills which started production of jute on the base of 50 tons a day of geotextiles with open mesh structures in June 1998, and in the recent time production and uses of these types of jute geotextiles have improved drastically. The possible market for jute geotextiles is bigger; though it is not yet being completely utilized. In order to get benefits of the prospects presented, the jute industries must carry a more market driven approach, reacting to the requirement of customers and must develop standards for its products. Jute, one of the oldest surviving agro-industries in India, has been traditionally in use for flexible packaging, especially sacks. Its special physical attributes have opened up new avenues for diversification promoted mostly as a result of global concerns for environment. Jute geotextile is one such diversified product of jute which has proved to be highly effective in addressing a number of soil-related problems in civil engineering Functionally, Jute geotextile does not have any dissimilarity with man-made geotextiles commonly known as synthetic geotextiles - made of artificial fibers with various petro-chemical derivatives as their source.

The functions are - separation, filtration, drainage and initial reinforcement. Besides, biodegradablility of geotextile helps in quick growth of vegetation by coalescing with the soil, increasing its permeability, retaining the appropriate humidity as "mulch" and creating a micro-climate that is conductive to vegetative growth. In fact, geotextile is the most acclaimed natural fabric that provides biotechnical solutions to vulnerable exposed soil. Biodegradability is considered by some as a disadvantage. This is to be borne in mind that all geotextiles act as catalyst in the process of improving engineering properties of soil. An effective life span of two season-cycles is found to be sufficient for natural consolidation of soil known as "filter cake" formation from extensive laboratory tests by leading academics and field trials. Biodegradability of geotextile is, therefore, not a discouraging.

1.4.2. Flax/Coir

Coir is a natural insulation material produced from flax fibers, intertwined together into non-woven matting, which can then be set in lofts or put into wall cavities. Coir geotextiles are applied in areas of erosion control, soil conservation, and other civil and bioengineering applications. It also has the appropriate strength and toughness to protect the slopes from erosion while permitting vegetation to flourish. They can dissolve the energy of flowing water and absorb the extra solar radiation.

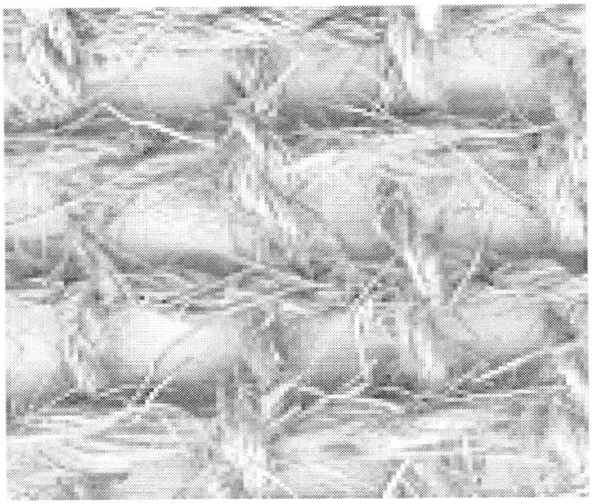

Figure 3. Coir geotextile.

1.4.3. Coconut (Coir) Matting

Used or recycled goods of the coconut fibers from the post industrial waste can also be applied as the economical applicators. It gives low impact and reasonable result to the problems of soil erosion and land sliding on manmade slopes such as motorway and railway embankments.

1.4.4. Cotton

Cotton is mainly made from old clothes in the form recycled cotton or from the post consumer waste stream, is available in a broad variety of colorways, each having the feature of speckling due to the lack of discrimination during the recycling and sorting stages of production.

1.4.5. Hemp

These are available in a broad woven variety:

1.4.6. Straw

Produced by a mixer of straw and a loosely woven net of biodegradable string, gives a minimal influence and realistic answer to the problems of soil erosion and land sliding on manmade slopes such as motorway and railway embankments.

Table 1. Mechanical properties of different natural and conventional fibers

Fiber	Density (g/cm^3)	Young's modulus (GPa)	Tensile strength (MPa)	Elongation at break (%)
Cotton	1.5-1.6	5.5-12.6	287-597	7-8
Jute	1.3	26.5	393-773	1.5-1.8
Flax	1.5	27.6-46.9	345-1035	2.7-3.2
Ramie	-	61.4-128	400-938	3.6-3.8
Sisal	1.5	9.4-22	511-635	2-2.5
Softwood kraft (spruce)	1.5	18-40	1000	-
Hardwood kraft (birch)	1.2	37.9	-	-
E-glass	2.5	70	2000-3500	2.5
Aramide	1.4	63-67	3000-3150	3.3-3.7
Carbon	1.4	230-240	4000	1.4-1.8

When we talk about natural geofiber composites we mean a composite material that is reinforced with geofibers from natural or renewable resources, in contrast to for example carbon fibers that have to be synthesized (with crude oil as origin). Natural geofibers may come from plants, animals or minerals. However, in the following discussion only cellulose based fibers from plants will be considered. The use of natural geofibers and natural geofiber composites are certainly not new to mankind. Bricks made from clay reinforced with straws have been used for thousands of years as building material. Textiles and ropes made from flax and hemp have been around for very long time and are still used today. Paper and cotton sheets impregnated with phenol- or melamine-formaldehyde resin were introduced in the early 1900 for electronic purposes. Natural geofibers in composites offer some interesting properties. Geofibers such as flax have relatively good specific properties, similar to those of glass fibers. Natural geofibers are environmental friendly and together with a thermoplastic matrix they are also recyclable. Economical aspects are of course important. The ever rising oil price has lead to an increased interest in the use of natural geofibers over the recent years. Drawbacks of natural fibers are sensitivity to moisture, low compatibility to non-polar polymers used as matrix and lack of well defined mechanical properties. The latter is because many parameters (geographic location, climate, soil condition etc.) affect the growth of plants and consequently their properties (may even vary a lot from different locations within the plant). Some typical mechanical data for different natural fibers are shown in table 1 in comparison to synthetic man-made fibers.

1.5. Composite/Synthetic Materials for Geotextiles

1.5.1. Kevlar

Woven Kevlar textiles possess the structural dynamics and works better than steel, but with a fraction of its weight. Whether it is working in a plant, or handling sharp edge materials kelwar bring together high strength (5 times stronger than steel) with light weight, and comfort with protection. Kevlar is also made at low temperatures. They are applied as the reinforcement of cement concrete in the form of chopped fibers both for ordinary and autoclave cure types. Kevlar can also be used with other materials as a protective coating.

1.5.2. Polyester

Made from the post consumer waste such as bottles, fabrics, etc. in the composition of polyester ethylene terephthalate, and because of its superior strength and elasticity it is applicable to numbers of geotechnical applications.

1.5.3. Polypropylene

Polypropylene is a famous alternative for geo-grids or geo-matrices, as it is strong and chemically static. Polypropylene is accepted to carry out poor long-term creep behavior but is economical and light in weight.

1.5.4. Phenolic/Jute Composite

This composite is a strong and abundantly durable, made by the mixing of jute fibers with a phenol based resin binder. Bio-composites such as this one give substitute of making durable, high performance materials whilst applying as few non-renewable as probable. This material is extensively applicable within a range of furniture design applications.

Besides these, many other types of materials may be applied to make geotextiles, according to its applications. Geotextiles are more and more applied in civil engineering works and so their production has new openings to the textile industries. New and various types of geotextiles are made worldwide and are applied in a variety of applications. The majority of geo-fabrics are made from synthetic materials, but there are few uses that require biodegradable, natural materials such as coir or jute. Unlike synthetics, they can absorb and store moisture, their natural flexibility permits them to match directly to soil profile and their bulk provides great productive cover/weight ratio. As natural material, these are well-matched with surrounding vegetation and are easily available with an economical rate. Natural fibers are available in a large quantity in India. They are comparatively economical and can give a cost-effective solution to short span projects. In the coming years, all these factors show bright prospects for geotextiles worldwide.

In the most common reinforcement application, the geotextile interacts with soil through frictional or adhesion forces to resist tensile or shear forces. To provide reinforcement, a geotextile must have sufficient strength and embedment length to resist the tensile forces generated, and the strength must be developed at sufficiently small strains (i.e. high modulus) to prevent excessive movement of the reinforced structure. To reinforce embankments and retaining structures, a woven geotextile is recommended because it can provide high strength at small strains. Geotextiles can be used successfully in pavement rehabilitation projects.

Geotextile interlayers are used in two different capacities-the full-width and strip methods. The full-width method involves sealing cracks and joints and placing a nonwoven material across the entire width of the existing pavement.

Geotextiles have been used in construction of gravel roads and airfields over soft soils to solve these problems and either increase the life of the pavement or reduce the initial cost. The placement of a permeable geotextile between the soft subgrade and the granular material may provide one or more of the following functions, (1) a filter to allow water but not soil to pass through it, (2) a separator to prevent the mixing of the soft soil and the granular material, and (3) a reinforcement layer to resist the development of rutting. The reinforcement application is primarily for gravel surfaced pavements. The required thicknesses of gravel surfaced roads and airfields have been reduced because of the presence of the geotextile. There is no established criterion for designing gravel surfaced airfields containing a geotextile.

1.6. Advantages

Geosynthetics, including geotextiles, geomembranes, geonets, geogrids, geocomposites and geosynthetic clay liners, often used in combination with conventional materials, offer the following advantages over traditional materials:

- Space Savings - Sheet-like, geosynthetics take up much less space in a landfill than do comparable soil and aggregate layers.
- Material Quality Control - Soil and aggregate are generally heterogeneous materials that may vary significantly across a site or borrow area. Geosynthetics on the other hand are relatively homogeneous because they are manufactured under tightly controlled conditions in a factory. They undergo rigorous quality control to minimize material variation.
- Construction Quality Control - Geosynthetics are manufactured and often factory "prefabricated" into large sheets. This minimizes the required number of field connections, or seams. Both factory and field seams are made and tested by trained technicians. Conversely, soil and aggregate layers are constructed in place and are subject to variations caused by weather, handling and placement.
- Cost Savings - Geosynthetic materials are generally less costly to purchase, transport and install than soils and aggregates.
- Technical Superiority - Geosynthetics have been engineered for optimal performance in the desired application.
- Construction Timing - Geosynthetics can be installed quickly, providing the flexibility to construct during short construction seasons, breaks in inclement weather, or without the need to demobilize and remobilize the earthwork contractor.
- Material Deployment - Layers of geosynthetics are deployed sequentially, but with a minimum of stagger between layers, allowing a single crew to efficiently deploy multiple geosynthetic layers.
- Material Availability - Numerous suppliers of most geosynthetics and ease of shipping insure competitive pricing and ready availability of materials.

- Environmental Sensitivity – Geosynthetic systems reduce the use of natural resources and the environmental damage associated quarrying, trucking, and other material handling activities.

1.7. Geotextile Polymers

Almost all geotextiles available in the United States are manufactured from either polyester or polypropylene. Polypropylene is lighter than water (specific gravity of 0.9), strong and very durable. Polypropylene filaments and staple fibers are used in manufacturing woven yarns and nonwoven geotextiles. High tenacity polyester fibers and yarns are also used in the manufacturing of geotextiles. Polyester is heavier than water, has excellent strength and creep properties, and is compatible with most common soil environments.

1.8. Geotextile Structures

There are two principal geotextile types, or structures: wovens and nonwovens. Other manufacturing techniques, for example knitting and stitch bonding, are occasionally used in the manufacture of specialty products. *Nonwovens:*

Nonwoven geotextiles are manufactured from either staple fibers (staple fibers are short, usually 1 to 4 inches in length) or continuous filaments randomly distributed in layers onto a moving belt to form a felt-like "web". The web then passes through a needle loom and/or other bonding machine interlocking the fibers/filaments. Nonwoven geotextiles are highly desirable for subsurface drainage and erosion control applications as well as for road stabilization over wet moisture sensitive soils.

Wovens

Weaving is a process of interlacing yarns to make a fabric. Woven geotextiles are made from weaving monofilament, multifilament, or slit film yarns. Slit film yarns can be further subdivided into flat tapes and fibrillated (or spider web-like) yarns. There are two steps in this process of making a woven geotextile: first, manufacture of the filaments or slitting the film to create yarns; and second, weaving the yarns to form the geotextile. Slit film fabrics are commonly used for sediment control, i.e. silt fence, and road stabilization applications but are poor choices for subsurface drainage and erosion control applications. Though the flat tape slit film yarns are quite strong, they form a fabric that has relatively poor permeability. Alternatively, fabrics made with fibrillated tape yarns have better permeability and more uniform openings than flat tape products.

2. A Case Study on Application of Geofibers in Concrete

Concrete is the most widely used construction material in the world. Its low cost, ease of application and compressive strength are the principle reasons for its universal acceptance.

However it has few shortcomings most of which are attributable to the Portland cement binder. Shortcomings include poor tensile strength, high porosity freeze thaw deterioration, and destruction by corrosive chemicals etc [1].

Improving concrete properties such as strength and durability by adding fiber to the mix has become a common practice.

The development of new composite materials possessing increased strength and durability when compared with conventional types is a major requirement of applications in repairs and in the improvement of infrastructure materials used in the civil construction industry. Polymer concrete (PC) is an example of a relatively new material with such high performance [2].

The demand for lightweight concrete in many applications of modern construction is increasing, owing to the advantage that lower density results in a significant benefit in terms of load-bearing elements of smaller cross sections and a corresponding reduction in the size of the foundation [3].

Lightweight aggregate concrete, popular through the ages, was reported to have a comparable or some times better durability even in severe exposure conditions.

Lightweight aggregates are broadly classified in to two types—natural (pumice, diatomite, volcanic cinders, etc.) and artificial (perlite, expanded shale, clay, slate, sintered PFA, etc.). Expanded polystyrene (EPS) beads are a type of artificial ultra-lightweight, nonabsorbent aggregate [4, 5]. It can be used to produce low-density concretes required for building applications like cladding panels, curtain walls, composite flooring systems, and load-bearing concrete blocks [6, 7].

Expanded polystyrene (EPS) concrete is a lightweight concrete with good energy-absorbing characteristics, consisting a discrete air voids in a polymer matrix. However, polystyrene beads are extremely light, with a density of only 12 to 20 kg/m^3, which can easily cause segregation in mixing. Hence some chemical treatment of surface on this hydrophobic material is needed. Other investigators also reported that EPS tends to float and can result in a poor mix distribution and segregation, necessitating the use of admixtures [8,9].

2.1. Scope

Four key points of this investigation are:

1. To convert carpet waste into useful product.
2. To consume carpet wastes which would otherwise gone to landfill.
3. Protection of Environment from being heavily contaminated.
4. Using recycled Polyamide (PA 66) fibers could be a cost-effective way to improve the performance for lightweight EPS concrete.

2.2. Experimental

Materials

Ordinary Portland cement (OPC) was used. The main chemical composition of (OPC) used in this study are listed in table 2.

Table 2. Chemical and physical compositions of cement

SiO_2	20.40%
Al_2O_3	6.12%
Fe_2O_3	3.051%
CaO	63.16%
MgO	2.32%
SO_3	2.40%
Specific gravity	3.13

Two types of commercially available spherical EPS beads were used (table 3). As indicated in table 3, the grading of EPS shows that Type B has mostly 3.0-mm-size beads and Type C has mostly 6.0-mm-size beads.

Table 3. EPS beads characteristics

Type	Size mm	Bulk density kg/m^3	Specific gravity
B	3	20	0.029
C	6	16	0.014

The physical properties of PA 66 used in this study are summarized in table 4.

Table 4. Some physical properties of PA 66

Specific gravity	1.16
Tensile strength	965 MPa
Elastic modulus	5.17 GPa
Ultimate elongation	20%

Admixture

A naphthalene based super plasticizer (that complies with ASTM C 494) was used to produce the mixes a flow able or highly workable nature, to suit the hand compaction adopted. This admixture is well known in the market and has a good history in international concrete works worldwide.

Specimen Preparation

A number of standard test specimens of different sizes were selected for investigating the various parameters. For studying the compressive strengths at 3, 7, and 28 days and also for investigating the absorption tests cubes of 100 mm size were used. Meanwhile, the split tensile strength test was conducted on 100 mm diameter × 200 mm cylinders at 28 days.

Lightweight EPS concrete were made with PA and without PA reinforcement. Initially, the EPS beads were wetted with 25% of the mixing water and super plasticizer before adding the remaining materials. It is expected that by addition of EPS in place of normal aggregate the weight of concrete can be reduced significantly and the interaction of Polyamide 66 fibers with EPS concrete can improve the strength. In this study, the estimated component additions were measured by volume in order to simplify mixing process. The non-absorbent, hydrophobic and closed cellular aggregates (expanded polystyrene beads) were mixed in a planetary mixer in compliance with the recommendation of ASTM C 305. To obtain a very uniform and flowing mixture, the mixing operation was continued on regular biases. The molds are then filled with fresh EPS concrete and then with PA reinforced EPS concrete. Afterward, the mixture is firmly compacted by hand.

Based on ASTM C 143-98 we measured the slump values of the fresh concrete. Afterward, the compressive strength test was carried out in a testing machine with capacity of 2000 KN. The loading rate used during this test was 2.5 KN/s. According to ASTM C 496-89 we conducted the split tensile strength test on cylinders at 28 days. The absorption test was then carried out as per ASTM C 642-82.

2.3. Experimental Results and Discussion

The results of this investigation are explained in the following subsections.

Fresh Concrete

It appeared that the workability of the concrete in terms of the slump measurements were about 45–80 mm in non-reinforced cases and around 40-75 for polyamide EPS concretes. In addition, it should be noted that the mixes having the higher percentage silica fume can present higher slump values. During the experimental study, all the concretes specimens shown an interesting flexibility and provided a very easy condition to work with. Meanwhile, the specimens could be compacted using just hand compaction and they could be easily finished. In order to be able to solve the problems associated with the hydrophobic nature of the EPS beads, the silica fume and the super plasticizer were added. This will help to improve the cohesiveness of the mix significantly.

Development of Compressive Strength with the Age

Figure 4 displays the development of compressive strength with the age for reinforced EPS concrete. This illustration clearly states that the compressive strength of reinforced EPS concretes in almost all mixes show a continuous increase with age. It is seen that the rate of strength development was greater initially and decreased as the age increased. Comparison of strengths at 7 days revealed that non-reinforced concretes developed almost 40–50% of its 28-day strength. While it is seen that reinforced concretes developed almost 20–40% of the corresponding 28-day strength.

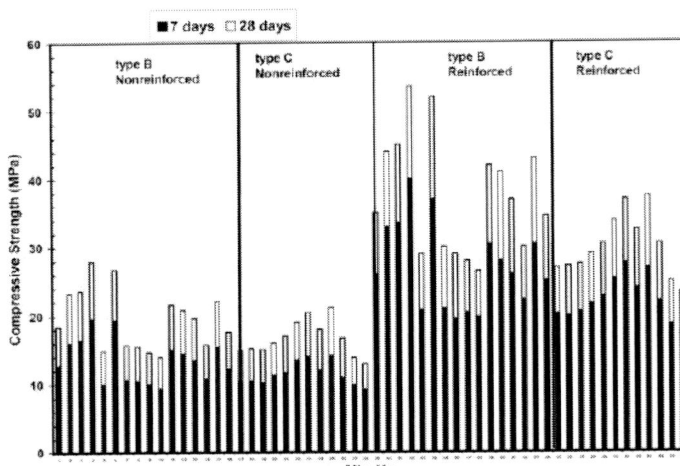

Figure 4. Effects of age on compressive strength.

Effect of Density and EPS Volume

Figures 5 and 6 illustrate the compressive strengths of EPS concretes with different plastic densities for reinforced and non-reinforced concrete and the volume content of EPS.

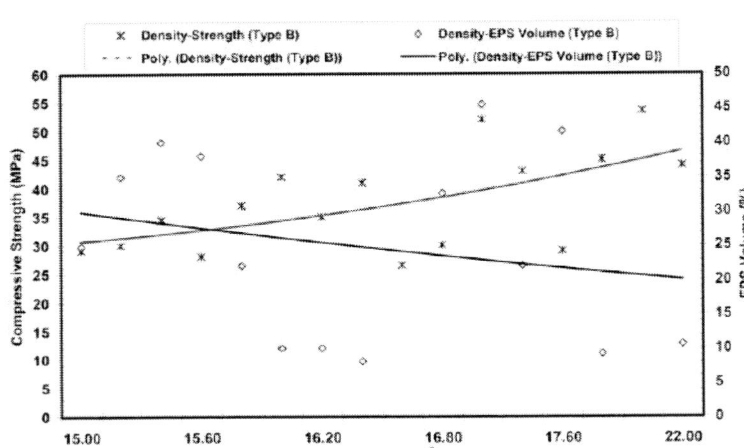

Figure 5. Variation of composite density with EPS volume and Compressive strength (type B).

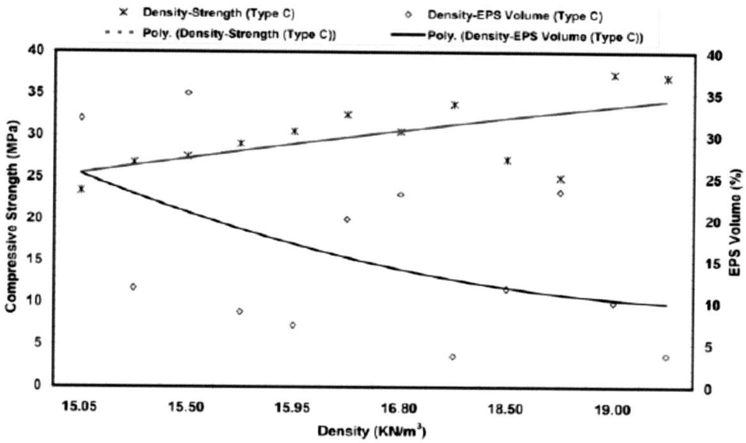

Figure 6. Variation of composite density with EPS volume and Compressive strength (type C).

It is appeared that the strength of non-reinforced and reinforced EPS concrete to increase linearly with a decrease in the EPS volume. Furthermore, it is noted that the density of EPS concretes decreased significantly with an increase in the EPS volume. It is clearly observed that the polyamide EPS concrete have high strength in compression, comparing to that non-reinforced concretes.

In some earlier studies reported on EPS concrete, the densities varied from 15 to 20 KN/m^3. At 28 days, it is reported a compressive strength of 10 to 21 MPa for these concretes. In the present study, we obtained the strength of (23.5 to 37.5) and (26.5 to 54.0) with use of polyamide 66 yarns.

Effect of EPS Bead Size

The plot represented in figure 7 show an inverse relationship between the concrete strength and the bead size. That means the strength in EPS concrete increased with a decrease in the EPS bead size for the same mix proportions.

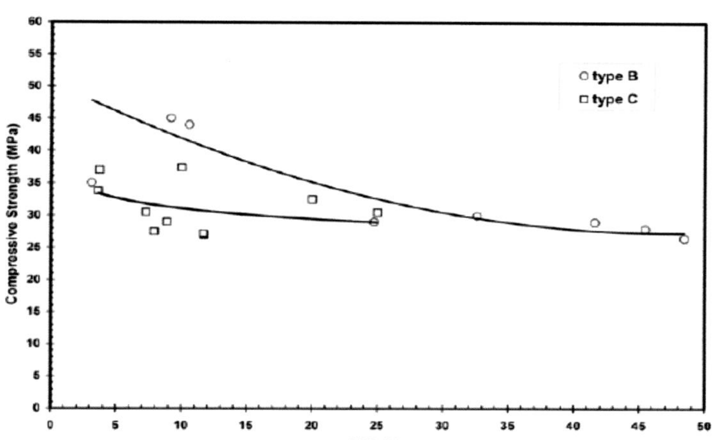

Figure 7. Effects of EPS bead size on compressive strength.

Effect of Silica Fume and Polyamide 66 Fibers

From figure 8 it can be clearly appreciated that in the mixes content fine silica fume the compressive strength can increase significantly (i.e.: mixes no. 4, 6,…). This represents the effects of fine silica fume and polyamide 66 on the compressive strength of EPS concrete. This is clearly visible that fine silica fume can improve the dispersion of EPS in the cement paste and interfacial bonding between EPS and cement paste. Moreover, the compressive strength of polyamide EPS concrete is almost 1.5 to 2 times non-reinforced EPS concretes

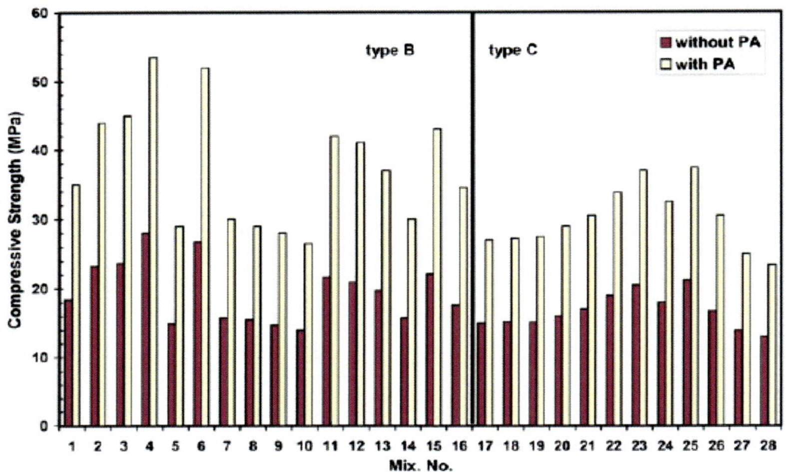

Figure 8. Effects of silica fume and polyamide on compressive strength.

Failure Mode

The failure modes of different specimens are presented in figure 9. Figure (9a) shows the EPS concrete specimen before loading. The failure mode observed in figure (9b) is rather gradual (more compressible), and the specimen is capable of retaining the load after failure, without full disintegration. Figure (9c) illustrates failure mode in polymer concretes containing EPS with polyamide yarns. This is attributed to the fact that the failure mode of this specimen is near to a typical brittle failure.

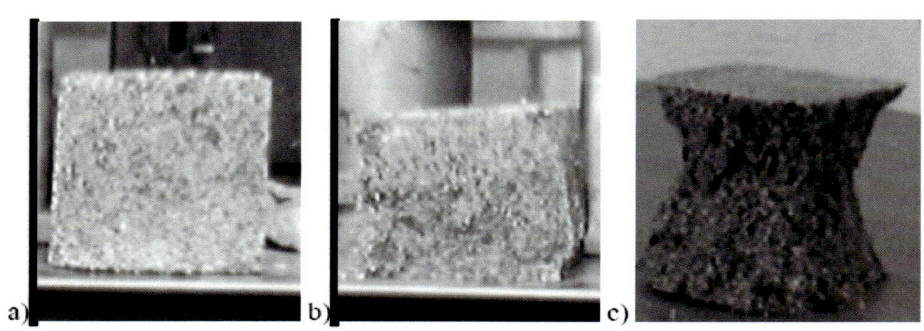

Figure 9. The EPS concrete samples with and without polyamide 66; (a) Compressive samples before failure, (b) Observed failure mode in polymer concretes containing EPS without polyamide yarns and c) Observed failure mode in polymer concretes containing EPS with polyamide yarns.

Effect of EPS Volume

Figure 10 illustrates the effects of volume content of EPS on the tensile strength of polyamide EPS concretes. This attribute to the fact that there is an inverse relationship between the concrete tensile strength and the EPS volume. This means that the tensile strength of EPS concrete appeared to increase with a decrease in the EPS volume.

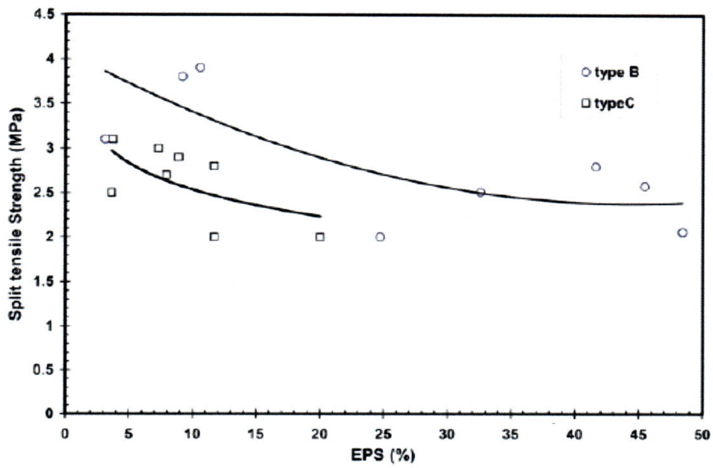

Figure 10. Variation of tensile strength with EPS volume.

Effect of EPS Bead Size

From figure 11, the tensile strength of polyamide EPS concrete (type B) is 4 MPa whereas, the maximum tensile strength for polyamide EPS concrete (type C) is only 3.1 MPa. But in general, we can expect an increase of 70% in the tensile strength for the lower bead sizes.

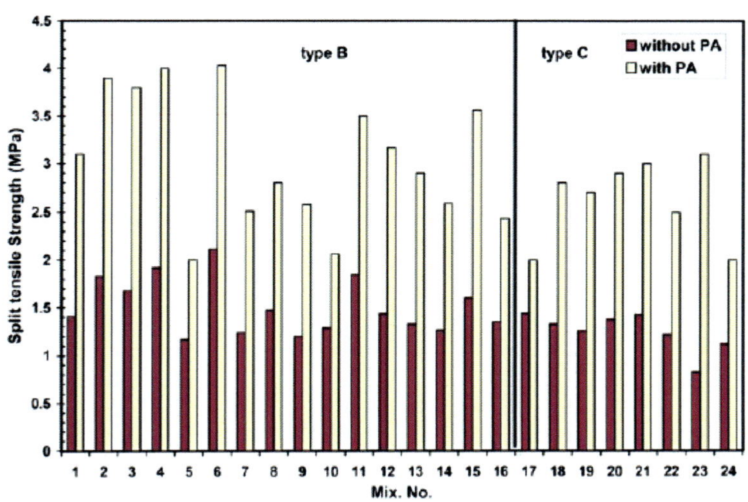

Figure 11. Effects of EPS bead size on tensile strength.

Effect of Polyamide 66 Fibers

The effect of polyamide 66 yarns on the tensile strength of EPS concrete is presented in figure 11. The tensile strength of Polyamide-EPS concrete demonstrated a higher value with respect to EPS concrete. This incremental ratio of strengths is about 60 to 100%. Furthermore it is observed that the Polyamide-EPS concrete containing smaller beads have higher tensile strength.

Variation of Tensile Strength Versus Compressive Strength

Efforts were made to study the variation of tensile strength with the compressive strength (figure 12). It appears that the tensile strength increased with an increase in compressive strength. The rate of strength increasing in polyamide-EPS concrete containing small beads (type B) is higher than other.

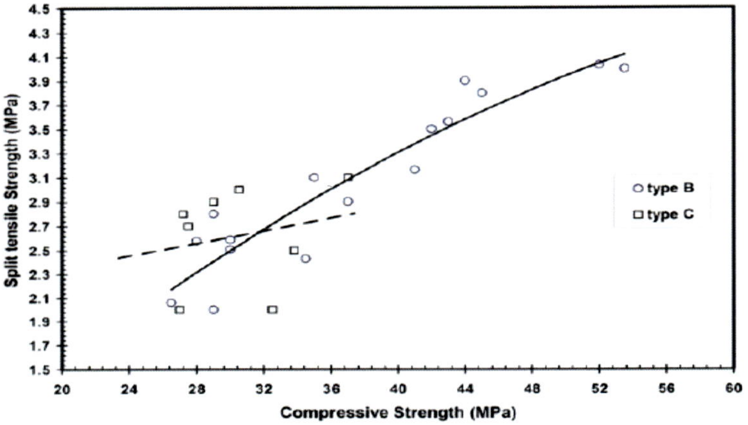

Figure 12. Variation of tensile strength with compressive strength.

Failure Mode

The splitting failure mode of the concrete specimens containing non-reinforced EPS aggregates also did not exhibit the typical brittle failure principally expected in conventional concrete as in compressive strength. Consequently, the tensile failure mode observed in the polymer concrete containing polyamide 66 yarns was rather a typical brittle failure which is normally resembles to conventional concrete (figure 13).

Figure 13. The failure mode in split tensile test.

Shrinkage

The most significant factor affecting the shrinkage of concrete is the degree of restraint by the aggregate (i.e. its elastic properties and the volumetric proportion of the paste in the mix). It should be noted that the EPS beads principally offer little hindrance to the shrinkage of the paste. Therefore it is expected that as the volumetric proportion of the EPS is increased, the shrinkage would increase as well. This is clearly represented in figure 14. In this study, the drying shrinkage of normal concrete at 90 days was 610 micro strains. Inversely, when the volume content of EPS in the concrete specimen was 20%, the drying shrinkage at 90 days raised to 1000 micro strain, which represents a disadvantage for the application of EPS.

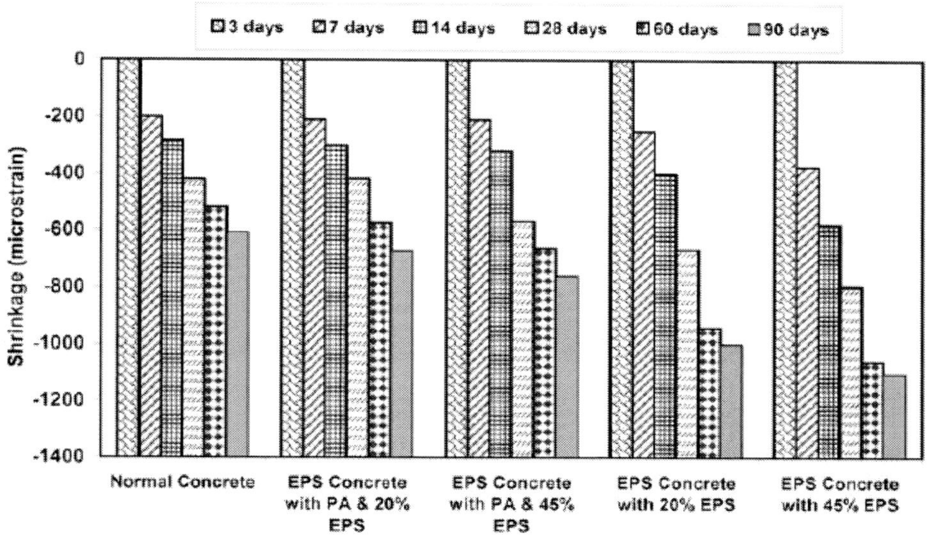

Figure 14. Relationship between drying shrinkage strain of EPS concrete (with polyamide and without polyamide) and reference concrete with age.

In a view of the above, it can be seen that polyamide yarns reduced the drying shrinkage of EPS concrete significantly. Even for the case of EPS concrete which contained 20% of its volume by EPS, the drying shrinkage at 90 days was 670 micro strains, which seems to be rather close to normal concrete range.

2.4. Conclusion

The use of recycled materials in concrete has been on the increase throughout the world due to conserving resources. The results of this study have shown that use of this locally available low-cost PA waste may provide an economical and effective alternative to concrete aggregates. The chief innovation of this work is the significant ability of polyamide 66 fibers to improve the behavior of lightweight EPS concrete. It is concluded that:

1. The interaction of Polyamide 66 fibers with EPS concrete did significantly influence the strength. For a 28 days lightweight Polyamide-EPS concrete, we demonstrated a strength of around 50 MPa while the densities varied from 15 to 20 KN/m^3.
2. The tensile strength of Polyamide-EPS concrete demonstrated a value of about 4 MPa. This incremental ratio of strengths represents an improvement of about 60 to 100%.
3. Final absorption of Polyamide-EPS concrete demonstrated a value of 4.5%. This value for EPS concrete was about 8.2%.
4. The drying shrinkage can be considered as the weakness of EPS concretes. This can be resolved by addition of Polyamide 66 fibers in EPS concrete. In this study, Polyamide 66 fibers improved the drying shrinkage of EPS concrete significantly. Addition of Polyamide 66 fibers to those EPS concretes containing 20% of EPS demonstrated a drying shrinkage equal to normal concrete. But the presence of polyamide 66 yarns in EPS concretes containing 45% of EPS reduced the drying shrinkage by 30%.
5. The failure mode of Polyamide-EPS concrete demonstrated a plastic behavior. While this characteristics was brittle in EPS concrete.

REFERENCES

[1] M. Muthukumar and D. Mohsen, "Studies on polymer concrete based on optimized aggregate mix proportion", Europe*an Polymer Journal,* Volume 40, 2004, pp. 2167-2177.

[2] J. P. Gorninski, D. C. Dal Mobin and C. S. Kazmierczak, Study of the modulus of elasticity of polymer concrete compounds and comparative assessment of polymer concrete and Portland cement concrete, *Cement and Concrete Research*, Volume 34, 2004, pp. 2091-2095.

[3] ACI Committee 213 R-87. Guide for Structural Lightweight Aggregate Concrete, ACI Manual of Concrete Practice, Part 1, American Concrete Institute, Farmington Hills (1987).

[4] A. Short and W. Kinniburgh, *Lightweight Concrete.* (3rd ed.), Applied Science Publishers, London (1978).

[5] V. Sussman, Lightweight plastic aggregate concrete. *ACI J.* (1975 (July)), pp. 321–323.

[6] C. Bagon and S. Frondistou-Yannas, Marine floating concrete made with polystyrene expanded beads. *Mag. Concr. Res.* 28 (1976), pp. 225–229

[7] K. G. Babu and D. S. Babu, Behaviour of lightweight expanded polystyrene concrete containing silica fume, *Cement and Concrete Research*, Volume 33, Issue 5 , May 2003, pp. 755-762.

[8] A.A. Al-Manaseer and T.R. Dalal, Concrete containing plastic aggregates. *Concr. Int.* (1997 (August)), pp. 47–52.

[9] CEB-FIP. Diagnosis and assessment of concrete structures—*State of the art report,* CEB Bulletin 1989.

[10] R. L. Roy, E. Parant and C. Boulay, Taking into account the inclusions' size in lightweight concrete compressive strength prediction, *Cement and Concrete Research,* 2004.

[11] B. Chen and J. Liu, Properties of lightweight expanded polystyrene concrete reinforced with steel fiber, *Cement and Concrete Research,* Volume 34, 2004, pp. 1259-1263.

[12] P.H. Bischoff, Polystyrene aggregate concrete subjected to hard impact, *Proc. Inst. Civ. Eng., Part 2* 6 (1990) 225– 239.

In: Progress in Chemistry and Biochemistry
Editors: E. M. Pearce and G. E. Zaikov

ISBN: 978-1-60692-344-3

Chapter 16

CONVECTIVE DRYING OF TEXTILES: A DETAILED REVIEW ON THEORETICAL AND EXPERIMENTAL ASPECTS

A.K. Haghi
University of Guilan, P. O. Box 3756, Rasht, Iran
Haghi@Guilan.ac.ir

ABSTRACT

The objective of any drying process is to produce a dried product of desired quality at minimum cost and maximum throughput possible. A very common method of removing water from textiles is convective drying. Hot air is used as the heat transfer medium and is exhausted to remove vaporized water. Considerable thermal energy is required to heat make-up air as the hot air is exhausted. In this chapter, it is shown that the effect of humidity on the drying rate of textiles depends on both gas stream temperature and flow rate, and are much larger in the warm-up and constant rate periods than in the falling rate period.

1. INTRODUCTION

When faced with a drying problem on an industrial scale, many factors have to be taken into account in selecting the most suitable type of dryer to install and the problem must be analysed from several standpoints. Even an initial analysis of the possibilities must be backed up by pilot-scale tests unless previous experience has indicated the type most likely to be suitable. The accent today, due to high labour costs, is on continuously operating unit equipment, to what extent possible automatically controlled. In any event, the selection of a suitable dryer should be made in two stages, a preliminary selection based on the general nature of the problem and the textile material to be handled, followed by a final selection based on pilot-scale tests or previous experience combined with economic considerations.

Textile manufacturing involves a crucial energy-intensive drying stage at the end of the process to remove moisture left from dye setting. Determining drying characteristics for textiles, such as temperature levels, transition times, total drying times, and evaporation rates, is vitally important so as to optimize the drying stage. Meanwhile, a textile material undergoes some physical and chemical changes that can affect the final textile quality.

The stenter is the universal machine for treating textile fabric in web form. It can be used for drying, heat-setting, curing, permanent finishing and even thermosolling, as well as many other processes. In the course of these operations the fabric can be stretched or shrunk in longitudinal and cross directions. On the one hand, the stenter should be able to treat highly elastic, light-weight and extremely delicate fabric types with care and, on the other, be suitable for carrying out drying or coating processes with heavy materials, right to felts and carpets. The sensitivity of a material to temperature will have an important bearing on the choice of dryer.

2. Effect of Humidity on the Drying Rate

A typical drying curve showing moisture content versus time is illustrated in figure 1. It should be noted that the slope of this curve is the drying rate, at which moisture is being removed. The curve begins with a warm-up period, where the material is heated and the drying rate is usually low. As the material heats up, the rate of drying increases to a peak rate that is maintained for a period of time known as the constant rate period. Eventually, the moisture content of the material drops to a level, known as the critical moisture content, where the high rate of evaporation cannot be maintained. This is the beginning of the falling rate period. During falling rate period, the moisture flow to the surface is insufficient to maintain saturation at the surface. This period can be divided into the first and second falling rate periods. The first falling rate period is a transition between the constant rate period and the second falling rate period. In the constant rate period, external variables such as gas stream humidity, temperature, and flow rate dominate. In the second falling rate period, internal factors such as moisture and energy transport in the textile material dominate.

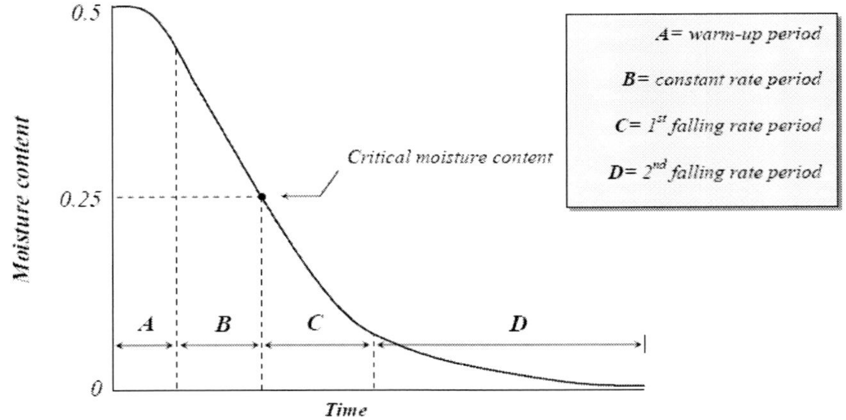

Figure 1. A typical moisture content profile for textile material.

Although much of the water is removed in the constant rate period of drying, the time required to reduce the moisture in the product to the desired value can depend on the falling rate period. If the target moisture content is significantly lower than the critical moisture content, the drying rates in the falling rate period become important.

The drying process can also be represented by a plot of drying rate versus moisture content, as illustrated if figure 2 In this plot, time proceeds from right to left. The warm-up period is on the far right, and the constant rate period corresponds to the plateau region. The falling rate period is the section between the plateau region and the origin.

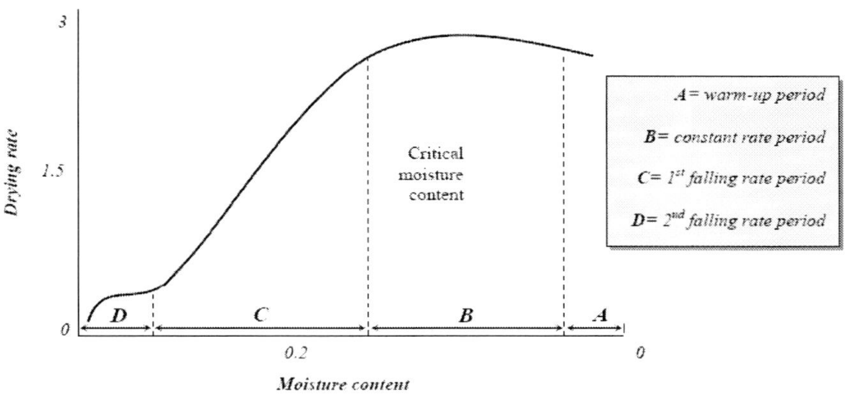

Figure 2. A typical drying rate profile for textile material.

3. Constant Rate Period and Falling Rate Period

During the constant rate period, evaporation is taking place from the fabric surface. The rate of drying is essentially that of the evaporation of the liquid component under the conditions of temperature and air flow during the process. High air velocities will reduce the thickness of the stationary gas film on the surface of textile material and hence increase the heat and mass transfer coefficients. In commercial forced convection dryers the effects of heat transfer by conduction and radiation may be appreciable, due to the fact that the material surface temperatures are higher than the wet-bulb temperature of the drying air. Thus in stenter dryer, the drying rate in the constant rate period may be higher than that calculated values through the effect of radiation.

As has been stated earlier, the period of constant rate of evaporation from wet textile material is followed by a period during which the rate of drying progressively decreases, the transition from one period to the other taking place at the point of critical moisture content of the textile material. During the constant rate drying period, surface of the exposed textile material is completely wetted, at the change to falling rate period some of the fabric surfaces will be still wet and some dry depending largely on the physical form of the textile material being dried. The rate of evaporation of the less moist surfaces will be lower than that of the completely wetted portions, the net result being a falling off in the rate of drying as drying proceeds when compared with the rate during the constant rate period. The rate of drying in this part of the drying curve will still be affected by factors which influence the constant rate drying period as discussed earlier. Once all the exposed surfaces of the textile material cease to be wetted, however, the rate of drying will be a function of the rate at which moisture or moisture vapour can move physically by diffusion and capillarity from within the fabric to its surface.

4. Convective Heat Transfer Rate

Many investigators have attempted to explain the effects of humidity on drying rates and the existence of inversion temperatures. The explanations are usually based on changes that occur in convective heat transfer, radiative heat transfer, and mass transfer as the humidity and temperature of the gas stream change.

At a given gas stream temperature, convective heat transfer rates can change as the humidity in the gas stream is varied, because product temperature and fluid properties vary with humidity. These effects can be explained using the following relationship for the convective heat transfer rate:

$$q / A = h(T_{\infty} - T_s) = h\Delta T \tag{1}$$

here q/A is the convective heat transfer rate per unit surface area A, h is the heat transfer coefficient,

T_{∞} is the free stream temperature of driving medium, T_s is the surface temperature of textile material being dried, and $\Delta T = T_{\infty} - T_s$ is the temperature difference between the drying medium and the textile material being dried

Since product temperature is dependent on humidity, clearly ΔT is also dependent. Further, the heat transfer coefficient *h* is a function of both product temperature and fluid properties. Thus the convective heat transfer rate changes with humidity, as does the drying rate of a textile material.

5. Equilibrium Moisture Content

In considering a drying problem, it is important to establish at the earliest stage, the final or residual moisture content of the textile material which can be accepted. This is important in many hygroscopic materials and if dried below a certain moisture content they will absorb or "regain" moisture from the surrounding atmosphere depending upon its moisture and humidity. The material will establish a condition in equilibrium with this atmosphere and the moisture content of the material under this condition is termed the equilibrium moisture content. Equilibrium moisture content is not greatly affected at the lower end of the atmospheric scale but as this temperature increases the equilibrium moisture content figure decreases, which explains why materials can in fact be dried in the presence of superheated moisture vapour. Meanwhile, drying medium temperatures and humidities assume considerable importance in the operation of direct dryers.

In the more common drying operations met with in practice, the equilibrium moisture content of a material is important as drying may be carried out unnecessarily far, resulting in a reduction in the capacity of a given drying installation and an unjustifiably high cost of drying. Thus, if the equilibrium moisture content of wool hanks in contact with normal ambient air is of the order of 13 to 14% on the wet-weight basis there would be no point in drying to a much lower moisture content. Table 1 gives examples of the equilibrium moisture content in contact with air at different percentage relative humidities at 16 °C.

Table 1. Equilibrium moisture content (percentage on dry-weight baises)

Material	20% r.h.a	30%r.h.a	40%r.h.a	50%r.h.a	60%r.h.a
Wool (worsted)	9.5	13.0	16.5	18.5	21.5
Cotton cloth	3.5	4.5	6.0	7.0	7.5
Egyptian cotton	3.5	4.5	5.5	6.0	7.25
Linen	2.75	3.5	4.5	5.1	6.0

r.h.a = relative humidity of air.

It should be noted that two processes occur simultaneously during the thermal process of drying a wet textile material, namely, heat transfer in order to raise temperature of the wet textile material and to evaporate its moisture content together with mass transfer of moisture to the surface of the textile material and its evaporation from the surface to the surrounding atmosphere which, in convection dryers, is the drying medium. The quantity of air required to remove the moisture as liberated, as distinct from the quantity of air which will release the required amount of heat through a drop in its temperature in the course of drying, however, has to be determined from the known capacity of air to pick up moisture at a given temperature in relation to its initial content of moisture. For most practical purposes, moisture is in the form of water vapour but the same principles apply, with different values and humidity charts, for other volatile components.

6. INVERSION TEMPERATURE

Drying air will always have an advantage over drying in steam because ΔT in equation 1 is larger for drying in air; this is a consequence of T_s being very nearly the wet bulb temperature. The wet bulb temperature is lowest for dry air, increases with increasing humidity, and reaches the saturation temperature of water for a pure steam environment. Thus ΔT_{AIR} will be larger than ΔT_{STEAM} , but $\Delta T_{AIR} / \Delta T_{STEAM}$ decreases with increasing T_∞. Further, the heat transfer coefficient increases with humidity. Apparently, the net effect of the changes in h and ΔT is the convective heat transfer rate increases faster for steam than for air with increasing temperature. Some authors indicated that this is the reason inversion temperature exist. If the dominant heat transfer mechanism is convection, this explanation is plausible.

The variation of gas stream properties with humidity has been reported to explain the existence of inversion temperatures by several investigators [1, 2, 3, 4].

For example, Sheikholeslami and Watkinson [4] explain the existence of inversion temperatures based on the total transferable heat in the drying gas. This depends on the specific heat of the gas, the mass flow rate of the gas, and ΔT between the gas and the drying textile material. Further, Chung and Chow [1] use a dimensionless correlation for the convective heat transfer coefficient in parallel flow to explain the existence of inversion temperatures. Large Renold's numbers and prandl numbers favor higher drying rates in steam, while the high thermal conductivity of air tends to offset these effects. This line of reasoning may be flawed according to Sheikholeslami and Watkinson [4] , who claim that convective heat transfer is altered in the presence of liquid evaporation. Thus, the correlations developed for heat transfer in the absence of drying may not be applicable to heat transfer with liquid evaporation.

Workers have proposed that in version temperatures are due to the higher radiative heat transfer in steam [5] .

Differences in radiative heat transfer are grounded in the Stefan-Boltzmann law. Radiative heat transfer increases with the forth power of temperature, while convective heat transfer varies linearly with temperature. Thus, the relative importance of radiative heat transfer increases with temperature. Since the emissivity of steam is higher than that of air, radiative heat transfer from steam is greater at high temperatures. When the temperature increases sufficiently, the radiative heat transfer effects cause the drying rates in steam to equal those in air.

7. Mass Transfer

In a specialized sense, the term "mass transfer" is the transport of a substance that is involved as a component (constituent, species) in a fluid mixture. An example is the transport of salt in saltine water. Moreover, convective mass transfer is analogous to convective heat transfer.

Consider a batch of fluid of volume V and mass *m*. Let the subscript *i* refer to the *i*th component of the mixture. The total mass is equal to the sum of the individual masses m_i so $m = \sum m_i$. Hence if the concentration of component *i* is defined as $C_i = m_i / V$ then the aggregate density ρ of the mixture must be the sum of all the individual concentrations, $\rho = \sum C_i$. Clearly the unit of concentration is kgm^{-3}. Instead of C_i the alternative notation ρ_i is appropriate if we think of each component spread out over the total volume V.

When chemical reactions are of interest it is convenient to work in terms of an alternative description, one involving the concept of mole. By definition, a mole is the amount of substance that contains as many molecules as there are in 12 grams of carbon 12. That number of entities is 6.022×10^{23} (Avogadro's constant). The molar mass of a substance is the mass of one mole of that substance. Hence, if there are *n* moles in a mixture of molar mass M and mass *m*, then $n = m / M$. Similarly the number of moles n_i in a mixture is the mass of that component divided by its molar mass $M_i, n_i = m_i / M_i$. The mass fraction of component *i* is $\phi_i = m_i / m$ so clearly $\sum \phi_i = 1$. Similarly the mole fraction of component *i* is $x_i = n_i / n$ and $\sum x_i = 1$. To summarize, we have three alternative ways to deal with composition-a dimensional concept (concentration) and two dimensionless ratios (mass fraction and mole fraction). These quantities are related by $C_i = \rho\phi_i = \rho(M_i / M)x_i$, where the equivalent molar mass (M) of the mixture is given by $M = \sum M_i x_i$. If, for example, the mixture can be modeled as an ideal gas, then its equation of state is $PV = mR_mT$ or $PV = nRT$, where the gas constant of the mixture (R_m) and the universal gas constant (R) are related by $R_m = n/m$, $R = R_m / M$. The partial pressure P_i of component *i* is the pressure we would measure if component *i* alone were to fill the mixture volume V at the same temperature T as the mixture. Thus $P_iV = m_iR_mT$ or $P_iV = n_iRT$. Summing these equations over *i*, we obtain Dalton's law, $P = \sum P_i$, which states that the pressure of a mixture of gases at a specified volume and temperature is equal to the sum of the partial pressures of the components. Note that $P_i / P = x_i$, and so using Equation $C_i = \rho\phi_i = \rho(M_i / M)x_i$ and $M = \sum M_i x_i$ we can relate C_i to P_i.

Mass transfer, as used here, is the transfer of moisture from the wet material to the gas stream. One explanation of the existence of inversion temperatures is based on the theory that the driving potential for mass transfer into an environment of steam is different from the driving potential into air [6, 7] .

The driving potential for mass transfer in convective drying in air is commonly considered to be differences of vapour concentration across a boundary layer. An alternative view is that mass transfer in steam occurs by bulk flow due to a pressure difference. The vapour pressure at the surface where evaporation occurs is thought to be slightly higher than the free stream pressure, and causes bulk flow of vapour into the gas stream. Surface temperatures slightly higher than the saturation temperature have been measured [8].

Only a slight elevation in surface temperature is required to produce a pressure difference capable of generating a large bulk flow of vapour into the gas stream.

8. Dry Air and Superheated Steam

A number of investigators have focused on the two humidity extremes, dry air and superheated steam. Comparing drying in superheated steam with drying in air, six observations, excluding economic considerations, warrant attention. First, the drying rate of the constant rate period in air at temperatures above the inversion temperature [9, 10].

At saturation temperature, the drying rate in superheated steam is zero, but as the temperature increases, drying rates in superheated steam increases faster than those in air. At the inversion temperature, drying rates in superheated steam increase faster than those in air. Second, the critical moisture content decreases with increasing humidity of the drying gas. Third, in the falling rate period, drying rates in superheated steam roughly equal those in air, even at temperatures below the inversion temperature. Fourth, the equilibrium moisture content of a material dried in superheated steam at high temperatures (usually as high as the inversion temperature) is often lower than the equilibrium moisture content of a textile material dried in air at the same temperature [11, 12].

As it was mentioned earlier, equilibrium moisture content is the moisture content of a material when it comes to equilibrium with its environmental conditions, primarily the conditions of humidity and temperature. Fifth, more even moisture distribution in the textile material being dried occurs in superheated steam. When drying in air, heterogeneous surface wetting typically occurs below the critical moisture content. Sixth, material porosity and pliability are sustained better when dried in superheated steam than when dried in air.

9. HEAT SETTING PROCESS

The main aim of the heat setting process is to ensure that fabrics do not alter their dimensions during use. This is particularly important for uses such as timing and driving belts, where stretching of the belt could cause serious problems. It is important to examine the causes of this loss in stability so that a full understanding can be obtained of the effects that heat and mechanical forces have on the stability of fabrics. All fabrics have constraints place on them by their construction and method of manufacture, but it is the heat-setting mechanism that occurs within the fiber that will ultimately influence fabric dimensions. Heat setting on modern stenter is often carried out by first drying and then heat setting in one passage through stenter. The temperature and time of setting must be carefully monitored and controlled to ensure the consistent fabric properties are achieved. During heat setting, the segmental motion of the chain molecules of the amorphous regions of the fiber are generally increased leading to structural relaxation within the fiber structure. During cooling, the temperature is decreased below the fiber glass transition temperature (T_g) and the new fiber structure is established. Because the polymer chain molecules have vibrated and moved into new equilibrium positions at a high temperature in heat setting, subsequent heat treatments at lower temperatures do not cause heat-set fiber to relax and shrink, so that the fabric dimensional stability is high. Presetting of fabric prior to dyeing alters the polymer chain molecular arrangement within the fibers, and hence can alter the rate of dye uptake during dyeing. Process variations (e.g. temperature, time or tension differences) during heat setting may thus give rise to dyeability variations that become apparent after dyeing. Fabric postsetting after coloration can lead to the diffusion of dyes such as disperse dyes to the fiber surface and sublimation, thermomigration and blooming problems, all of which can alter the colour and markedly decrease the colour fastness to washing and rubbing of technical textiles containing polyester fibers.

A mixture of superheated steam and air is necessary for the rapid heat setting process. In the steam/air mixture used as the setting medium, the total heat capacity is increased considerably and thus it is possible to apply much more energy to the fabric within a given time and it is possible to heat-set the synthetic fiber fabrics in a shorter time.

An explanation for this increased energy transfer is that the specific heat of steam/air mixture is almost twice of that of air alone. Also by using steam/air mixture, heat exchange at the exterior of the fiber is increased, but the heat conduction from the surface to the interior of the fiber remains unchanged. Thus warp knitted fabrics and light weight wovens like taffeta and georgette require 3-4 seconds with steam/air mixture, compared with 15 seconds if hot air alone is used. Heavier fabrics, especially wovens require 9 seconds with rapid heat setting (with steam/air mixture) and 30 seconds with hot air.

Drawn polyester filaments, which have not been set have good tenacity and elasticity properties, but lack dimensional stability when subjected to the action of heat. Also, twisted or doubled filaments have a tendency to curl, which adversely affects further processing. Planar structures produced from unset continuous filaments or fibers exhibit creasing behaviour when in use. It is necessary to impart dimensional stability to polyester fabrics so that the garments made from them retain their shape on being subject to washing and ironing – therefore it is necessary to set the drawn, twisted filament or fabric in this state in order to attain resistance against shrinkage, and dimensional curling and crease resistance. This can be achieved by a process called heat setting in which they are subjected to the action of heat in the presence or absence of swelling agents, with or without tension. In practice, hot water, saturated steam or dry heat is used.

The favorable effect of the heat setting process lies in the fact that it compensates the stresses in the filaments resulting from manufacturing processes (spinning, drawing, twisting). The process may be explained as follows:

The linkage between the molecular chains or crystallites which were "frozen in" under tension and the mechanical stresses can be balanced by heat, thereby giving rise to greater freedom to the chains to oscillate. This provides the bonds an opportunity to snap into the sites of least energy. Subsequent cooling results in a state of lower strain. Further, the action of heat induces crystallization and orientation processes.

Pin stenter is one of the machines used for heat setting synthetic fiber fabrics and operates at temperature up to 230 °C using hot air. Polyester fabrics have been set at 230 °C and then dyed to a good standard of quality. Even dyeing obtained indicates uniformity of heating provided by the hot which impinges on both sides of the fabric and rapidly equilibriates both stenter pins and the fabric to the same temperature as they pass through the hot zone.

In summery, only by heat-setting the polyester fibers acquire the dimensional stability, crease-resistance and resilience desired in use. It should be noted that heat-setting is usually indispensable for ensuring satisfactory behaviour of the material during other finishing processes and is one of the most important finishing processes employed with materials containing polyester fibers or their mixtures with other fibers.

10. Convective Heat and Mass Transfer Coefficients

The convective heat and mass transfer coefficients at the surface of textile materials are important parameters in drying processes; they are functions of velocity and physical properties of the drying medium [13], and in general, can be expressed in the form of

$$Nu = a\,\mathrm{Re}^{b}\,\mathrm{Pr}^{c} \tag{2}$$

$$Sh = a'\,\mathrm{Re}^{b'}\,Sc^{c'} \tag{3}$$

where Nu is the Nusselt number, Re, is the Reynolds number, Pr, is the Prandtl number, S is pore saturation and *a, b, c,* are constants.

It should be noted that for a fully wetted surface, the areas for heat and mass transfer are virtually the same, so that the surface temperature is close to the wet bulb temperature; for partly wetted surface, the effective area for mass transfer decreases with the surface moisture content.

Suzuki and Maeda [14] proposed a model for convective mass transfer coefficient which assumed that evaporation take places from discontinuous wet surface consisting of dry and wet patches. The ratio of a wet area to the total surface area decreases with decreasing moisture content. However, it is not clear how the fraction of wet area varies with the surface moisture content. Moreover for hygroscopic textile materials, when the fraction of wet area at the surface approaches zero or the surface moisture content is equal to the maximum sorptive value, the evaporation rate at the fabric surface may not be equal to zero.

11. Convective Drying of Textile Material: Simple Case

In this section we focus on the equations which considers all the major internal moisture transfer mechanisms and the properties of the textile material to be dried, including whether it is hygroscopic or non-hygroscopic. The convective heat and mass transfer coefficients are assumed to vary with the surface moisture content.

As it was mentioned earlier, drying of textile materials involves simultaneous heat and mass transfer in a multiphase system. The drying textile materials may be classified into hygroscopic and non-hygroscopic . For non-hygroscopic textile materials, pores of different sizes form a complex network of capillary paths. The water inside the pores that contribute to flow is called free water. However, the water inside very fine capillaries is difficult to replace by air. This portion of water is known as the irreducible water content. Here, it is defined as bound water. The voids in textile materials are interconnected and filled with air and certain amount of free water.

When a textile material is exposed to convective surface condition, three main mechanisms of internal moisture transfer is assumed to prevail;

- capillary flow of free water,
- movement of bound water and
- vapour transfer.

If the initial moisture content of the textile material is high enough, the surface is covered with a continuous layer of free water and evaporation takes place mainly at the surface. Internal moisture transfer is mainly attributable to capillary flow of free water through the pores. Therefore, the drying rate is determined by external conditions only, i.e. the temperature, humidity and flow rate of the convective medium, and a constant drying rate period will be observed. As drying proceeds, the fraction of wet area decreases with decreasing surface moisture content, so that the mass transfer coefficient decreases. In order to predict how the wet area fraction varies with surface moisture content, it is necessary to introduce percolation theory [15].

According to percolation theory, when water passes through randomly distributed paths in a medium, there exist a percolation threshold, which usually corresponds to critical free water movement content. When the free water content is greater than the critical, the water phase is continuous. For a two-dimensional porous medium, this critical value is about 50% of the saturated free water content and for a three-dimensional porous medium it is around 30%.Regardless of the rate of internal moisture transfer, so long as the free water content at the surface is less than the critical, the surface will form discontinuous wet patches. Thus, the mass transfer coefficient decreases with the surface free water content and the first falling rate periods starts. In the first falling rate period, a new energy balance will be reached at the surface, the 'dry' pitches still contain bound water, and the vapour pressure at the surface is determined by the Clausius-Clapeyron equation. When the surface moisture content reaches its maximum value, no free water exists. The surface temperature will rise rapidly, signaling the start of the second falling rate period, during which receding evaporation front often appears, dividing the system into two regions, the wet region and the sorption regions. Inside the evaporation front, the material is wet, i.e. the voids contain free water and the main mechanism of moisture transfer is capillary flow. Outside the front, no free water exists. All water is in the sorptive or bound water state and the main mechanisms of moisture transfer are movement of bound water and vapour transfer. Evaporation takes place at the front as well as in the whole sorption region, while vapour flows through the sorption region to the surface. Therefore, based on the definitions of the constant rate, first falling rate and second falling rate periods, the characteristics of most drying processes can be described mathematically [16].

12. Capillary Flow of Free Water

In textile material, pores provide capillary paths for free water to flow. The driving force for capillary flow is tension gradient or pressure gradient. The pertinent expression for capillary flow of free water is given by [17,18]:

$$J_L = -\rho_W \frac{K}{\mu}(\nabla P_g - \nabla P_c - \rho_w g) \tag{4}$$

Here,

J_L $[kgm^{-2}s^{-1}]$ represent the free water flux, ρ_w $[kgm^{-3}]$is the density of water, K $[m^2]$ refers to the permeability, μ $[kgm^{-1}s^{-1}]$ is the viscosity, P_g $[Nm^{-2}]$ is the gas pressure, P_c $[Nm^{-2}]$ is the capillary pressure { $(P_g - P_L)[Nm^{-2}]$ } and P_L $[Nm^{-2}]$ is the liquid moisture or free water pressure

We can also assume that:

1. the textile material is macroscopically homogeneous;
2. the flow in the capillary paths is laminar;
3. there is no significant temperature gradient;

4. the effect of the gas phase pressure and gravity force are negligible.

Thus equation (4) can be simplified as :

$$J_L = \rho_w \frac{K_L}{\mu}\left(\nabla P_g - \nabla P_c - \rho_w g\right) \tag{5}$$

As pointed out by Miller and Miller [19], for homogeneous media and negligible gravity forces, tension is proportional to moisture content. It appears that Krischer and Kast's equation for liquid flow [20] may be valid;

$$J_L = -\rho_0 D_L \nabla U \tag{6}$$

where ρ_0 is the bulk density of dry material, D_L $[m^2 s^{-1}]$ is the capillary conductivity and U$[kgkg(solid)^{-1}]$ is the moisture content.

Since the permeability K_L depends on the pore structure of the material and the interaction between water and the textile material, it is difficult to find a theoretical form to relate K_L and the capillary conductivity D_L . However, the velocity of flow in capillaries may be assumed to follow the Hagen-Poiseuille law [21]:

$$v = \frac{r^2}{8\mu\tau}\nabla P_c \tag{7}$$

where v$[ms^{-1}]$ is the fluid velocity, r [m] is the radius and τ is the tortuosity factor of capillary paths.

The total mass flow rate of free water can then be expressed in terms of the local velocity and corresponding capillary radius r_c, which is defined as the radius of the largest capillary in which free water exist, i.e.

$$J_L = \frac{1}{A}\int \rho_w \frac{r^2}{8\mu\tau}\nabla P_c dA_c = \int_{r_{min}}^{r_c} \rho_w \frac{r^2}{8\mu\tau}\nabla P_c \gamma(r) dr \tag{8}$$

where $\gamma(r)$ is the pore volume density function, defined as

$$\gamma(r) = \frac{\rho_0}{\rho_w}\frac{\Delta U}{\Delta r}\bigg|_r = \varepsilon \frac{\Delta S}{\Delta r}\bigg|_r$$

here, ε is the porosity and S is the pore saturation.The relationship between the pore saturation and the pore density function is

$$S = \frac{\int_{r_{\min}}^{r_c} \gamma(r)dr}{\int_{r_{\min}}^{r_{\max}} \gamma(r)dr} = \frac{1}{\varepsilon} \int_{r_{\min}}^{r_c} \gamma(r)dr \tag{9}$$

where $r_{\max}$ is the overall largest capillary in the porous material. Comparing equations (5) and (6) with equation (8), K_L may be expressed as

$$K_L = -\frac{K_L}{\mu}\frac{\nabla P_c}{\nabla U} = \frac{2}{r_c^2 \gamma(r_c)}\frac{\sigma}{\mu} K_L \tag{10}$$

here σ is the surface tension.

The capillary conductivity D_L may be expressed as

$$D_L = -\frac{K_L}{\mu}\frac{\rho_w \nabla P_c}{\rho_0 \nabla U} = \frac{2}{r_c^2 \gamma(r_c)}\frac{\sigma}{\mu} K_L \tag{11}$$

The pore volume density function can be obtained from the relationship between capillary pressure and pore saturation of a porous medium. Solving equations (9) and (10) using the correlation of pore volume density obtained from the experimental results of Chatzis [22], a relationship between pore saturation and relative permeability can be obtained which is very close to the empirical correlation

$$K_r = \left(\frac{S - S_{ir}}{1 - S_{ir}} \right)^3 \tag{12}$$

Therefore, equation (12) can be used to predict the relative permeability, K_r . For water, σ / μ is a linear function of temperature [23], which can be expressed as ;

$$\frac{\sigma}{\mu} 1.064T - 394.3(ms^{-1}) \tag{13}$$

here T[K] is the temperature. The value of r in equation (11) might be constant or a function of free water content. Consequently, the following form is obtained to predict the capillary conductivity of free water for non-hygroscopic textile materials:

$$D_L = (1.604T - 394.3)\beta K_0 \left(\frac{S - S_{ir}}{1 - S_{ir}} \right)^3 \tag{14}$$

here K_0 $[m^2]$ is the single phase permeability of porous material and similarly, for hygroscopic materials

$$D_L = (1.604T - 394.3)\beta K_0 \left(\frac{U - U_{ms}}{U_0 - U_{ms}} \right)^3 \tag{15}$$

13. Movement of Bound Water

Movement of bound water, sometimes known as "liquid moisture transfer near dryness" or "sorption diffusion", has been studied by a number of investigators [24, 25, 26].

It has been shown that liquid moisture transfer still exists in the sorption region and is a strong function of free water content. Whitaker and Chou [27] studied

gas phase convective transport in the dry region which contains irreducible water and concluded that there could be a liquid moisture flux in the region. Movement of bound water however cannot be simply defined as a diffusion process, which often creates confusion in the analysis of liquid moisture transfer in drying processes. Moreover, the bound water conductivity measured is strongly influenced by moisture content. Therefore, movement of bound water may rather be due to flow along very fine capillaries or through cellular membranes.

In the sorption region, both movement of bound water and vapour transfer play important roles in moisture transfer. The transport equation for bound water may be expressed as

$$J_b = -\rho_w \frac{K_b}{\mu} \nabla P = -\rho_w \frac{K_b^{'}}{\mu} \nabla P_v \tag{16}$$

Since the bound water in the sorption region is in equilibrium with vapour in the gas phase, equation (16) may be written as

$$J_b = -\rho_w \frac{K_b^{'}}{\mu} P_v^* \frac{\partial \psi}{\partial U} \nabla U = -\rho_0 D_b \nabla U \tag{17}$$

where ψ is the relative humidity, D_b $[m^2 s^{-1}]$ is bound water conductivity. Thus the bound water conductivity may be written as

$$D_b = D_{bo} \left(\frac{U_b - U_{eq}}{U_{ms} - U_{eq}} \right)^3 \exp\left(\frac{E_d}{Rt} \right) \tag{18}$$

here t [s] is time, E_d is defined as the activation energy of movement of bound water.

14. VAPOUR FLOW

During a drying process, water vapour flows through the pores of the textile material by convection and diffusion. The equations of vapour flow and air flow may be written as;

$$J_v = \frac{P_v M_w}{RT}\frac{K_g}{\mu_g}\nabla P + \frac{D_v M_w}{RT}\nabla P_v \tag{19}$$

$$J_a = \frac{(P - P_v)M_a}{RT}\frac{K_g}{\mu_g}\nabla P + \frac{mD_v M_a}{RT}\nabla(P - P_v) \tag{20}$$

where M is the molecular weight, m is the ratio of air and vapour diffusion coefficients.

15. MACROSCOPIC EQUATIONS GOVERNING HEAT AND MASS TRANSFER IN TEXTILE MATERIAL

The theoretical formulation of heat and mass transfer in porous media is usually obtained by a change in scale. We can pass from a microscopic view where the size of the representative volume is small with regards to pores, to a microscopic view where the size of the representative volume ω is large with regard to the pores. Moreover, the heat and mass transfer equations can be deduced from Whitaker's theory. The macroscopic equations can be obtained by averaging the classical fluid mechanics, diffusion and transfer equations over the averaging volume ω $[m^3]$. The average of a function *f* is;

$$\bar{f} = \frac{1}{\omega}\int_{\omega} f d\omega \tag{21}$$

and the intrinsic average over a phase *i* is

$$\bar{f}^i = \frac{1}{\omega_i}\int_{\omega} f d\omega \tag{22}$$

16. GENERALIZED DARCY'S LAW

Darcy's law is extended by using relative permeabilities. For gaseous phase, since no gravitational effect is noted;

$$\overline{v}_g = -\frac{KK_g}{\mu_g}\frac{\partial}{\partial x}(\overline{P}_g^g) \tag{23}$$

where $\overline{v}_g$ is the speed of the gaseous phase, K $[m^2]$ the intrinsic permeability, K_g the relative permeability to the gaseous phase, $\overline{P}_g^g$ the average intrinsic pressure of the gaseous mixture, and μ_g the viscosity of the gaseous phase.

For the liquid phase

$$\overline{v}_1 = -\frac{KK_1}{\mu_1}\frac{\partial}{\partial x}(\overline{P}_g^g - \overline{P}_c^1 + \overline{\rho}_1^1 g) \tag{24}$$

where $\overline{v}_1$ is the speed of the liquid phase, K_1 the relative permeability to the liquid phase, P_c the capillarity pressure, μ_g μ the viscosity of the liquid, g the gravitational constant, and ρ_1 the density of liquid.

17. Mass Conservation Equations

For the liquid

$$\frac{\partial\overline{\rho}_1}{\partial t} + \frac{\partial}{\partial x}(\overline{\rho}_v^g \overline{v}_v) = \dot{m} \tag{25}$$

where $\dot{m}$ is the evaporated water in units of time and volume.

For the vapour

$$\frac{\partial\overline{\rho}_v}{\partial t} + \frac{\partial}{\partial x}(\overline{\rho}_v^g \overline{v}_v) = \dot{m} \tag{26}$$

where

$$\overline{\rho}_v^g \overline{v}_v = \overline{\rho}_v^g \overline{v}_g - \overline{\rho}_g^g D_{eff} \frac{\partial}{\partial x}(\overline{\rho}_v / \overline{\rho}_g) \tag{27}$$

P and P are the average densities of the water vapour and of the gaseous mixture, D the coefficient of the effective diffusion of vapour in the porous medium.

For the gaseous mixture

$$\frac{\partial \overline{\rho}_g}{\partial t} + \frac{\partial}{\partial x}(\overline{\rho}_g^g \overline{v}_g) = \dot{m} \tag{28}$$

18. Energy Conservation Equation

By assuming all the specific heats as constant and with aid of the mass conservation equations, the energy balance takes a form which is unusual but efficient in the calculations

$$\begin{aligned} &\frac{\partial}{\partial t}(\overline{\rho C}_p \overline{T}) + \frac{\partial}{\partial x}(\overline{\rho}_1^1 C_{p1} \overline{v}_1 \overline{T} + \sum_{j=v,a} \overline{\rho}_j^g C_{pj} \overline{v}_j \overline{T}) \\ &= \frac{\partial}{\partial x}(\lambda_{eff} \frac{\partial \overline{T}}{\partial x}) - \Delta h_{vap}^{\circ} \dot{m} \end{aligned} \tag{29}$$

here, Δh_{vap}° is a constant defined by

$$\Delta h_{vap}^{\circ} = \Delta h_{vap} + (C_{p1} - C_{pv})\overline{T} \tag{30}$$

λ_{eff} the effective thermal conductivity of the textile material, Δh_{vap} the enthalpy of vaporization, $\overline{\rho C}_p$ the constant pressure heat capacity of the textile material.

$$\overline{\rho C}_p = \overline{\rho} C_{ps} + \overline{\rho}_1 C_{p1} + \overline{\rho}_v C_{pv} + \overline{\rho}_a C_{pa} \tag{31}$$

19. Thermodynamic Relations

The partial pressure of the vapour is equal to its equilibrium pressure

$$\overline{P}_v^g = \overline{P}_{veq}(T, S) \tag{32}$$

The gaseous mixture is supported to be an ideal mixture of perfect gases

$$\overline{P}_j = \overline{\rho}_j R\overline{T} / M_j \text{ ; j =a, v} \tag{33}$$

$$\overline{P}_g = \sum_{j=a,v} \overline{P}_j \text{ ; } \overline{\rho}_g = \sum_{j=a,v} \overline{\rho}_j \tag{34}$$

20. Heat and Mass Transfer of Textile Fabrics in the Stenter

By modification of the mathematical model developed by Nordon and David [28] the transient temperature and moisture concentration distribution of a fabric in the stenter can be determined;

$$D\frac{\partial^2 C_A}{\partial x^2} = \frac{\partial C_F}{\partial t} + \varepsilon\frac{\partial C_A}{\partial t} \tag{35}$$

and

$$k\frac{\partial^2 T}{\partial x^2} = \rho C_p\frac{\partial T}{\partial t} - \lambda\frac{\partial C_F}{\partial t} \tag{36}$$

Here, D$[m^2/s]$ is the diffusion coefficient, $C_A[kg/m^3]$ is the moisture content of air in fabric pores, $C_F[kg/m^3]$ is the moisture content of fibers in a fabric, ε is the porosity, $k[W/mK]$ is the thermal conductivity, $\rho[kg/m^3]$ is the density, $C_p[kJ/kgK]$ is the specific heat, $\lambda[kj/kg]$ is the latent heat of evaporation.

The boundary conditions for convective heat transfer are

$$q = h_e(T_e - T) \tag{37}$$

and

$$\dot{m} = h_m(C_e - C_A) \tag{38}$$

here, $q[W/m^2]$ is the convective heat transfer rate, $h_e[W/m^2K]$ heat transfer coefficient, $T_e[K]$ is the external air temperature, $\dot{m}[kg/m^2s]$ is the mass transfer rate, $h_m[m/s]$ is the mass transfer coefficient, and $C_e[kg/m^3]$ is the moisture content of external air. The deriving force determining the rate mass transfer inside the fabric is the difference between the relative humidities of the air in the pores and the fibers in the fabric. The rate of moisture exchange may be considered as proportional to the relative humidity difference. Thus the rate equation for mass transfer is

$$\frac{1}{\rho(1-\varepsilon)}\frac{\partial C_F}{\partial t} = K(y_A - y_F) \tag{39}$$

where y_A is the relative humidity of air in pores of fabric and y_F is the relative humidity of fiber of fabric. Also, the relative humidities of air and fabric for polyester are assumed to be

$$y_A = \frac{C_A RT}{P_s} \tag{40}$$

and

$$y_F = \frac{C_P}{\rho(1-\varepsilon)} \tag{41}$$

here $R[kJ/kgK]$ is the gas constant and $P_s[kg/m^2]$ is the saturation pressure.

The rate constant K in Equation 2.39 is an unknown empirical constant. The transient fabric temperatures can be calculated assuming various values of the rate constant K. The value of the rate constant can be varied from 0.01 to 10 .When the rate constant is small, the evaporation rate is so small that the moisture content decreases very slowly. Initially, the surface temperature increases rapidly, but later this rate declines (figure 3).

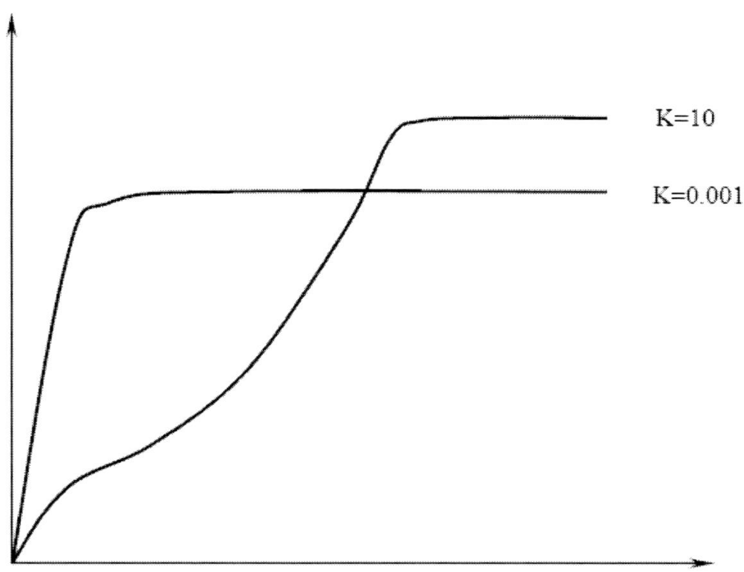

Figure 3. Effect of rate constant on the fabric surface temperature.

When K is greater than 1, however, the effect of the rate constant on the surface temperature distribution is not as significant. This indicates that when the rate is greater than 1, the evaporation rate is high and the diffusion mechanism inside the fabric. From figure (4) we see that the surface and center temperatures increase rapidly in the initial stage up to the saturation temperature, at which point the moisture in the fabric starts to evaporate.

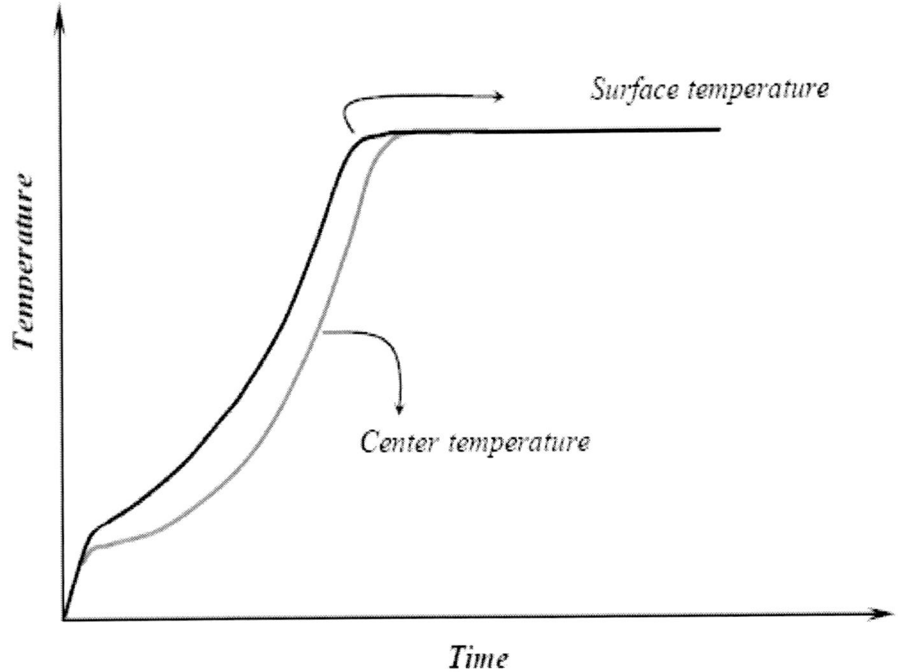

Figure 4. Temperature variation of surface and center fabric.

From that point, the difference between the surface temperature and the center temperature increases due to the different moisture contents of the surface and the center. In this stage the fabric stats to dry from the surface, and the moisture in the interior is transferred to the fabric surface. Then the moisture content decreases during drying of the fabric. Thereafter, the surface and center temperatures coverage to reach the external air temperature. The moisture variations of the surface and the center of fabric is shown in figure 5.

Initially, the surface moisture content decreases rapidly, but later this rate declines because moisture is transferred to the external air from the fabric surface. The center moisture content remains constant for a short time, and then decreases rapidly, because the moisture content difference between the surface and the interior of the fabric becomes large. After drying out, both center and surface moisture contents converge to reach the external air moisture content. When the initial moisture content is high, the temperature rise is relatively small and drying takes a long time. This may be because the higher moisture content needs much more heat for evaporation from the fabric. Also, the saturation temperature for higher moisture content is lower, and thus the temperature rise in the initial stage is comparatively small (figure 6).

Moreover, the effect of stenter air moisture content can be evaluated as shown in figure (7). When the moisture content is high, the initial temperature rise of the fabric also become high. This may be because the saturation temperature in the initial stage largely depends on the stenter air moisture content.

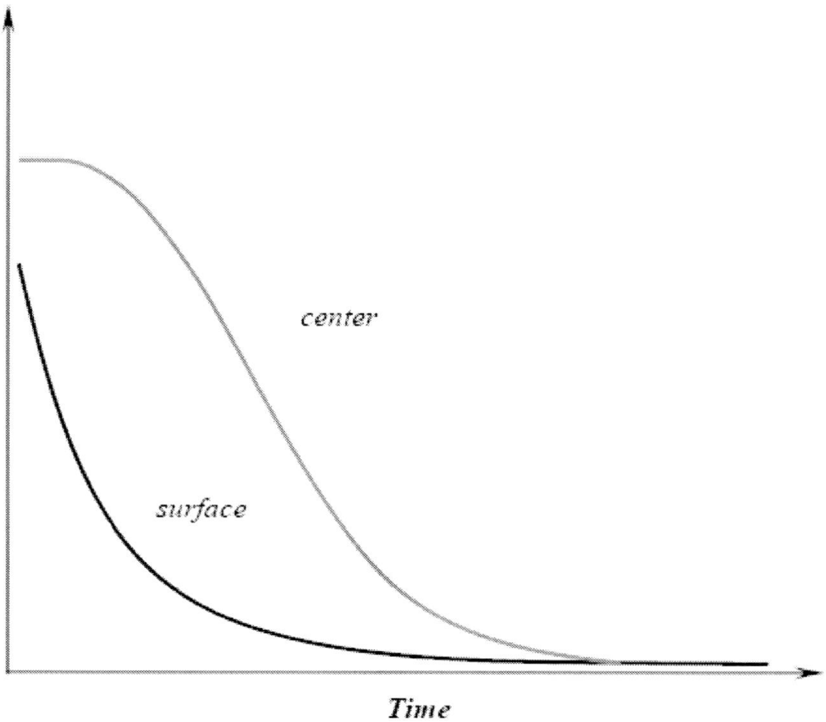

Figure 5. Moisture content variation of surface and center fabric.

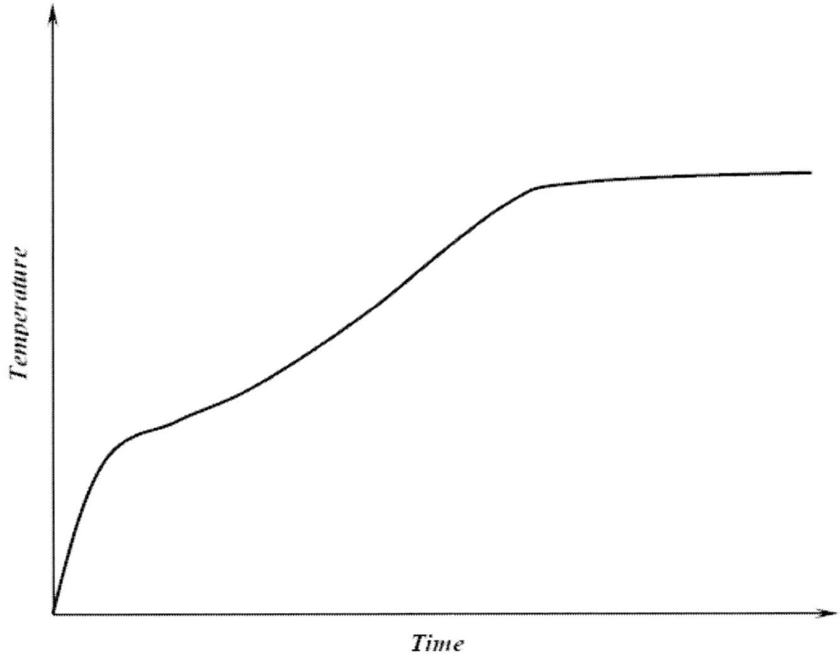

Figure 6. Effect of initial moisture content of fabric.

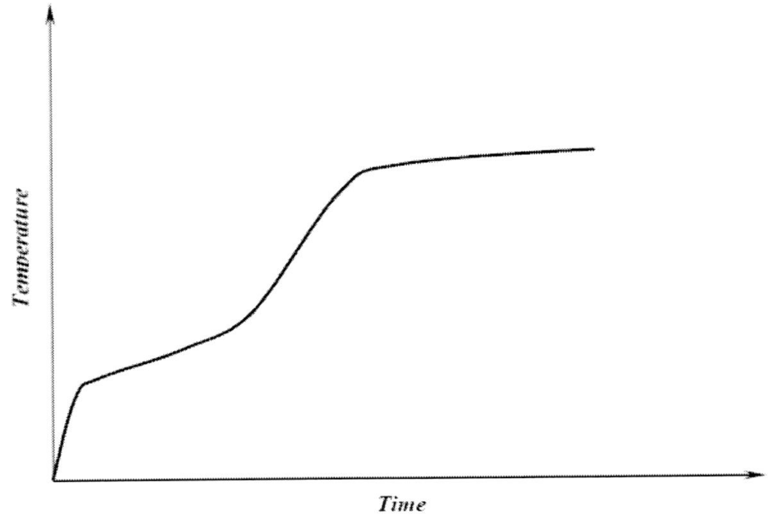

Figure 7. Effect of stenter air moisture content.

After the initial temperature rise, however, the temperature increase is relatively small, and thus the time required for complete drying is comparatively long. Meanwhile, when the stenter air temperature is high, the temperature rise of the fabric is great. Also, the average fabric temperature converges on the stenter air temperature converges on the stenter air temperature in a relatively short time (figure 8). This may be because the fabric dries more rapidly due to the large difference between its temperature and that of the stenter air [29].

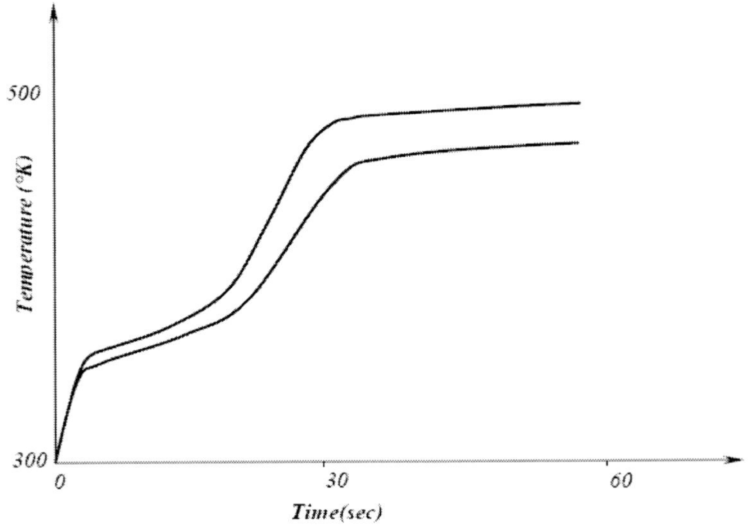

Figure 8. Effect of stenter air temperature.

REFERENCES

[13] Chow,L. C., and Chung, J. N., Evaporation of water into laminar stream of air and superheated stem, *Int. J. Heat Mass Trans.* 26(3), 373-380 (1983).

[14] Haji, M., and Chow, L.C., Experimental measurement of water evaporation rates into air and superheated steam, *J. Heat Trans.* 110 (2), 273-242 (1988).

[15] Lane, A.M., and Stern, S., Application of superheated vapor atmospheres to drying, *Mechan. Eng.* (5), 423-426 (1956).

[16] Sheikholeslami, R., and Watkinson, A.P., rate of evaporation of water into superheated steam and humidified air, *Int. J. Heat Mass Trans.* 35 (7), 1743-1751 (1992).

[17] Moyne, C., and Basilico, C., High temperature convective drying of softwood and hard wood: Drying Kinetics and Product quality interactions, *Drying* '85, 376-381 (1985).

[18] Beeby, C., and Potter, O. E., Steam Drying, *Drying.* '85, 41-58 (1985).

[19] Shi-Ruo, C., Jin-Young, C., and Mujumdar, A.S., Preliminary study on silkworm Cocoons dring, Drying. '91, 620-627 (1991).

[20] Wenzel, L. and White, R., Drying Granular solids in superheated steam, *Ind. Eng. Chem.* 43, 1823-1831, (1951).

[21] Nishimura, N., Nomura, T., and Ueda, S., Heat and Mass transfer with water evaporation into superheated steam stream, *Mem. Fac. Eng. Osaka city Univ.* 30, 1-10 (1989).

[22] Nomura, T., and Hyodo, T., Behavior of inversion point temperature and new applications of superheated vapor drying, *Drying*. '85, 517-522 (1985).

[23] Bond, J.F., An experimental study of the falling rate period of superheated steam impingement drying of paper, *Drying Technol.* 10(4), 961-977 (1992).

[24] Trommelen, A.M., and Crosby, E.J., Evaporation and drying of drops in superheated vapours, *AICHE J.* 16 (5), 857-867 (1970).

[25] R.B. Keey, *Drying Principles andPractice*, 2nd edition, Pergamon Press, Oxford (1978).

[26] M.Suzuki and S. Maeda, On the mechanism of drying of granular beds, mass transfer from discountinuous source, *J. Chem. Engng. Japan.* 1, 26-31 (1968).

[27] R.G.Larson, L.E.Scriven and H.T. Davis, Percolation theory of two-phase flow in porous media, *Chem. Engng. Sci.* 36, 57-73 (1981).

[28] P.Chen and D.C.Pei, A mathematical model of drying processes, *Int J. Heat Mass Transfer,* 32(2), 297-310, 1989.

[29] R.A. Greenkorn, *Flow phenomena in porous media*, Marcel Dekker, New York (1983).

[30] S. Whitaker, Flow in porous media II : the governing equations for immiscible, two-phase flow, *Transport porous media* 1, 105-125 (1986).

[31] E.E. Miller and R.D. Miller, Theory of capillary flow I, Practical implications, *Proc. Soil Sci. Soc. Am*. 19, 267-271 (1995).

[32] O. Krischer and W. Kast, Die Wissenschaftlichen Grundlagen der Trocknungstechnik, 3rd Edn, Springer, Berlin (1978).

[33] R.B. Keey, *Drying principles and practice*, 1st edition, Pergamon Press, Oxford (1972).

[34] I.Chatzis, A network approach to analyse and model capillary and transport phenomena in porous media, Ph.D. Thesis, University of waterloo, Canada (1980).

[35] P. Chen, Mathematical modeling of drying and freezing processes in the food industry, PhD. Thesis, University of Waterloo, Canada, (1987).

[36] E. Rotstein, Advances in transport phenomena and thermodynamics in the drying of cellular food systems. In Drying. '86 (Edited by A.S. Mujumdar), Vol. 1, pp. 1-11. Hemisphere, Washington, DC (1986).
[37] E. Rotstein and A.R.H. Cornish, Influence of cellular membrane permeability on drying behavior, *J. Food Sci.* 43, 926-934 (1978).
[38] G. Bramhall, Sorption diffusion in wood, *Wood Sci.* 12, 3-13 (1979).
[39] S. Whitaker and W.T. Chou, Drying granular porous media-theory and experiment, *Drying Tech. Int. J.* 1, 3 (1983).
[40] Nordon, P., and David, H. G., Coupled diffusion of moisture and heat in Hygroscopic textile materials, *Int. J. Heat Mass Trans.* 10, 853-866 (1967).
[41] S. Park and D. Baik, Heat and mass transfer analysis of fabric in tenter frame, *Textile Res.* J. 67(5), 311-316 (1997).

In: Progress in Chemistry and Biochemistry
Editors: E. M. Pearce and G. E. Zaikov

ISBN: 978-1-60692-344-3

Chapter 17

MICROWAVE HEATING OF TEXTILES: THEORETICAL AND EXPERIMENTAL ASPECTS

A. K. Haghi
University of Guilan, P. O. Box 3756, Rasht, Iran
Haghi@Guilan.ac.ir

ABSTRACT

Radiation is a form of electromagnetic energy transmission and takes place between all matter providing that it is at a temperature above absolute zero. Infra-red radiation forms just part of the overall electromagnetic spectrum. Radiation is energy emitted by the electrons vibrating in the molecules at the surface of a body. The amount of energy that can be transferred depends on the absolute temperature of the body and the radiant properties of the surface.

Electromagnetic radiation is a form of energy that propagates through a vacuum in the absence of any moving material. We observe electromagnetic radiation as light and use it as radio waves, X-rays, etc. In this chapter, we are mostly interested in a form of electromagnetic radiation called microwaves that can be used to heat and dry textile materials.

1. INTRODUCTION

The word *microwave* is not new to every walk of life as there are more than 60 million microwave ovens in households all over the world. On account of its great success in processing food, people believe that the microwave technology can also be wisely employed to process materials. Microwave characteristics that are not available in conventional processing of materials consists of : penetrating radiation, controllable electric field distribution, rapid heating, selective heating materials and self-limiting reactions. Single or in combination, these characteristics lead to benefits and opportunities that are not available in conventional processing methods.

Since World War II , there have been major developments in the use of microwaves for heating applications. After this time it was realized that microwaves had the potential to provide rapid, energy-efficient heating of materials. The main applications of microwave heating today include food processing, wood drying, plastic and rubber treating as well as curing and preheating of ceramics. Broadly speaking, microwave radiation is the term associated with any electromagnetic radiation in the microwave frequency range of 300 MHz-300 Ghz. Domestic and industrial microwave ovens generally operate at a frequency of 2.45 Ghz corresponding to a wavelength of 12.2 cm. However, not all materials can be heated rapidly by microwaves. Materials may be classified into three groups, *i.e.* conductors insulators and absorbers. Materials that absorb microwave radiation are called dielectrics, thus, microwave heating is also referred to as dielectric heating. Dielectrics have two important properties:

- They have very few charge carriers. When an external electric field is applied there is very little change carried through the material matrix.
- The molecules or atoms comprising the dielectric exhibit a dipole movement distance. An example of this is the stereochemistry of covalent bonds in a water molecule, giving the water molecule a dipole movement. Water is the typical case of non-symmetric molecule. Dipoles may be a natural feature of the dielectric or they may be induced. Distortion of the electron cloud around non-polar molecules or atoms through the presence of an external electric field can induce a temporary dipole movement. This movement generates friction inside the dielectric and the energy is dissipated subsequently as heat[1].

The interaction of dielectric materials with electromagnetic radiation in the microwave range results in energy absorbance. The ability of a material to absorb energy while in a microwave cavity is related to the loss tangent of the material.

This depends on the relaxation times of the molecules in the material, which, in turn, depends on the nature of the functional groups and the volume of the molecule. Generally, the dielectric properties of a material are related to temperature, moisture content, density and material geometry.

An important characteristic of microwave heating is the phenomenon of "hot spot" formation, whereby regions of very high temperature form due to non-uniform heating. This thermal instability arises because of the non-linear dependence of the electromagnetic and thermal properties of material on temperature. The formation of standing waves within the microwave cavity results in some regions being exposed to higher energy than others. This result in an increased rate of heating in these higher energy areas due to the non-linear dependence. Cavity design is an important factor in the control, or the utilization of this "hot spots" phenomenon.

Microwave energy is extremely efficient in the selective heating of materials as no energy is wasted in "bulk heating" the sample. This is a clear advantage that microwave heating has over conventional methods. Microwave heating processes are currently undergoing investigation for application in a number of fields where the advantages of microwave energy may lead to significant savings in energy consumption, process time and environmental remediation.

Compared with conventional heating techniques, microwave heating has the following additional advantages :

- higher heating rates;
- no direct contact between the heating source and the heated material;
- selective heating may be achieved;
- greater control of the heating or drying process;
- reduced equipment size and waste.

The benefit of microwave technology has been realized over the past decade with the growing acceptance of microwave ovens in the home. This, together with the gloomy outlook of a worldwide energy crises, has paved the way for extensive research into new and innovative heating and drying processes. The use of microwave drying cannot only greatly enhance the drying rates of textile materials, but it may also enhance the final product quality.

While cost present a major barrier to wider use of microwave in textile industry, an equally important barrier is the lack of understanding of how microwaves interact with materials during heating and drying. The design of suitable process equipment is further confounded by the constraint that geometry places on the prediction of field patterns and hence heating rates within the materials. Effects such as resonance within the material can occur as well as large variations in field patterns at the textile material surface.

The phenomenon of drying has been investigated at considerable length and treated in various texts. However in general, there is only a very small section of this literature devoted to microwave to microwave drying of textile materials.

One of the main features which distinguishes microwave drying from conventional drying processes is that because liquids such as water absorb the bulk of the electromagnetic energy at microwave frequencies, the energy is transmitted directly to the wet material. The process does not rely on conduction of heat from the surface of the textile material and thus increased heat transfer occurs, speeding up the drying process. This has the advantage of eliminating case hardening of textile material which is usually associated with convective hot air drying operations. Another feature is the large increase in the dielectric loss factor with moisture content. This can be used with great effect to produce a moisture leveling phenomenon during the drying process since the electromagnetic energy will selectively or preferentially dry the wettest regions of the solid [2].

Meanwhile, infrared heating on textile lines has been in use for many years on dyeing lines to set the dyes prior to the tenter oven and to predry a host of fabric finishes or topical coatings on fabrics. The renewed interest in infrared predrying is due in large part to the need for ever-increasing line speeds and the availiability of improved infrared hardware. Infrared predrying of the dyed or finished fabric rapidly preheats and predries wetted fabrics far faster than the typical convection tenter dryer. Typically an air dryer requires 20-25% of its length just to preheat the wetted fabric to a temperature where water is freely evaporated. The infrared preheater/predryer section takes over this function in a fraction of the length required in the convection dryer. For dyed fabrics, infrared predryers are typically vertical in configuration, and are generally mounted on the line prior to the tenter frame. The systems consist of arrays of electric infrared emitters positioned on both sides of the fabric. The emitters are typically controlled from the fabric temperature. The evaporative load on the predryer dictates how much energy is required and how many vertical sections the predryer must be. With today's more efficient and higher powered emitters most predryers are one or two passes. In applications where two-sided heatinf is not required, such as latex backcoatings, an infrared predryer can be enclosed around pin and clip tenter frames immediately prior to the tenter oven. As a result, line speeds are increased as the added energy accelerates the heating the heating or drying process that has previously taken place only inside the oven. Heatsetting operations can benefit from preheating as well.

It should be noted that controlling shade variations and shade shifts in dyed fabrics has typically been problematic for manufacturing engineers. Without predrying, the likelihood of shade variation from one side of the fabric to the other increases. Dyestuffs tend to migrate to the heated side of the fabric as it passes through the oven. The migration is due partly to gravity, and partly to fluid dynamics. Dyed fabrics come onto the tenter frame at usually 50% to 80% wet pickup. Optimum product quality requires that wet pickup be reduced to the 30% to 60% range with equal water removal from both sides of the fabric. The predryed fabric is then presented to the horizontal tenter oven with the dyes "locked in" to position. Additional quality benefits can be realized on topical finishes or coatings. Rapid heating with infrared immediately after coating applications tends to keep the coating from deeply wicking into the fabric. For example, the infrared predrying of foamed on fluorochemical finishes for stain resistance tend to keep the coating more towards the surface of the fabric where they do the most good [3].

2. BACKGROUND

This section reviews the basic principles of physics pertaining to microwave heating.

-*Energy* : energy is the capacitance to do work, and work is defined as the product of a force acting over a distance, that is,

$$E = W = (F)(x) \tag{1}$$

where E = energy, W = the equivalent work, F = force that performs the work, x = distance a mass is moved by the force.

- *Atomic particles* : all matter are composed of atoms. Atoms, in turn, consist of a nuclei surrounded by orbiting electrons. The nucleus consists of positively charged protons and unchanged neutrons. The surrounding electrons are negatively charged. In neutral atoms, the number of protons in the nucleus equals the number of electrons, resulting in a 0 net charge.
- *Electrostatic forces* : if some electrons are removed from a piece of material, the protons will outnumber the electrons and the material will take on a positive charge. Similarly, if some electrons are added to a piece of material, the material will take a negative charge. If two positively charged objects are brought near to each other, they will each feel a force pushing them apart. Similarly, if two negatively charged objects are brought together, they will each experience a force pushing them apart. On the other hand, if a

negatively charged object is brought near a positively charged object, each will experience a force pulling them together.

Columb's law : if two charges of magnitude q_1 and q_2 are separated by a distance r as shown if figure 1, each will feel a force magnitude ;

$$F = k\frac{q_1 q_2}{r^2} \tag{2}$$

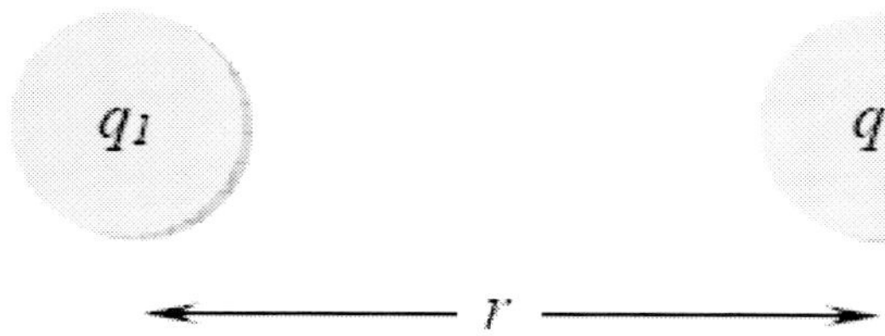

Figure 1. Principle of Columb's law.

It is clear from Equation (2) that the force is proportional to the magnitude of each charge and inversely proportional to the square of the distance between them. If, for example, we double the charge on either object, the force will double. On the other hand, if we double the distance between them, the force will be reduced to 1/4 of its previous value.

Electric fields : Electrostatic force is defined as "force at a distance" (Equation 1). If we have a charge Q, and a test charge q is placed a distance R away from it, Q will push on q across that distance as shown in figure 2. The magnitude of push will depend on the magnitudes of Q, q, and r as given in Equation (2).

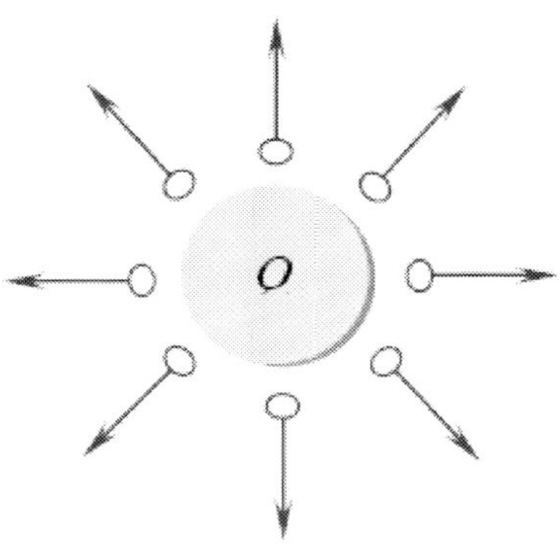

Figure 2. Forces around charge Q.

Another way to look at this is to say that Q creates a field in the space that surrounds it. At any point in that space , the field will have a strength E that depends on Q and r. If a test charge is placed at some point in the space, the field at that point will push on it with a force depends on the field strength E at that point and on q. To make these two explanations mathematically equivalent, we separate Equation (2) into two parts; thus

$$F = k\frac{Qq}{r^2}\left(k\frac{Q}{r^2}\right)(q) \tag{3}$$

The second part is simply the charge of the second particle. The first part we call E, the field strength at distance r away from Q:

$$E = \left(k\frac{Q}{r^2}\right) \tag{4}$$

Now the force on q can be defined in terms of the field strength times the magnitude of q:

$$F = E.q \tag{5}$$

A microwave oven consists of three major parts:

- The magnetron is the device that generates the microwaves.
- Wave guides direct these waves to the oven cavity.
- The oven cavity holds the material to be heated so that microwaves can impinge on them.

Magnetron : It generates microwaves and consists of the following parts:

a. Central cathode. The cathode is a metal cylinder at the center of the magnetron that is coated with an electron-emitting material. In operation, the cathode is heated to a temperature high enough to cause electrons to boil off the coating.
b. Outer anode. There is a metal ring called an anode around the magnetron that is maintained at a large positive potential (voltage) relative to the cathode. This sets up an electrostatic field between the cathode and anode that accelerates the electrons toward the anode.

Magnetic field: a strong magnetic field is placed next to the anode and cathode in such an orientation that it produces a magnetic field at right angles to the electrostatic field. This field has the effect of bending the path of the electrons so that, instead of rushing to the anode, they begin to circle in the space between the cathode and anode in a high-energy swarm.

Resonant cavities: they have been built into the anode. Random noise in the electron swarm causes occasional electrons to strike these cavities are such that most radiation frequencies die out. Microwave frequencies, on the other hand, bounce around the cavities and tend to grow, thus getting their energy from the magnetron, passes through the wave guides, and enters the cavity.

However, not all materials can be heated rapidly by microwaves. Materials are reflected from the surface and therefore do not heat metals. Metals in general have high conductivity and are classed as conductors. Conductors are often used as conduits (waveguide) for microwaves. Materials which are transparent to microwaves are classed as insulators. Insulators are often used in microwave ovens to support the material to be heated. Materials which are excellent absorbers of microwave energy are easily used and are classed as dielectric. Figure 3 shows these properties.

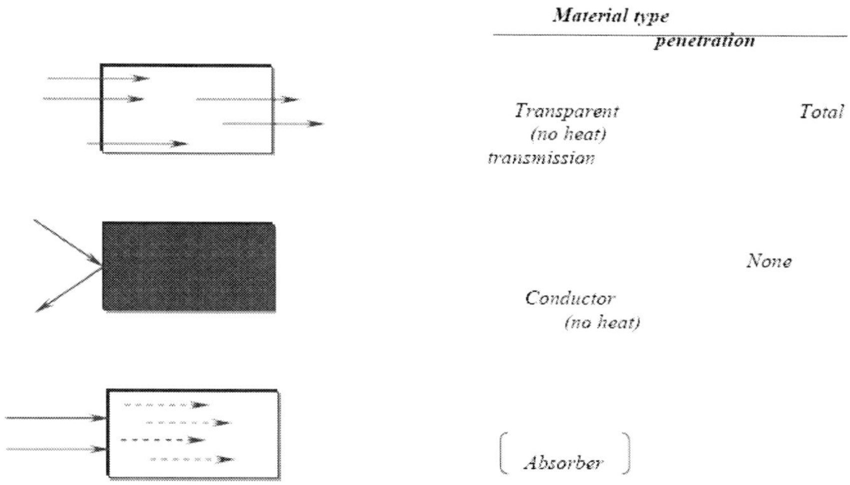

Figure 3. Interaction of microwave with materials.

Microwaves from part of a continuous electromagnetic spectrum that extends from low frequency alternating currents to cosmic rays shown in table 1.

In this continuum, the radio-frequency range is divided into bands as depicted in table.2. Radio-frequency (r.f.) energy has several possible benefits in textile processing. Substitution of conventional heating methods by radio-frequency techniques may result in quicker and more uniform heating, more compact processing machinery requiring less space, and less material in-process at a particular time. Radio-frequency energy has been used for many years to heat bulk materials such as spools of yarn. Bands 9, 10, and 11 constitute the microwave range that is limited on the frequency side by HF and on the high frequency side by the infrared. These microwaves propagate through empty space through empty space at the velocity of light. The frequency ranges from 300 MHz to 300 GHz.

Table 1. The electromagnetic spectrum

Region	Frequencies (Hz)	Wavelength
Audio frequencies	$30\text{-}30\times10^{3}$	10mm-10km
Radio frequencies	$30\times10^{3}\text{-}30\times10^{11}$	10km-1m
Infrared	$30\times10^{11}\text{-}4\times10^{14}$	1m-730nm
Visible	$4\times10^{14}\text{-}7.5\times10^{14}$	730nm-0.3nm
Ultraviolet	$7.5\times10^{14}\text{-}1\times10^{18}$	400nm-0.3nm
X-rays	$>1\times10^{17}$	< 3nm
Gamma rays	$>1\times10^{20}$	< 3nm
Cosmic rays	$>1\times10^{20}$	< 3nm

Table 2. Frequency bands

Band	Designation	Frequency limits
4	Very low frequency (VLF)	3-30 kHz
5	Low frequency (LF)	30-300 kHz
6	Medium frequency (MF)	300 kHz- 3MHz
7	High frequency (HF)	3-300 MHz
8	Very high frequency (VHF)	30-300 MHz
9	Ultra high frequency (UHF)	300-3 GHz
10	Super high frequency (SHF)	3-30 GHz
11	Extremely high frequency (EHF)	30-300 GHz

- Pertinent electromagnetic parameters governing the microwave heating:

The loss tangent can be derived from material's complex permittivity. The real component of the permittivity is called the dielectric constant whilst the imaginary component is referred to as the loss factor. The ratio of the loss factor to the dielectric constant is the loss tangent. The complex dielectric constant is given by:

$$\varepsilon = \varepsilon' - j\varepsilon'' \tag{6}$$

where ε is the complex permittivity, ε' is the real part of dielectric constant; ε'' is the loss factor, and $\varepsilon'/\varepsilon'' = \tan\delta$ is the loss tangent.

Knowledge of a material's dielectric properties enables the prediction of its ability to absorb energy when exposed to microwave radiation. The average power absorbed by a given volume of material when heated dielectrically is given by the equation:

$$P_{av} = \varpi\varepsilon_0\varepsilon_{eff}'' E_{rms}{}^2 V \tag{7}$$

where P_{av} is the average power absorbed (W); ϖ is the angular frequency of the generator (rad/s); ε_0 is the permittivity of free space; ε_{eff}'' is the effective loss factor; E is the electric field strength (V/m); and V is the volume (m^3).

The effective loss factor ε_{eff}'' includes the effects of conductivity in addition to the losses due to polarization. It provides an adequate measure of total loss, since the mechanisms contributing to losses are usually difficult to isolate in most circumstances.

Another important factor in dielectric heating is the depth of penetration of the radiation because an even field distribution in a material is essential for the uniform heating. The properties that most strongly influence the penetration depth are the dielectric properties of the material. These may vary with the free space wavelength and frequency of the propagating wave. For low loss dielectrics such as plastics (ε''<<1) the penetration depth is given approximately by :

$$D_P = \frac{\lambda_0 \sqrt{\varepsilon'}}{2\pi\varepsilon''_{eff}} \tag{8}$$

where D_P is the penetration depth; λ_0 is the free space wavelength; ε' is the dielectric constant; and ε''_{eff} is the effective loss factor.

The penetration depth increases linearly with respect to the wavelength, and also increases as the loss factor decreases. Despite this, however, penetration is not influenced significantly when increasing frequencies are used because the loss factor also drops away maintaining a reasonable balance in the above equation

As the material is heated, its moisture content decreases leading to a decrease in the loss factor. It can be seen from equation (8) that the decrease in loss factor causes in the penetration depth of radiation.

Microwaves cause molecular motion by migration of ionic species and/or rotation of dipolar species. Microwave heating a material depends to a great extent on its "dissipation" factor, which is the ratio of dielectric loss or "loss" factor to dielectric constant of the material. The dielectric constant is a measure of the ability of the material to retard microwave energy as it passes through; the loss factor is a measure of the ability of the material to dissipate the energy. In other words "loss" factor represents the amount of input microwave energy that is lost in the material by being dissipated as heat. Therefore, a material with high "loss" factor is easily heated by microwave energy. In fact, ionic conduction and dipolar rotation are the two important mechanisms of the microwave energy loss (i.e., energy dissipation in the material). Non-homogeneous material (in terms of dielectric property) may not heat uniformly, that is, some parts of the materials heat faster than others. This phenomenon is often referred to as thermal runway.

Continuous temperature measurement during microwave irradiation is a major problem. Luxtron fluoroptic or accufiber can be employed to measure temperature up to 400°C but are too fragile for most industrial applications. An optical pyrometer and thermocouple can be employed to measure higher temperatures. Optical pyrometers, such as thermovision infrared camera, only records surface temperature, which is invariably much lower than the interior sample temperature. When a thermocouple (metallic probe) is employed for temperature measurements, arcing between the sample and the thermocouple can occur leading to temperature measurements, arcing between the sample and thermocouple can occur leading to failure in thermocouple performance. A recent development is the ultrasonic temperature probe, which covers temperature up to 1500 °C.

In summery, microwave heating is unique and offers a number of advantages over conventional heating such as:

-non-contact heating;
-energy transfer, not heat transfer;
-rapid heating;
-material selective heating;

- volumetric heating;
- quick start-up and stopping;
- heating starts from interior of the material body;

- higher level of safety and automation.

Some glossaries of microwave heating system are shown in table 4.3[4].

Table 3. Some glossaries of microwave heating system

Applicator or cavity	A closed space where a material is exposed to microwaves for heating
Choke	Barriers placed at entrance and exit of the applicator to prevent leakage of microwaves.
Circulator	A three port ferrite device allowing transmission of energy in one direction but directing reflected energy into water load (dummy load) connected at the third port.
Coupling	The transfer of energy from one portion of a circuit to another.
Dielectric	It is a measure of a sample's ability to retard microwave energy as it passes through.
Dielectric loss or loss factor	It is a measure of a sample's ability to dissipate microwave energy.
Hertz (Hz)	1 Hz = 1 cycle/s.
Magnetron	An electronic tube for generating microwaves.
Single mode applicator	Dimension of applicator or cavity is comparable with the wave length of microwave.
Multimode applicator	An applicator dimension is large in relation to the wave legth of incident microwaves.

3. Basic Concepts of Microwave Heating

As it was mentioned earlier, microwaves are electromagnetic waves having a frequency ranging from 300 MHz and 0.3 THz. Most of the existing apparatuses, however, operate between 400 MHz and 60 GHz, using well defined frequencies, allocated for industrial, Scientific and Medical (ISM) applications. Among them, the 2.45 GHz is widely used for heating applications, since it is allowed word-wide and it presents some advantages in terms of costs and penetration depth.

It was also mentioned earlier, that quantitative information regarding the microwave-material interaction can be deduced by measuring the dielectric properties of the material, in particular of the real and imaginary part of the relative complex permittivity, $\varepsilon = \varepsilon^{'} - j\varepsilon''_{eff}$, where the term $\varepsilon^{''}_{eff}$ includes conduction losses, as well as dielectric losses. The relative permeability is not a constant and strictly depends on frequency and temperature. A different and more practical way to express the degree of interaction between microwaves and materials is given by two parameters; the power penetration depth (D_P) and the power density dissipated in the material (P), as defined earlier in a simplified version as follows:

$$D_P = (\lambda_0 \sqrt{\varepsilon'}) / 2\pi\varepsilon'' \quad P = 2\pi f \varepsilon_0 \varepsilon^{''}_{eff} E^2_{rms} \qquad (9)$$

where $\varepsilon = \varepsilon' - j\varepsilon''_{eff}$ is the complex permittivity of the material under treatment, λ_0 is the wavelength of the radiation, f is its frequency, $\varepsilon_0 = 8.854.10^{-12}$ F/m is the permittivity of empty space and E_{rms} is the electric field strength inside the material itself. It should be noted that P and D_P can only give quantitative and often misleading information, especially when it is critical to determine the temperature profiles inside the material. Others are the variables involved, however from this two parameters can be deduced most of the peculiarities which make the microwave heating a unique process[5].

First of all, it can be noticed the existence of temperature profile inversion with respect to conventional heating techniques. The air in proximity of the materials during the heat treatment, in fact, is not a good microwave absorber so that it can be considered that the atmosphere surrounding the material is essentially at low temperature.

Viceversa, the material under treatment, interacting in a stronger way with the electromagnetic field, heats up and reaches higher temperature. The result is that, in most cases, the surface temperature of the sample is lower than inside the material itself. This effect is more pronounced for poor heat conducting materials.

Since the given formulation for D_P and P show a strong dependence upon the real and imaginary part of the material permittivity, for multiphase systems having components with quite different permittivities, it is expected a strong selectivity of the microwave heating process. Power, in fact, is transferred preferentially to lossy materials (high ε''_{eff}) so that it can be possible to rise the temperature of just a single phase or component, or to spatially limit the heat treatment to the material, without involving the surrounding environment. This peculiarities can be particularly useful when treating composite materials.

The rapid variations of the permittivity as a function of temperature is responsible for a not always desirable phenomenon, the thermal runaway, that is to say the rapid and uncontrollable overheating of parts of the material under processing. Considering a low thermal conductivity material, whose permittivity increases as the temperature rises, in particular ε" increasing the temperature growing, it will be subject to gradient of temperature, being colder in the regions where heat is rapidly dissipated or the field strength is lower, and hotter in the remaining zones. These zones, presenting higher values of ε", and thus of P, will start absorbing microwaves more than the cold ones, further rising their temperature and consequently the local value of ε", strengthening the phenomenon.

Finally, dielectric heating is penetrating, depending on the operating wavelength, and permits to directly heat treat the surface and the core of the body, without waiting for the heat to reach the core of the sample by means of conduction, particularly time-taking for low thermal conductivity materials, like most polymers are. In these materials, the penetration depth is high, of the order of some tens of centimeters, thus facilitating the processing of large bodies, too[6].

4. Heat and Mass Transfer Classical Equations

The conservation of mass and energy for a textile material give the following equations:

$$\frac{\partial X}{\partial t} = D_n \nabla^2 X \tag{10}$$

$$C_{pn}\rho_n \frac{\partial T}{\partial t} = \nabla(K_n \nabla T) + Q(r,z,t) \tag{11}$$

where n =1 and 2 refers to the inner and outer layer of material, and D is diffusivity (m^2/s); X the moisture content (kg/kg dry basis); k the thermal conductivity (W/m K); ρ the density (kg/m^3); C_p the heat capacity (J/kg K); and Q is the microwave source term (W/m^3).

The empirical model for calculating moisture diffusivity as a function of moisture and temperature;

$$D = \frac{1}{1+X} D_0 \exp\left[-\frac{E_0}{R}\left(\frac{1}{T}-\frac{1}{T_r}\right)\right] + \frac{1}{1+X} D_i \exp\left[-\frac{E_i}{R}\left(\frac{1}{T}-\frac{1}{T_r}\right)\right] \tag{12}$$

where D (m^2/s) is the moisture diffusivity; X the moisture content (kg/kg dry basis); T (°C) the material temperature; T_r a reference temperature, and R=0.0083143 kj/mol K is the ideal gas constant; $D_0(m^2/s)$ the diffusivity at moisture X= 0 and temperature $T = T_r$; D_i (m^2/s) the diffusivity at moisture X=∞ and temperature $T = T_r$; E_0 (kJ/mol) the activation energy of diffusion in dry material at X=0 and E_i (kj/mol) is the activation energy of diffusion in wet material at X=∞. The proposed model may uses the estimated parameters in table 1.

- Initial conditions: At time t =0 : $T = T_0(r,z)$, $X = X_0(r,z)$
- Boundary conditions: $\left.\frac{\partial X}{\partial t}\right|_{(r=0,H/2=0,t)} = 0$, $\left.\frac{\partial T}{\partial t}\right|_{(r=0,H/2=0,t)} = 0$

Figure 4 shows that there is an increase in drying rate because of the microwave power density.

Table 3. Numerical values for wool (based on data from various authors)

Diffusion coefficient of water vapor - 1st stage :
$(1.04 + 68.20W_c - 1342.59W_c^2)10^{-14}$, t < 540s
Diffusion coefficient of water vapor - 2nd stage :
$1.6164\{1 - \exp[-18.163\exp(-28.0W_c)]\}10^{-14}$, t ≥ 540s
Diffusion coefficient in the air : $2.5e^{-5}$
Volumetric heat capacity of fiber : $373.3 + 4661.0W_c + 4.221T$
Thermal conductivity of fiber :
$(38.49 - 0.720W_c + 0.113W_c^2 - 0.002W_c^3)10^{-3}$
Heat of sorption: $1602.5\exp(-11.72W_c) + 2522.0$
Porosity of fiber : 0.92
Density of fiber ; 1300 kg/m^3
Radius of fiber ; 1.03 e^{-5} m
Mass transfer coefficient : 0.137 m/s
Heat transfer coefficient : 99.4 $W/m^2\ K$

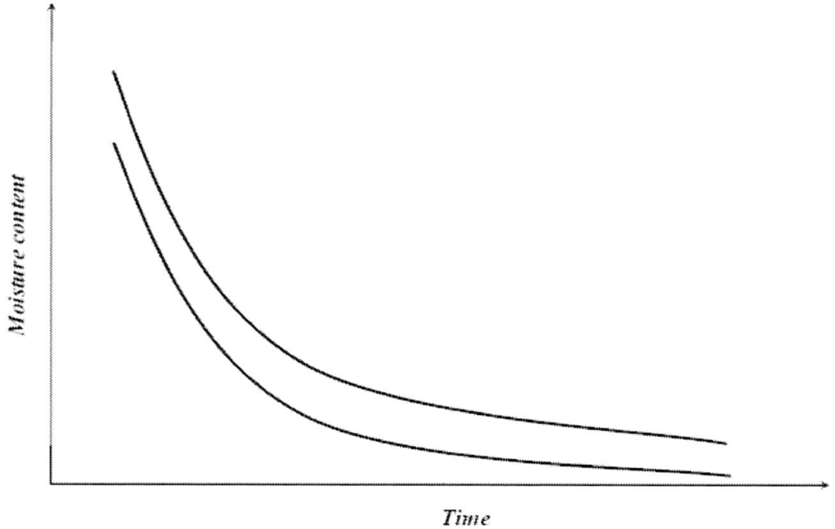

Figure 4. Comparison of conventional and microwave heating on average moisture content.

This can be attributed to the effect of microwave on moisture by rapidly increasing the moisture migration to the surface and increased evaporation. A comparison of these drying curves demonstrates improvement in drying times, under microwave heating. Nevertheless the results show significant improvement in average drying times over the conventional heating method.

5. Heat and Mass Transfer Exponential Model

It has been recognized that microwaves could perform a useful function in textile drying in the leveling out of moisture profiles across a wet sample. This is not surprising because water is more reactive than any other material to dielectric heating so that water removal is accelerated. An exponential model presented here.[7] can be used to describe the drying curves.

$$X = (a - X_{eq}).\exp(-b.t^{d}) + X_{eq} \tag{13}$$

and its derivative form:

$$(-dX / dt) = b.d.t^{(d-1)}.X \tag{14}$$

Parameters *a, b, d* can be determined by regression by the least square method. The quantities *b* and *d* vary with the experimental conditions and they are drying coefficients. X is the moisture content of the drying material, *dX/dt* is the drying rate and *t* is the drying time.. Parameter *a* represents the initial moisture content.

From figure 5 it is seen that the incident power strongly influenced the drying kinetics of a textile sample, reducing the drying time by raising the microwave heating power.

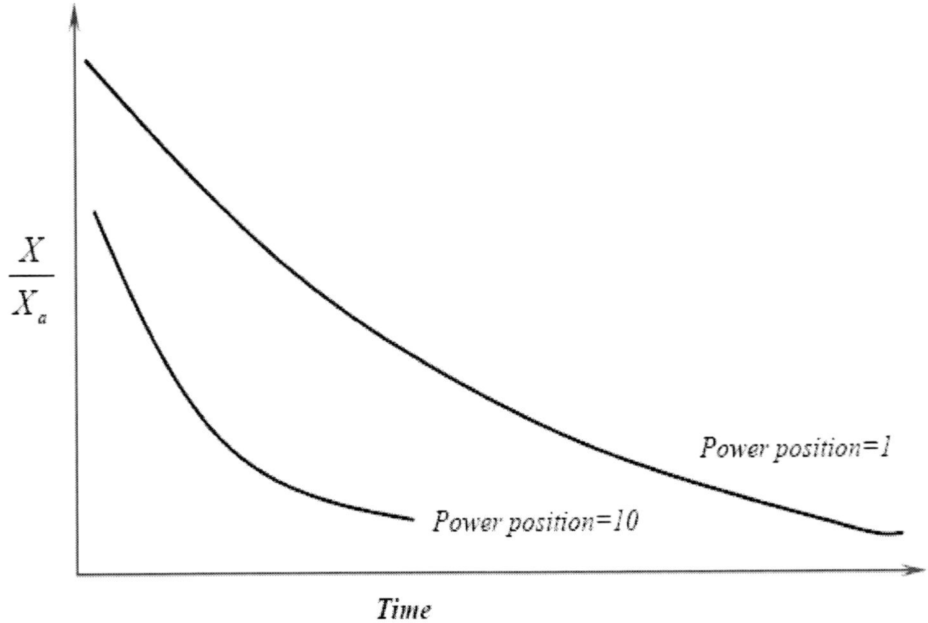

Figure 5. Normalized moisture content for two power of microwave heating.

6. Combined Microwave and Convective Drying of Tufted Textile Material

It should be noted that because of the higher temperature and pressure gradients generated during combined microwave and convective drying, greater care must be taken not to damage the textile material to be dried, whilst still taking advantage of the increased drying rates provided by the microwave environment. To fully understand the heat and mass transfer phenomenon occurring within the material during combined microwave and convective drying, it is required to analyse the moisture, temperature and pressure distributions generated throughout the process

It was shown by Ilic and Turner [8] that a theory based on a continuum approach led to the following equations of motion governing the drying of a slab of material :

Total mass:

$$\frac{\partial}{\partial t}\left(\phi S_g \rho_g + \phi S_W \rho_W\right) + \nabla.\left(\chi_g \rho_g V_g + \chi_W \rho_W V_W\right) = 0 \qquad (15)$$

Total liquid:

$$\frac{\partial}{\partial t}\left(\phi S_g \rho_{gv} + \phi S_W \rho_W\right) + \nabla.\left(\chi_g \rho_{gv} V_{gv} + \chi_W \rho_W V_W\right) = 0 \qquad (16)$$

Here, S is the volume saturation, ϕ is the porosity, $\rho[\mathrm{kg}\, m^{-3}]$ is the density of the fibers χ is the surface porosity, ϕ is the porosity

Total enthalpy:

$$\frac{\partial}{\partial t}\left(\phi S_g \rho_{gv} h_{gv} + \phi S_g \rho_{ga} h_{ga} + \phi S_g \rho_{ga} h_{ga} + \phi S_W \rho_W h_W + (1-\phi)\rho_S h_S - \phi \rho_W \int_0^{S_W} \Delta h_W(S) dS\right)$$

$$+ \nabla.\left(\chi_g \rho_{gv} V_{gv} h_{gv} + \chi_g \rho_{ga} V_{ga} h_{ga} + \chi_W \rho_W V_W h_W\right)$$

$$= \nabla.\left((K_g X_g + K_W \chi_W + K_S (1-\chi))\nabla T\right) + \phi \qquad (17)$$

where ϕ is the internal microwave power dissipated per unit volume, $\mathrm{K}[m^2]$ is permeability, and h $[Jkg^{-1}]$ is the averaged enthalpy. In equation (16) the effects of viscous dissipation and compressional work have been omitted.

The equations (15, 16 and 17) are augmented with the usual thermodynamic relations and the following relations:

Flux Expressions Are Given As Follows

Gas flux :

$$\chi_g \rho_g V_g = -\frac{KK_g(S_W)\rho_g}{\mu_g(T)}\left[\nabla P_g - \rho_g g\right] \tag{17a}$$

here, g $[ms^{-2}]$ is the gravitational constant and K_g is the relative permeability of gas.

Liquid flux :

$$\chi_W \rho_W V_W = -\frac{KK_W(S_W)\rho_W}{\mu_W(T)}\left[\nabla\left(P_g - P_C(S_W, T)\right) - \rho_W g\right] \tag{17b}$$

where, K_W is the relative permeability of water, and μ $[Hm^{-1}]$ is the permeability of free spaces.

Vapour flux :

$$\chi_g \rho_{gv} V_{gv} = \chi_g \rho_{gv} V_g - \frac{\chi_g \rho_g D(T, P_g) M_a M_v}{M^2} \nabla\left(\frac{P_{qv}}{P_g}\right) \tag{17c}$$

here, V $[ms^{-1}]$ is the averages velocity and M $[kgmol^{-1}]$ is the molar mass.

Air flux :

$$\chi_g \rho_{ga} V_{ga} = \chi_g \rho_g V_g - \chi_g \rho_g V_{gv} \tag{17d}$$

Relative Humidity (Kelvin Effect)

$$\psi(S_W, T) = \frac{P_{gv}}{P_{gvs}(T)} = \exp\left(\frac{2\sigma(T) M_v}{r(S_W)\rho_W RT}\right) \tag{18}$$

where ψ is the relative humidity and $P_{gvs}(T)$ is the saturated vapour pressure given by the Clausius-Clapeyron equation.

Differential Heat of Sorption

$$\Delta h_W = R_v T^2 \frac{\partial(\ln \psi)}{\partial T} \tag{19}$$

Enthalpy-Temperature Relations

$$h_{ga} = C_{pa}(T - T_R) \tag{20}$$

$$h_{gv} = h^0_{vap} + C_{pv}\left(T - T_R\right) \tag{21}$$

$$h_W = C_{pW}\left(T - T_R\right) \tag{22}$$

$$h_s = C_{ps}(T - T_R) \tag{23}$$

The expressions for K_g , K_W are those given by Turner and Ilic [8], and μ_g , μ_W have had functional fits according to the data by Holman [9]. The diffusivity $D\left(T, P_g\right)$ given by Quintard and Puiggali [10] and the latent heat of evaporation given by,

$$h_{vap}\left(T\right) = h_{gv} - h_W \tag{24}$$

After some mathematical manipulations, the one-dimensional system of three non-linear coupled partial differential equations which model the drying process in a thermal equilibrium environment are given by :

$$a_{s1}\frac{\partial S_W}{\partial t} + a_{s2}\frac{\partial T}{\partial t} = \frac{\partial}{\partial Z}\left[K_{S1}\frac{\partial S_W}{\partial Z} + K_{T1}\frac{\partial T}{\partial Z} + K_{T1}\frac{\partial T}{\partial Z} + K_{P1}\frac{\partial P_g}{\partial Z} + K_{gr1}\right] \tag{25}$$

$$a_{T1}\frac{\partial S_W}{\partial t} + a_{T2}\frac{\partial T}{\partial t} = \frac{\partial}{\partial Z}\left(K_e\frac{\partial T}{\partial Z}\right) - \phi\rho_W h_{vap}\frac{\partial}{\partial Z}\left[K_S\frac{\partial S_W}{\partial Z} + K_T\frac{\partial T}{\partial Z} + K_P\frac{\partial P_g}{\partial Z}K_{gr}\right]$$
$$+\left[\phi\rho_W C_{pW}\left(K_{S2}\frac{\partial S_W}{\partial Z} + K_{T2}\frac{\partial T}{\partial Z} + K_{P2}\frac{\partial P_g}{\partial Z} + K_{gr2}\right)\right]\frac{\partial T}{\partial Z} + \Phi\left(S_W, T\right) \tag{26}$$

$$a_{P1}\frac{\partial S_W}{\partial T} + a_{P2}\frac{\partial T}{\partial t} + a_{P3}\frac{\partial P_g}{\partial t} = \frac{\partial}{\partial Z}\left[K_S\frac{\partial S_W}{\partial Z} + K_T\frac{\partial T}{\partial Z} + K_{P3}\frac{\partial P_g}{\partial Z} + K_{gr3}\right] \tag{27}$$

The capacity coefficients a_{S1} , a_{T1} , a_{p1} and the kinetic coefficients K_{S1} , K_{T1} , K_{P1}, K_{gr1} all depend on the independent variables : Saturation S_W , Temperature T and total pressure P_g . The boundary conditions are written in one dimension as :

At z=0 (Drying surface) :

$$K_{S1}\frac{\partial S_W}{\partial Z}+K_{T1}\frac{\partial T}{\partial Z}+K_{P1}\frac{\partial P_g}{\partial Z}+K_{gr1}=\frac{K_m M_V}{R\phi\rho_W}\left(\frac{P_{gV}}{T}-\frac{P_{gV0}}{T_0}\right) \tag{27a}$$

$$K_e\frac{\partial T}{\partial Z}-\phi\rho_W h_{Vap}\left(K_S\frac{\partial S_W}{\partial Z}+K_T\frac{\partial T}{\partial Z}+K_P\frac{\partial P_g}{\partial Z}+K_{gr}\right)=Q(T-T_0) \tag{27b}$$

$$P_g=P_o \tag{27c}$$

At z=L (Impermeable surface):

$$K_{S1}\frac{\partial S_W}{\partial Z}+K_{T1}\frac{\partial T}{\partial Z}+K_{P1}\frac{\partial P_g}{\partial Z}+K_{gr1}=0 \tag{28a}$$

$$K_e\frac{\partial T}{\partial Z}-\phi\rho_W h_{Vap}\left(K_S\frac{\partial S_W}{\partial Z}+K_T\frac{\partial T}{\partial Z}+K_P\frac{\partial P_g}{\partial Z}+K_{gr}\right)=0 \tag{28b}$$

$$\left(K_{S1}-K_S\right)\frac{\partial S_W}{\partial Z}+\left(K_{T1}-K_T\right)\frac{\partial T}{\partial Z}+\left(K_{P1}-K_{P3}\right)\frac{\partial P_g}{\partial Z}+\left(K_{gr1}-K_{gr3}\right)=0 \tag{28c}$$

Initially:

$$T(z,0)=T_1 \tag{29a}$$

$$P_g(z.0)=P_0 \tag{29b}$$

$$\frac{\partial P_c}{\partial Z}=-\rho_W g \tag{29c}$$

Figures 6 and 7 show a comparison of convective drying with or without microwaves. Whilst for convective drying there are definite constant rate and falling rate periods, when microwaves are added the form of the curves change [2].

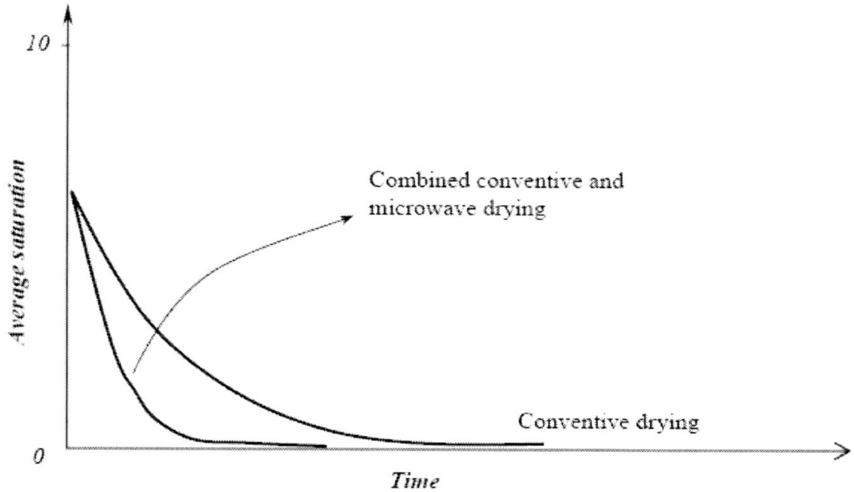

Figure 6. Average saturation profiles in time for drying with or without microwaves.

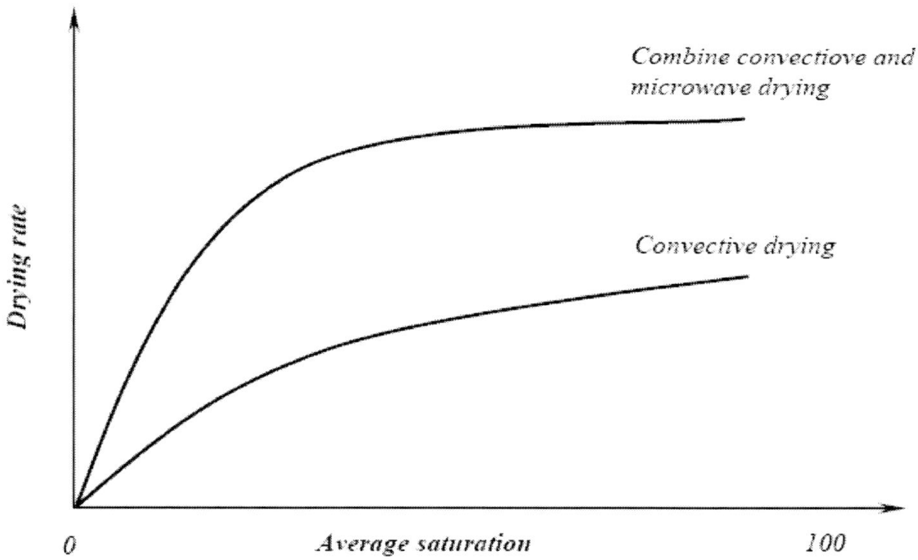

Figure 7. Drying rate curves corresponding to profiles plotted in Figure 6.

REFERENCES

[1] D.A.Jones, T.P. Lelyveld, S.D. Mavrofidis, S.V. Kingman, and N. J. Miles, Microwave heating application in environmental engineering, *Resources, Conservation and Recycling,* 34, 75-90, 2002.

[2] I.W. Turner and P.G. Jolly, Combined microwave and conventional drying of a porous material, *Drying technology,* 9(5), 1209-1269, 1991.

[3] T.V. Denend, Infrared predrying yields significant benefits, *American Dyestuff Reporter,* 45-51, December 1998.

[4] K. E. Haque, Microwave energy for mineral treatment processes, *Int. J. Miner. Process,* 57 , 1-24, 1999.
[5] A.C. Metaxas and R.J. Meredith, Industrial microwave heating, Peter Peregrinus Ltd, London 1993, pp. 7-24.
[6] P. Veronesi, C. Leonelli, G. Pellacani and A. Boccaccini, Unique microstructure of glass-metal composites obtained by microwave assisted heat-treatment, *J. Thermal Anal. Cal.,* 72, 1141-1149, 2003.
[7] D. Skansi, Z. Bajza an A. Arapovic, Experimental evaluation of the microwave drying of leather, *J. of the society of Leather Thechnologies and chemists,* 79, 171-177.
[8] Ilic, M. and Turner, I.W.: Convective drying of a consolidated slab of wet porous material, *Int.J.Heat Mass Transfer*, , 32 (12),(1989).
[9] Holman, J. P., 1989, Heat Transfer, McGraw-Hill Book Company.
[10] Quintard M., and Puiggali J.R., 1986, Numerical modeling of transport processes during the drying of granular porous medium, *J. Heat and Technology*, 4, 2.

In: Progress in Chemistry and Biochemistry
Editors: E. M. Pearce and G. E. Zaikov

ISBN: 978-1-60692-344-3

Chapter 18

BIODEGRADATION OF OIL PRODUCTS IN POLLUTED SOIL BY USING THE MIXTURE OF THE POLYMER COMPLEX [CHITOSAN – LACTOSERUM PROTEIN]–ACTIVE SILT

Y.P. Ioshchenko*, V.F. Kablov* and G.E. Zaikov**

* Volzhsky Politechnical Instititute (branch) of Volgograd
State Technical University
42a Engels St., Volzhsky 404121, Volgograd Region
**N.M. Emanuel Institute of Biochemical Physics,
Russian Academy of Science
4 Kosygin St., Moscow 119991

ABSTRACT

Biodegradation processes of oil products in polluted soil were studied by using the mixture of the polymer complex [chitosan – lactoserum protein]–active silt. It is shown, that all these processes take place due to the synergism of the components which form part of the mixture; the use of the developed biomixture ensures the effective biodegradation of complicated oil hydrocarbons, a high degree of cleaning soil from pollutants, except the possible secondary pollution of the environment. As a result of biocomposting, the cleaned ground contains biogenic substances, which fit in the natural cycle and can be used for realization of agricultural work.

Keywords: biodegradation, petroleum products, chitosan–lactoserum protein polymer complex, active silt, biomixture.

INTRODUCTION

During oil outflows the vital functions of biota are repressed to a considerable degree [1, 2]. That's why the problem of recultivation and biodegradation of oil polluted soils becomes that actual [3, 4]. Unfortunately, the visible success in the practice of oil polluted soils biodegradation has not been achieved yet.

Today the most progressive technology in the oil polluted soils cleaning is considered to be the usage of introductive microorganisms [5, 6], strains of fungus, actinomycetes [7-9], bacteria [10-12]. At the same time the microorganisms are able to destroy only the upper layers (15-20 cm) of oil pollutions, to have a selective effect on the definite fractions of oil products; the strains don't solve the problems of recultivation, the possibilities of their transformation also have their limits.

To a considerable degree the difficulty of biodegradation is caused by the high concentration of oil products in soil that results in obstacles in the biochemical processes of oxidation and biodestruction. Sorbing oil products on sorbents' particles and macromolecules allows decreasing the concentration in the local systems. It is important, that a sorbent is being degradated by itself in the course of time. Biopolymers are very perspective for sorbing of oil products.

Chitosan is a biopolymer of the future; the interest in it is connected with unique physiological and ecological prosperities such as biocompatibility, biodegradability, non-toxic physiological activity, and easy access to resources needed for its production. The polymer complexes based on chitosan in combination with other synthetic and biodegradable polymers is of a particular interest.

However, the use of the mixture of bio- and synthetic polymers and microorganisms in biodegradation processes of oil products in the polluted soil has not studied enough.

The aim of this work was the study of biodegradation processes by using the mixture of the polymer complex [chitosan – lactoserum protein (ChS-LP)]–active silt.

EXPERIMENTAL

The soil, used in this work, contains oil products left after oil outflow; it was taken in the industrial area of The Volzhsky Chemical Complex.

The mixture, which consists of ChS-LP polymer complex and active silt, was used as an agent, accelerating biodegradation processes.

The advantages of the polymer complex are accessibility of raw materials, non-toxicity, biodegradability, and ecological safety [13, 14].

The active silt from Volzhsky waste disposal plant was used as the test-cultures. It consists of the filiform bacteria, Bodo ovatus, Oikomonos mutabilis, Euglena visidis, Amotba limax, Arcella centropyxis, Arcella diskodes, Eoglipha leavis, Vorticella convallaria, Vorticella mikrostoma, Epistulis plikatilis, Carchesium polipinum, Opercularia coarktata, Fhabdostyla ovum, Aspidiska costata, Lionotus lamella, Amphileptus carchesi, Colpidium colpoda, Chilodonella uond, Lakrumaria pupula, Stulonicha mvtilis, Oxitricha pelionella, Euplotec patella, Prorodon teres, Tokohrya mollis, Podophrya fixa, Rabdofrya, Monostyla, Cathyrnia, Notommata ansata, Rotatria, Philodina roseola, Colurella, Nematoda, Aelosoma. The bacteria and microorganisms that form part of active silt are able to use for their nutrition different materials of natural and synthetic origin. Besides, the active silt is the source of biodegradants, nitrogen and the organic component of the soil.

In spring-summer period the oil polluted soil was cultivated on planed and balanced plot by the initial mixture of the polymer complex [chitosan–lactoserum protein]–active silt. After that the soil was loosened to 15-20 cm for the improvement of moisture and aeration. The optimal moisture of the soil essential for ensuring the favorable conditions of microorganisms' vital functions must be 50-60 %.

Biodegradation processes of oil products were studied for 90 days at the temperature of 10 °C, the optimal temperature range was 25-45 °C. A cover material was used to maintain the optimal temperature. Monthly the loosening was made to a depth of 50-55 cm, and the soil was watered whenever its humidity decreased to 50 %. Besides, the periodic soil loosening prevented the moisture evaporation. The decreasing of index oilsaturation (oil destruction) indicated the efficacy of cultivation of the worked soil. In autumn to provide a complete soil recultivation it was exposed to the phytoremediation - the soil's sowing by annual and biennial seeds.

The optimal correlation of the mixture "oil polluted soil/sorbent/active silt" was found by the formula describing the dependence of the degree of oil product biodegradation on content of sorbent in the mixture, received by data handling, using the following dependence [15]:

$$S(n) = k \cdot (1 - e^{-n/T}),$$

S – degree of oil product biodegradation, %;

n – correlation of soil's quantity (volume unit) to the sorbent's quantity (volume unit);

k – scale coefficient, used for calculation of oil product biodegradation with introduced sorbent depending on scale of coordinate;

T – time constant of oil product biodegradation with introduced sorbent;

$e \approx 2{,}718$,

for all that, the optimal correlation of components was chosen in the point of intersection function graph *S(n)* with horizontal line, which is smaller then steady-state value to 3-5%.

The value of the optimal doze of active silt was determined according the formula describing the dependence of the degree of oil product biodegradation on doze of active silt, received by data handling, using the following dependence [15]:

$$S(d) = k_1 \cdot (1 - e^{-d/Q}),$$

d – doze of active silt, mg/kg soil;

k_1 – scale coefficient, used for calculation of oil product biodegradation with introduced active silt depending on scale of coordinate;

Q - time constant of oil product biodegradation with introduced active silt,

for all that, the optimal doze of active silt was chosen in the point of intersection of function graph *S(d)* with horizontal line, which is smaller then steady-state value to 3-5%.

In this work for biodegradation of oil polluted soil for each 1 m^2 of the soil was used 50 g of the developed biomixture; the optimal correlation bringing in the soil is following:

polymer complex : active silt=1:4.

The index of residual of oil products in the soil was used for valuation of decomposition's intensity. The definition was carried out using a spectra-photometrical method – photo-electro-calorimeter "KFK-2" and a weight method [16].

The number of the main group's of microorganisms, which take part in biodegradation of oil products was defined by sewing the soil suspension, using the dilution technique on the dense nutrient medium [17].

The measurement of mass concentrations of the oil products in waste waters samples was carried out using the fluorescent method by the means of the liquid analyzer "Fluorat-02" [18].

The quantitative and structural-group composition of residual oil products was carried out with the help of methods of the quantitative NMR-spectroscopy [19] and liquid chromatography [20], using the apparatuses "Mercury-300" and "Lumachrom" accordingly.

RESULTS AND DISCUSSION

The proposed method of oil product biodegradation is based on the bringing biological active mixture of ChS-LP polymer complex-active silt in the soil. The given mixture is used for decreasing or full degradation of different structured oil hydrocarbons, which take place in processes of oxidative destruction and mineralization during the vital functions of hydrocarbons oxidizing microorganisms.

The use in biodegradation processes of oil products in the polluted soil of biomixture ChS-LP polymer complex-active silt is very perspective.

ChS-LP polymer complex (figure 1) represents the microsubstances with cavities of nanosizes, fixed between each other with hydrogen links, in which the sorbtial processes happen. The complex has good sorbtial, protective and reconstruction properties [21]. Due to the different nature of its functionally-active groups, chitosan withholds the lactoserum protein in the complex, preventing its washing-out under the action of different natural factors.

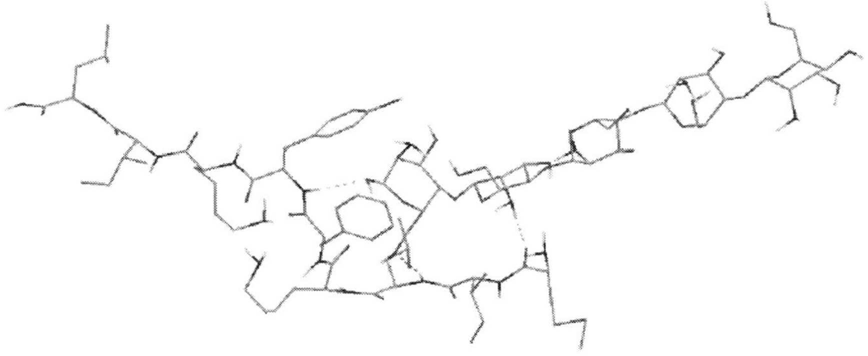

Figure 1. The structure of the fragment of ChS-LP polymer complex.

Biodegradation processes of oil products in the polluted soil takes place due to the synergism of the components which form part of the mixture: ChS–LP polymer complex and active silt. The processes of biodegradation procced with the detachment of CO_2, water, biomass' formation and partially oxide biologically inert by-products. At that time the transformation processes of organic substances are accompanied by the humification processes. The completeness of oil hydrocarbons biodegradation with increased resistance and with the low content (less 3 weight %) of oil products in the polluted soil is provided by the use of the mixture of ChS-LP polymer complex-active silt in the proposed method.

Chitosan is a base of the polymer complex and a sorbent that accelerates biodegradation processes of oil products; lactoserum protein is the active nutrient medium for bacteria and microorganisms of active silt. The model of the structure of ChS-LP polymer complex with sorbed oil products (OP) in its cavities and with active silt is represented in the figure 2. Besides, due to the most activity not only of the polymer complex, but also of the active silt, the whole mixture has an increased activity towards the oil products biodegradation. The bringing of such a mixture in the polluted soil exerts influence over the habitat of natural microflora, causing the consistent continuity of the microorganisms' complex in the soil.

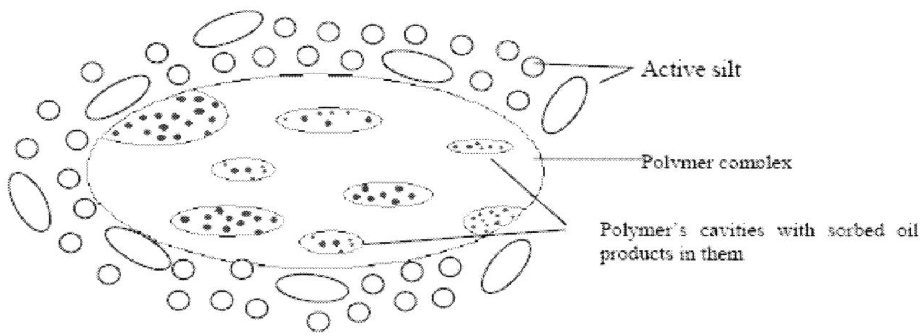

Figure 2. The model of the structure of ChS-LP polymer complex with sorbed oil products in its cavities and with active silt.

The dynamics of changing number of microorganisms in soil is represented in the table 1. As is shown in the table, the number of microorganisms in the polluted soil has increased in 12 times, at the same time the number of microorganisms in the processed soil by the developed mixture has increased more than in 1500 times. The considerable augmentation of the number of microorganisms is explained by the high activity of added mixture of the polymer complex-active silt. The mixture activates the soil' microbiological activity; the biomass of active silt plays a role of an effective fertilizer, which stimulates the development of microbiological biota in oil polluted soil.

Table 1. The dynamics of changing number of microorganisms in soil

Compositions	The dynamics of changing number of microorganisms in soil (in 1 g), 10^{-6}				
	before processing	after 2 weeks	after 1 month	after 2 month	after 3 month
The soil before processing	6,1	6,5	7,3	7,9	8,4
The soil with OP (control)	0,005	0,008	0,01	0,03	0,06
The soil with OP and with mixture of ChS-LP polymer complex-active silt	0,005	0,09	2,1	5,5	8,1

The combination of chitosan and lactoserum protein in the polymer complex ensures a comparatively larger efficiency of a degree of the oil biodegradation in the soil, which can be explained by the presence of a great number of active complex-forming centers due to the chelatogenic protein's group, which are able to hold the sorbed particles of oil products firmly and fixedly and to block their repeated return in soil.

The data in table 2 shows, that the use of the mixture of ChS-LP polymer complex-active silt in biodegradation process of oil products allows to decrease the content of oil hydrocarbons on 98,9 %.

As it is shown in the table 2, nature is not able to ensure a rather complete cleaning of oil polluted soils, the time of biodegradation are lengthening too much, and the degree of biodegradation achieves only 71,4 %, while using the developed mixture the degree of biodegradation achieves 98,9 %.

Table 2. The dynamics of changing content of OP in soil

Compositions	The degree of OP biodegradation in soil, %				
	before processing	after 2 weeks	after 1 month	after 2 month	after 3 month
The soil before processing	-	-	-	-	-
The soil with OP (control)	0,02	8,9	18,6	47,8	71,4
The soil with OP and with mixture of ChS-LP polymer complex-active silt	0,02	24,8	65,2	81,3	98,9

The process of changing of a relative concentration in time of oil products in soil was studied in this work (figure 3). It is shown, that under the action of the developed mixture the considerable changes of concentration of oil products occur for a short period of time.

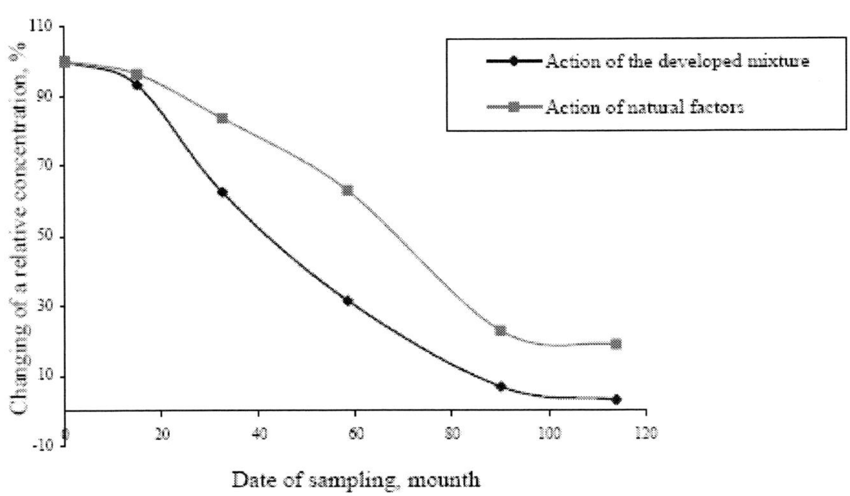

Figure 3. Changing of a relative concentration in time of oil products in soil.

A character of the developed mixture of ChS-LP polymer complex-active silt used for biodegradation processes of oil products in the polluted soil is the high degree of decomposition of complicated oil hydrocarbons: resinous and asphalt fractions (table 3). Due to its functional activity the polymer complex "prepares" asphaltenes for entering in humus layer of soil and allows achieving the high cleaning degree of the soil from pollutions.

Table 3. The quantitative and structural-group composition of residual OP

Compositions	Time of biodegradation, days	Gross content of residual OP, weight %	The quantitative and structural-group composition of residual OP, weight %		
			paraffin	resins	asphaltenes
The soil before processing	-	12,8	80,1	15,2	6,3
The soil with OP (control)	30	11,3	78,1	13,6	5,9
	60	9,8	74,5	13,0	5,2
	90	7,7	71,2	12,8	5,0
The soil with OP and with mixture of ChS-LP polymer complex-active silt	30	6,8	70,6	12,5	5,2
	60	3,0	55,8	8,2	4,1
	90	0,6	49,2	5,9	3,6

The experiments show that the use of the developed biomixture allows achieving a fuller decomposition of resinous and asphalting oils fractions, which can be explained by the presence of microorganisms and enzymes growth substratum in the mixture; all these alive organic beings take part in oxidation of complicated polycyclic oil hydrocarbons.

So, the bringing of the mixture of the polymer complex-active silt in the polluted soil extends the sphere of microorganisms, which are drawn into the process of hydrocarbons' decomposition, which helps to use wider potential possibilities of microcenosis. All these processes allow shortening the time of full biodegradation in the polluted soil with different degree of pollution and achieving a higher degree of oil hydrocarbons' transformation into the harmless substances.

As a result of biocomposting the cleaned ground contains biogenic substances, fit well in natural cycle and can be used for realization agricultural works. Besides, the used mixture allows not only to clean the soil from pollution, but also to receive the soil material which contains precious organic substances, ensures the nutrition for wide specter of plants and microorganisms.

The use of chitosan and lactoserum protein in the mixture allows to obtain the biologically active material enriched with nitrogen and to decrease the amount of the mineral nitric fertilizer, which are necessary for composting of oil polluted soil, and to except the possible secondary pollution of environment.

Conclusions

1. Biodegradation of oil products in the polluted soil is carried out by using the biological active mixture, which consists of ChS-LP polymer complex and active silt. The advantages of

this biomixture are accessibility of raw materials, non-toxicity, biodegradability, and ecological safety.

2. Biodegradation of oil products in the polluted soil takes place due to the synergism of the components which form part of the mixture: ChS–LP polymer complex and active silt. As a result of it, expenses on the production of oil-oxidizing bacteria and microorganisms decrease, and expenses on the utilization of active silt disappear.

3. The use of the biomixture polymer complex-active silt ensures an effective biodegradation of complicated oil hydrocarbons possesses to achieve a high degree of cleaning soil from pollutions. The given mixture allows obtaining the final biologically active soil material which contains precious organic substances and nitrogen, ensures the nutrition for wide specter of plants and microorganisms and allows to decrease the amount of mineral nitric fertilizer, which are necessary for composting of oil polluted soil and to except the possible secondary pollution of environment.

4. As a result of biocomposting the cleaned ground contains biogenic substances, fit well in natural cycle and can be used for realization agricultural works.

References

[1] V.S. Guzev, S.V. Levin, G.I. Seletckiy, et al. The role of soil microbiota in recultivation of oil polluted soil // Edited by D.G. Zvyagetcev. – M.: Microorganisms and soil protection: Chap. 4. 1989. P. 129-150.

[2] S.A. Buzmakov, I.V. Ladigin. // International scientific conferences' report theses. Part 1. Perm'. 1993. P. 201-205.

[3] V.F. Kablov, Y.P. Ioshchenko. // Povolzhsky Ecological Herald, Volgograd. Vol. 11. 2005. P. 49-54.

[4] C.T. Chen. Understanding the fete of petroleum hydrocarbons in the subsurface environment // J. Chem. Educ. 69, 5. 1992. P. 357-361.

[5] E.M. Halimov, S.V. Lenin, V.S. Guzev. // Moscow University News. Series "Soil Science". № 2. 1996. P. 59-64.

[6] Y.P. Ioshchenko, D.A. Kondrutsky. // IX regional conference of young researchers in Volgograd region report theses. Volgograd. 2005. P. 30-32.

[7] Patent 3813316 U.S., Int. Cl. C12B/00. A.M. Chakrabarty. Microorganisms having multiple compatible degradative energy-generating plasmids and preparation thereof.

[8] Patent 2280013 RU, Int. Cl. C02F3/34. V.V. Chakov, E.A. Karetnikova. Preparation for cleaning soil and water surfaces polluted by petroleum and petroleum derivatives.

[9] Patent 2081854 RU, Int. Cl. C02F3/34. L.E. Tolstokorova, V.P. Shchipanov, T.N. Morozova, L.S. Podenko. Bioreagent for water and soil treatment from petroleum pollution.

[10] E.V. Stabnikova, M.B. Selezneva, O.N. Reva, V.N. Ivanov. // Applied chemistry and microbiology. - T. 31, № 5. 1995. P. 534-539.

[11] Patent 2077397 RU, Int. Cl. B09C1/10. R.K. Andreson, F.K. Khaziev, V.S. Deshura, F.Y. Bagautdinov, T.F. Bojko, E.I. Novoselova. Method of restoration of soils contaminated with petroleum and its derivatives.

[12] Patent 2077579 RU, Int. Cl. C12N1/20. N.A. Zhirkova, V.S. Kobelev, V.P. Kholodenko. Strain of bacterium Acinetobacter species used for water and soil treatment from oil and petroleum products.

[13] V.F. Kablov, Y.P. Ioshchenko, D.A. Kondrutsky. // Review MITHT. Vol. 1. №. 5. 2006. P. 49-53.

[14] V.F. Kablov, Y.P. Ioshchenko, D.A. Kondrutsky. // New High Technologies. №. 4. 2004. P. 87-88.

[15] A.N. Kudrin, G.T. Ponomoreva. The use of mathematics in experimental and clinical medicine. M.: Medizdat. №3. 1967. 138 p.

[16] R.A. Gruzdyakova. // Hygiene and sanitation. № 3. 1993. P. 73-74.

[17] D.G. Zvyagintcev. Methods of soil microbiology and biochemistry. M.: MSU. 1980. 223 p.

[18] TU 4321-001-20506233. Methodology for carrying-out of measurements of mass concentrations of oil products in waste waters samples by means of the liquid analyzer "Fluorat-02".

[19] G.A. Kalabin, K.V. Kanitcskaya, D.F. Kushnarev. Quantitative NMR-spectroscopy of natural organic raw materials and products of their processing. M.: Khimiya. 2000. 408 p.

[20] V.I. Sokolova, M.A. Kolbin. Liquid chromatography of oil products. M.: Khimiya. 1984. 144 p.

[21] Y.P. Ioshchenko. Preparation and study of chitosan polymer complexes with proteins and hydroxylous polymers // Thesis for a Degree of Candidate of Technical Science. – Volgograd, 2006. 119 p.

In: Progress in Chemistry and Biochemistry
Editors: E. M. Pearce and G. E. Zaikov
ISBN: 978-1-60692-344-3

Chapter 19

TRANSPORT PROPERTIES IN DRYING OF WOODS: A REVIEW ON THEORETICAL APPROACHES

Sh. Rahrovan and A. K. Haghi*
University of Guilan, P. O. Box 3756
Rasht, Iran

ABSTRACT

High temperature heat treatment of wood is a complex process involving simultaneous heat, mass and momentum transfer phenomena and the effective models are necessary for process design, optimization, energy integration, and control. In this paper transport properties of wood drying is reviewed in detail.

NOMENCLATURE

A	Area,
a	Half of medium thickness,
a_m	Moisture diffusion coefficient,
a_q	Thermal diffusion coefficient,
Bi_m	Mass transfer Biot number,
Bi_q	Heat transfer Biot number,
C	Concentration,
C_m	Moisture capacity,
C_p	Specific heat,
C_{PS}	Specific heat capacity of moist wood,

* Haghi@Guilan.ac.ir

C_q	Heat capacity,
D	Diffusion coefficient,
D_V	Binary bulk diffusivity,
D_V^{eff}	Effective diffusivity,
d	Width of the wood sample,
E	Fraction of total concentration in the medium in sorption,
E'	Fraction of total concentration in the medium in desorption,
E_b	Activation energy,
$\overline{E}$	Moisture ratio,
F_0	Fourier number,
G	Specific gravity,
g	Gravitational acceleration,
H	Heat of vaporization,
H_m	Coefficients defined in Eq. (23),
h	Heat transfer coefficient ,
h'	Heat enthalpy,
h_m	Convective mass transfer coefficient,
h_q	Convective heat transfer coefficient,
I	Number of exposed surfaces,
j	Total water flux,
K	Permeability,
K_r	Relative permeability,
K_0	Kossovitch number,
k	Thermal conductivity,
k_m	Moisture conductivity,
k_{ma}	Mass transfer coefficient based on absolute humidity difference,
k_{mp}	Mass transfer coefficient based on partial pressure difference,
L	Characteristic length,
L'	Magnitude of the internal and external resistances to the moisture loss from the wood to air or airflow,
Lu	Luikov number,
M	Total concentration in the medium,
MC	Moisture content,
$\overline{MC}$	Average moisture content,
MC_0	Initial moisture content,
MC_e	Equilibrium moisture content,
mc	Dimensionless moisture content,

m_V	Molar mass,
$\dot{m}_{ij}$	Transition from phases i and j,
N	Flux,
Nu	Nusselt number,
n	Number of constants,
n'	Number of observation,
$\overline{n}$	Average of the data points,
n_V	Number of moles,
P	Total pressure,
Pr	Prandtl number,
p	Partial pressure,
p_c	Capillary pressure,
p_{SV}	Saturation water vapor pressure in wood,
Pn	Posnov number,
Q	Heat transferred,
q	Volumetric heat generation,
R	Ideal gas constant,
r	Correlation coefficient,
Re	Reynolds number,
S	Surface emission coefficient,
S'	Saturation,
S''	Standard error,
s	Dimensionless time from the inception of the falling-rate period,
T	Temperature in Kelvin,
T_c	Critical temperature,
t	Time,
U	Moisture potential,
V	Volume of wood,
v	Velocity,
X	Space distance,
x	Dimensionless space coordinate,
Y	Dimensionless distance,
y	Distance measured in a direction normal to the drying surface.

Greek Letters

α	Drying rate per unit area during the constant-rate period,
β	Coefficient of thermal expansion,
β_n	Positive roots,
f	Frequency,

δ	Thermal gradient coefficient,
δ'	Diagonal matrix,
ℓ	Density ratio of water vapor to drying air,
ε	Positive constant close to zero,
ε''	Dielectric loss factor,
η	Ratio of vapor diffusion coefficient to coefficient of total moisture diffusion,
θ	Dimensionless moisture concentration,
ϑ	Dimensionless temperature,
λ	Latent heat,
γ	Drying intensity,
μ	Viscosity,
ρ	Density,
ρ_0	Initial moisture concentration,
ρ_d	Dry body density,
ρ_{eq}	Equilibrium moisture concentration,
ρ_m	Moisture concentration,
Θ	Dimensionless total quantity of moisture during the constant-rate period or falling-rate period,
Φ	Dimensionless moisture content,
ϕ	Volumetric fraction,
ζ	Attenuation factor in the effective diffusivity,
Ψ	Coefficient for partial pressure of water vapor in wood,
ψ	Relative humidity,
Ξ	Intensity of the applied field,
Ω	Boundary surface of the computational domain,
τ	Dimensionless time,
τ_{cr}	Dimensionless critical time,
ω	Absolute humidity,
ϖ	Mole fraction.

Subscripts and Superscripts

a	Ambient condition,
air	Air properties b bound,
b	Bound water,
cr	Critical,
cw	Cell wall,
d	Dry body,
e	Heat loss,

eq	Equilibrium,
exp	Experimental,
f	Free water,
final	Final,
FSP	Fiber saturation points,
g	Gas,
in	Inside,
l	Liquid,
m	Moisture,
ma	Mass transfer coefficient based on absolute humidity difference,
mp	Mass transfer coefficient based on partial pressure difference,
out	Outside,
pred	Predicted,
r	Relative,
S	Saturated,
s	Solid skeleton of wood,
surf	Surface of board,
v	Vapor,
w	Free water,
wv	Water vapor,
max	Maximum,
0	Initial,
∞	Bulk,
$\_$	Average.

1. DIFFUSION

1.1. Moisture Diffusion in Microwave Drying

Classical convective drying of wood is energy intensive and a slow process, as it relies on heat conduction from the outer layer toward the interior of the board. Meanwhile, moisture content gradient is needed for the moisture to transfer from the interior to the wood surface. Microwave heating offers an opportunity to increase the rate of evaporation, because the energy is absorbed throughout the volume, and the heat is generated directly within the material being dried. Understanding the mechanism of moisture evaporation and movement during drying of wood is particularly important to correctly interpret moisture transport phenomena. Moisture diffusion theory (Fick's second law) was widely utilized to predict the moisture transfer in conventional kiln drying of wood, especially at low moisture content [1, 2]. An effective diffusion coefficient, combining the vapor and bound water diffusion, is used to determine the rate of moisture movement. D. Jia and M. T. Afzal [3] developed a one-dimensional mathematical model for simulation of microwave drying of red maple and white oak. They modeled the moisture transfer during microwave drying of hard wood based on the diffusion theory. An energy conservation and heat transfer equation was adopted to predict the temperature change. The model was verified at low moisture contents (below FSP) and low power levels (wood temperature below 100°C). To solve the model, MATLAB software was used for finite element analysis. Results show that the diffusion model is effective at low moisture content and can reflect the main mechanism of moisture transfer during the microwave drying. Diffusion coefficients of red maple and white oak obtained experimentally at different power levels were observed to increase exponentially with drying temperature. In the microwave drying of wood, internal temperature was higher than the boundary temperature.

1.1.1. Mathematical Model

A one-dimensional unsteady-state moisture diffusion model based on the Fick's second law is expressed as:

$$\frac{\partial (MC)}{\partial t} = \frac{\partial}{\partial X}\left(D\frac{\partial (MC)}{\partial X}\right) \tag{1}$$

In the microwave drying of wood, energy balance can be described as:

$$\rho C_p \frac{\partial T}{\partial t} = \frac{\partial}{\partial X}\left(k\frac{\partial T}{\partial X}\right) + q \tag{2}$$

where MC is the moisture content, t is the time, X is the space distance and D is the effective diffusion coefficient which combines the bound water and water vapor diffusion; k is the thermal conductivity, ρ is the density, C_p is the specific heat and q is the volumetric heat generation.

Thermal conductivity of wood with moisture content below 40% can be approximated by the following empirical equation:

$$k = G(0.2 + 0.004(MC)) + 0.024 \tag{3}$$

where G is the specific gravity based on volume at moisture content MC.

When wood contains water, the specific heat increases because the specific heat of water is larger than that of dry wood [4].

- For MC< 30%:

$$C_p = (1176 + (MC)/100.5859)/(1 + (MC)/100) \quad (J/kg.K) \tag{4}$$

- For MC > 30%:

$$C_p = (1678 + (MC)/100.4185)/(1 + (MC)/100) \quad (J/kg.K) \tag{5}$$

At moisture content higher than the fiber saturation point, the moisture mainly exists in two states: free (liquid) water and bound water. The energy associated with removing the bound water in drying is greater than the latent heat of liquid water vaporization.

1.1.2. Microwave Power Absorption

An important step in the finite element modeling of microwave drying lies in the formulation to determine the heat generation term "q" for absorbed heat from the microwave by the wood. The heat generated in a given medium depends primarily on three variables: the intensity of the applied field Ξ (V/m), its frequency f (1/s), and the dielectric loss factor ε'' of the medium [5]:

$$q \propto f\,\Xi^2 \varepsilon'' \tag{6}$$

The loss factor of the wood varies with a number of physical parameters such as moisture content, density, grain direction, mineral content, and wood temperature. Among these parameters, moisture content is the most significant, as water has a high value of dielectric permittivity substance attenuating microwave.

To determine the exact strength of heat buildup, Antti [6] provided a method by using the temperature rise during first few minutes,

$$\frac{\delta q}{\delta t} = \frac{\rho_d C_p \delta T}{\delta t} \tag{7}$$

where ρ_d is the dry density (dry wood mass/moist volume). During this period, all the microwave energy supplied to the board was used to increase the specimen temperature. The drying process only took place once the temperature was high enough for the water evaporation to take place in the wood and the vapor was transferred from the board surface.

1.1.3. Moisture Diffusion Coefficients

Jia and Afzal [7] developed an analytical approach to determine the moisture diffusion coefficient for slab shaped wood subjected to microwave drying. The logarithm of moisture ratio against time is regressed by means of the least squares curve fitting method in the following form:

$$\ln(\overline{E}) \approx \ln\left(\frac{2L'^2}{\beta_n^2\left(\beta_n^2 + L'^2 + L'\right)}\right) - \frac{\beta_n^2 D}{a^2} t, \quad L' < 5.0 \tag{8}$$

where the moisture ratio is defined as:

$$\overline{E} = \frac{\overline{MC} - MC_e}{MC_0 - MC_e}, \; L' = S.a / D, \; and \; \beta_n \tan \beta_n = L' \tag{9}$$

$\overline{MC}$ is the average moisture content across the specimen thickness, S is the surface emission coefficient, MC_e is the equilibrium moisture content, MC_0 is initial moisture content, a is the half-thickness of the specimen and L' indicates a magnitude of the internal and external resistances to the moisture loss from the wood to air or airflow during drying. The moisture diffusion coefficient can be obtained from the regression of the drying curve.

The diffusion coefficient is temperature dependent, which is, in turn, affected by the power level. This relationship can be expressed by Arrhenius equation [8]:

$$D = B\exp\left(\frac{-E_b}{R.T}\right) \quad (10)$$

where B is constant in m^2/s, R is gas constant, E_b is the activation energy at a moisture content MC for bound water and vapor diffusion, and T is temperature in Kelvin.

Two regressed equations of diffusion coefficients for red maple were obtained as a function of moisture content and power level:

- For longitudinal direction:

$$D = 3.60\times10^{-4}\exp\left(\frac{-4066}{T}\right).(MC)^{1.17} \quad (11)$$

- For transverse direction:

$$D = 1.17\times10^{-5}\exp\left(\frac{-4066}{T}\right).(MC)^{1.53} \quad (12)$$

Similarly, the diffusion coefficient of white oak was proposed as:

- For longitudinal direction:

$$D = 2.12\times10^{-3}\exp\left(\frac{-4155}{T}\right).(MC)^{0.49} \quad (13)$$

- For transverse direction:

$$D = 8.8\times10^{-5}\exp\left(\frac{-4155}{T}\right).(MC)^{0.8} \quad (14)$$

These equations work satisfactorily for low power levels, with maximum temperature of specimen below 100°C.

1.2. Diffusion Model During Falling Rate Period

For small and large values of γ, Sherwood [9] provided models for the falling-rate period over 75 years ago. Where γ is the drying intensity. Sherwood's models for the falling-rate period do not cover intermediate values of the drying rate, when the moisture distribution at the beginning of the falling-rate period is neither uniform nor parabolic. Y.T. Puyate and C.J. Lawrence [10] presented an exact diffusion model for the moisture distribution during the falling rate period of drying of solids that is valid at all drying rates. The exact model is valid for all values of γ and hence provides a missing link between Sherwood's models at moderate drying intensity.

The equation proposed by Sherwood to govern mass transport during drying of solids is given in dimensionless form as [11, 12]:

$$\frac{\partial \theta}{\partial \tau} = \frac{\partial^2 \theta}{\partial Y^2} \tag{15}$$

The dimensionless variables are given by:

$$\theta = \frac{\rho_m - \rho_{eq}}{\rho_0 - \rho_{eq}}, \quad Y = \frac{y}{(a)}, \quad \tau = \frac{tD}{(a)^2}, \tag{16}$$

where ρ_m is the moisture concentration (in mass units), ρ_0 is the initial uniform moisture concentration, ρ_{eq} is the equilibrium moisture concentration, D is the constant moisture diffusion coefficient, y is the distance measured in a direction normal to the drying surface, a is the thickness of a flat plate that is dried from only one exposed face (or half thickness for a slab that is dried from both sides), and t is time.

1.2.1. Constant Rate Period

With the assumption of a uniform moisture distribution at the start of drying, Gilliland and Sherwood [13] obtained the solution to equation 15 for the constant-rate period as:

$$\theta_1(Y,\tau) = 1 - \gamma\left[\tau + \frac{Y^2}{2} - \frac{1}{6} - \frac{2}{\pi^2}\sum_{n=1}^{\infty}\frac{(-1)^n}{n^2}\exp\left(-n^2\pi^2\tau\right)\cos(n\pi Y)\right], \tag{17}$$

which they applied to the drying of hemlock wood.

The parameter $\gamma = \dfrac{\alpha(a)}{(\rho_0 - \rho_{eq})D} = \dfrac{((a)^2/D)}{(a(\rho_0 - \rho_{eq})/\alpha)}$ = (time for diffusion)/ (time for evaporation) is the drying intensity (dimensionless drying rate) which relates the characteristic times for evaporation and liquid diffusion, where α is the rate of drying per unit area in the constant-rate period.

Sherwood [9] presented a model that describes the moisture distribution during drying of an infinite slab on the condition that no constant-rate period occurs, and the falling rate period is controlling from the start with the initial uniform moisture concentration of the body. This model is also appropriate if the drying rate is very fast such that the constant-rate period is negligible, and the moisture distribution at the beginning of the falling-rate period is approximately the initial uniform moisture concentration of the body. In some materials such as heartwood whose moisture content is normally close to the irreducible moisture content where capillary-driven flow ceases due to discontinuity in the liquid flow path, the constant-rate period is very short and may be neglected [14].

1.2.2. The End of Constant Rate Period

It is assumed that the constant-rate period ends at a dimensionless critical time $\tau = \tau_{cr}$ when the moisture content at the drying surface has dropped to the equilibrium value corresponding to the humidity of the atmosphere [13, 15]. This assumption is justified by the fact that the evaporation rate can remain constant even when dry patches appear on the surface of a body [16]. Equation 17 provides a relationship between γ and τ_{cr} in the form:

$$\gamma = \left[\tau_{cr} + \frac{1}{3} - \frac{2}{\pi^2} \sum_{n=1}^{\infty} n^{-2} \exp(-n^2 \pi^2 \tau_{cr}) \right]^{-1}, \tag{18}$$

which may be approximated when γ is large by [15]:

$$\tau_{cr} \approx \pi / \left(4\gamma^2\right) \tag{19}$$

Or for small γ by:

$$\tau_{cr} \approx \frac{1}{\gamma} - \frac{1}{3} \tag{20}$$

The total quantity of moisture, Θ, in the body at any time during the constant-rate period is obtained by integrating equation 17 across the thickness as:

$$\Theta_1(\tau) = 1 - \gamma\tau \tag{21}$$

1.2.2. The Falling Rate Period

Equation 15 can be solved to obtain the exact model for the moisture distribution during the falling-rate period, which is valid for all values of γ, in dimensionless form as:

$$\theta_2(Y, s) = \frac{4}{\pi} \sum_{m=1}^{\infty} H_m \exp\left[-(2m-1)^2 \pi^2 s / 4\right] \times \cos\left[(2m-1)\pi Y / 2\right], \tag{22}$$

where the coefficients H_m are given by

$$H_m = \frac{(-1)^m}{(2m-1)}\left\{\gamma\left[\tau_{cr} + \frac{1}{3} - \frac{4}{\pi^2(2m-1)^2} - \frac{2(2m-1)^2}{\pi^2}\sum_{n=1}^{\infty}\frac{\exp(-n^2\pi^2\tau_{cr})}{n^2\left[(2m-1)^2 - 4n^2\right]}\right] - 1\right\} \quad (23)$$

and $s = \tau - \tau_{cr}$ is the dimensionless time from the inception of the falling-rate period. The total quantity of moisture in the body at any time during the falling-rate period is obtained by integrating equation 22 across the thickness as:

$$\Theta_2(s) = -\frac{8}{\pi^2}\sum_{m=1}^{\infty}\frac{(-1)^m}{(2m-1)}H_m \exp\left[-(2m-1)^2\pi^2 s/4\right] \quad (24)$$

1.2.3. Sherwood's Models for the Fallinf Rate Period

- High drying rate

When the drying rate is very fast $(\gamma \succ\succ 1)$, the constant-rate period is negligible $(\tau_{cr} \to 0)$, and the moisture distribution at the beginning of the falling-rate period is approximately the initial uniform moisture concentration of the body (i.e. $\theta_1(Y,\tau_{cr}) \approx 1$). Equation 15 can then be solved using the initial condition for the constant-rate period and the boundary conditions for the falling-rate period, to obtain Sherwood's [35] high drying rate model in dimensionless form as:

$$\Theta_2(\tau) = \frac{8}{\pi^2}\sum_{m=1}^{\infty}(2m-1)^{-2}\exp\left[-(2m-1)^2\pi^2\tau/4\right] \quad (25)$$

- Low drying rate

When the drying rate is slow ($\gamma \prec\prec 1$), the constant-rate period is not negligible because τ_{cr} is large, and the series in equation 17 may be neglected after a short initial transient to obtain the parabolic concentration profile assumed by Sherwood to exist at the beginning of the falling-rate period. Accordingly, the exponential terms in equation 23 approach zero for large τ_{cr} , and the resulting expression for H_m can be combined with equations 20 and 24 to obtain the dimensionless form of Sherwood's [9] low drying rate model as:

$$\Theta_2(s) = \frac{32\gamma}{\pi^4}\sum_{m=1}^{\infty}(2m-1)^{-4}\exp\left[-(2m-1)^2\pi^2 s/4\right] \quad (26)$$

1.3. Effect of Moisture Content on Diffusion and Surface Emission Coefficients

For moisture diffusion in wood, some authors have reported a strong moisture dependence of the diffusion coefficient [17-21] while others have taken the diffusion coefficient as constant [22-25]. The surface emission coefficient, defined as a constant of proportionality [26] has been reported to depend on the ambient or environmental conditions either completely [22, 27] or in combination with material surface conditions [19, 28-30]. In addition, some authors have chosen to set the surface emission coefficient equal to infinity in analyzing their test data [31, 32]. Jen Y. Liu and William T. Simpson [33] determined whether the transport coefficients are constant or not. The closed-form solutions of the diffusion equation with constant transport coefficients were used to derive the conditions under which the transport coefficients can be taken as constant. The diffusion coefficient was assumed to be finite, but the surface emission coefficient could be either finite or infinite.

1.3.1. Basic Equations and Analyses

In a one-dimensional formulation with the diffusing substance moving in the direction normal to a sheet of medium of thickness 2a, the diffusion equation can be written as:

$$\frac{\partial C}{\partial t} = \frac{\partial}{\partial X}\left(D\frac{\partial C}{\partial X}\right) \quad (0 < X < a, t > 0) \tag{27}$$

where C is concentration of the diffusing substance, t is time, D is diffusion coefficient assumed to be constant, and X is space coordinate measured from the center of the sheet.

The boundary condition is:

$$D = \frac{\partial C}{\partial X} = S(C_{eq} - C) \quad (X \approx a, t > 0) \tag{28}$$

Equation 27 applies to sorption and desorption. When sorption defines diffusing substance entering the medium and desorption defines diffusing substance leaving the medium [26].

$$\text{Let } c = \begin{cases} (C - C_0)/(C_{eq} - C_0) & sorption \\ (C - C_{eq})/(C_0 - C) & desorption \end{cases}, \quad \tau = Dt/a^2, and\ x = X/a \tag{29,30}$$

Equation 27 can be put in dimensionless form in the same order as previously presented:

$$\frac{\partial c}{\partial \tau} = \frac{\partial^2 c}{\partial x^2} \quad (0 < x < 1, \tau > 0) \tag{31}$$

1.3.2. Mathematical Development

- Sorption

Let

$$M(\tau)=\int_0^{1-\varepsilon} c(\tau,x)dx \quad (\varepsilon \to 0) \tag{32}$$

The fraction of total concentration in the medium, E, is [26]:

$$E=\begin{cases} \dfrac{M(\tau)}{M(\infty)}=1-\dfrac{8}{\pi^2}\sum_{n=0}^{\infty}\dfrac{1}{(2n+1)^2}\exp\left[-\left(n+\dfrac{1}{2}\right)^2\pi^2\tau\right] & S=\infty \\ 1-2L'^2\sum_{n=0}^{\infty}\dfrac{\exp(-\beta_n^2\tau)}{\beta_n^2(\beta_n^2+L'^2+L')} & S\neq\infty \end{cases} \tag{33,34}$$

In which β_n are the positive roots of

$$\beta_n \tan\beta_n = L' \tag{35}$$

Which can be expressed in closed form as follows [34, 35]:

$$\tau=\begin{cases} \dfrac{\pi E^2}{4} \quad E\le 0.52, & (36) \\ & S=\infty \\ \dfrac{4}{\pi^2}\ln\left[\dfrac{\pi^2(1-E)}{8}\right] \quad E>0.52, & (37) \\ & (38) \\ \dfrac{\pi E^2}{4}-\dfrac{1}{L'}\ln(1-E) \quad E\le 0.52, & \\ & S\neq\infty \\ -\dfrac{4}{\pi^2}\ln\left[\dfrac{\pi^2(1-E)}{8}\right]-\dfrac{1}{L'}\ln(1-E) \quad E>0.52 & (39) \end{cases}$$

- Desporption

From the dimensionless concentration defined for sorption and for desorption we obtain:

$$E+E'=1 \tag{40}$$

Where E is defined in equations 33, 34 for sorption and E' is the corresponding quantity for desorption. We therefore halve

$$E' = \begin{cases} = \dfrac{M(\tau)}{M(0)} = \dfrac{8}{\pi^2}\sum_{n=0}^{\infty}\dfrac{1}{(2n+1)^2}\exp\left[-\left(n+\dfrac{1}{2}\right)^2\pi^2\tau\right] & S=\infty \quad (41) \\ 2L'^2\sum_{n=1}^{\infty}\dfrac{\exp\left(-\beta_n^2\tau\right)}{\beta_n^2\left(\beta_n^2+L'^2+L'\right)} & S\neq\infty \quad (42) \end{cases}$$

with β_n defined the same as before.

Closed-form solution is:

$$\tau = \begin{cases} \dfrac{\pi(1-E')^2}{4} \quad E' \geq 0.48, & (43) \\ -\dfrac{4}{\pi^2}\ln\left[\dfrac{\pi^2 E'}{8}\right] \quad E' < 0.48, & S=\infty \quad (44) \\ \dfrac{\pi(1-E')^2}{4} - \dfrac{1}{L'}\ln E' \quad E' \geq 0.48, & (45) \\ -\dfrac{4}{\pi^2}\ln\left[\dfrac{\pi^2 E'}{8}\right] - \dfrac{1}{L'}\ln E' \quad E' < 0.48 & S\neq\infty \quad (46) \end{cases}$$

As demonstrated in figure 1 a strong dependence of S on C_e is found. It has been suggested that S may also be dependent on C_0. According to the present observation, if such a dependence does exist, it must be very weak. The results in figure 1 are in qualitative agreement with those reported by Avramidis and Siau [22].

The diffusion coefficient D appears to be dependent on C_0 as shown in figure 2, which is not unexpected in an isothermal desorption environment. As C_0 approaches the fiber saturation point, D tends to level off toward a constant value. For a relatively large equilibrium moisture content value (EMC), the close agreement could extend well below a moisture fraction of 50%. This suggests that the applicability of the diffusion equation with constant transport coefficients, D and *S*, depends very much on the magnitude of EMC and, to a somewhat lesser degree, on the initial moisture content, as can be seen from figures 1 and 2. Under the test conditions considered, the common range of applicability defined by the fraction of total moisture content in wood is about 50% and above.

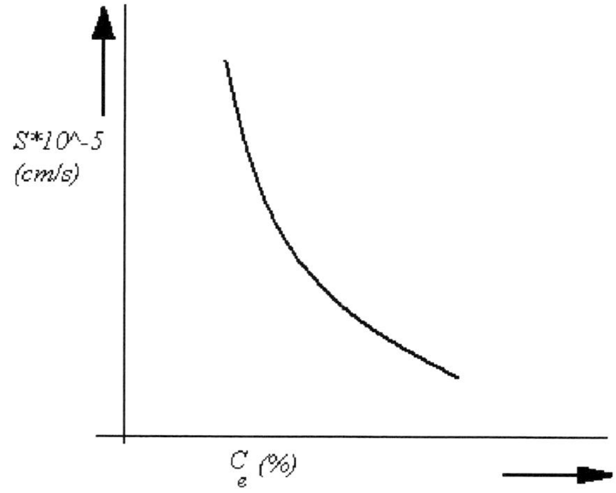

Figure 1. Dependence of surface emission coefficient on equilibrium moisture content.

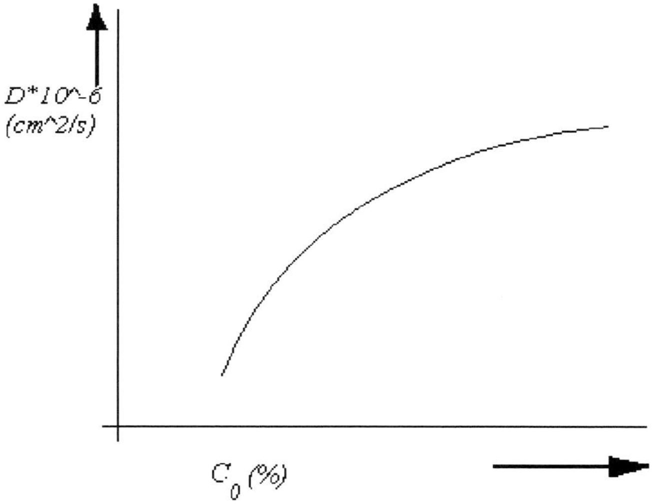

Figure 2. Dependence of diffusion coefficient on initial moisture content.

The desorption data of northern red oak showed that when the moisture fraction in wood is above a certain value, the transport coefficients can be taken as constant; below this value, they cannot. Where they can be taken as constant, the diffusion coefficient is found to be dependent on the initial moisture content in wood and the surface emission coefficient on the equilibrium moisture content in the environment. The closed-form solutions of the diffusion equation and the conditions derived there from for constant transport coefficients are valid for the whole range of diffusion, which renders unnecessary the half-time technique applied for several decades.

2. HIGH TEMPERATURE DRYING

2.1. Effect of Dimensionless Parameters on Heat and Moisture Transfer

High temperature heat treatment of wood is a complex process involving simultaneous heat, mass and momentum transfer phenomena and the effective models are necessary for process design, optimization, energy integration, and control. Younsi. R. et al. [36] presented numerical simulation of three dimensional heat and moisture transfer in wood. The physical problem involves a three-dimensional capillary porous sample (wood) in Cartesian coordinates, initially at uniform temperature and uniform moisture content. The boundaries are in contact with the hot surrounding gas, thus resulting in a convection boundary condition for both the temperature and moisture potential. The mathematical model is based on a set of coupled, heat and mass transfer equations proposed by Luikov [37]. Numerous series of calculations were carried out over the range of 0.01 < Lu < 0.1, 2 < K_0 < 30, 0.2 < Pn < 2, 10 < Bi_m < 50, 5 < Bi_q < 50. Where Lu, Pn, k_0, Bi_m *and* Bi_q are the Luikov , Posnov, Kossovitch, mass and heat transfer Biot numbers, respectively.

2.1.1. Governing Equation

The equations describing three-dimensional heat and moisture transfer in wood during heat treatment are given as follows [38, 39]:

- Heat transfer

$$\rho_d C_q \frac{\partial T}{\partial t} = \vec{\nabla}.\left[\left(k + \frac{\eta\lambda k_m \delta}{C_m}\right)\vec{\nabla}T + \eta\lambda k_m \vec{\nabla}U\right] \tag{47}$$

- Moisture transfer

$$\rho_d C_m \frac{\partial U}{\partial t} = \vec{\nabla}.\left[k_m \vec{\nabla}U + \left(\frac{k_m \delta}{C_m}\right)\vec{\nabla}T\right] \tag{48}$$

where the parameters C_q, k, η, λ, k_m, δ, C_m and U are the heat capacity, thermal conductivity coefficient, ratio of vapor diffusion coefficient to coefficient of total moisture diffusion, latent heat, moisture conductivity coefficient, thermo gradient coefficient, moisture capacity and moisture potential respectively.

The moisture potential U is related to the moisture content MC by:

$$MC = C_m.U \tag{49}$$

- Boundary conditions

$$-k\frac{\partial T}{\partial n}=h_q\left(T-T_a\right)+\left(1-\eta\right)\lambda h_m\left(U-U_a\right)\quad at\ \Omega \tag{50}$$

$$-k_m\frac{\partial U}{\partial n}=\left(\frac{k_m\delta}{C_m}\right)\frac{\partial T}{\partial n}+h_m\left(U-U_a\right)\quad at\ \Omega \tag{51}$$

where h_q is convective heat transfer coefficient, h_m is convective mass transfer coefficient, T_a is temperature at ambient condition, U_a is moisture potential, Ω is boundary surface of the computational domain.

2.1.2. Dimensionless Formulation

First some parameters should be defined:

- Fourier number $F_0=\frac{a_q t}{d^2}$
- Dimensionless temperature $\vartheta=\frac{T-T_0}{T_a-T_0}$
- Dimensionless moisture content $\Phi=\frac{\left(U_0-U\right)}{\left(U_0-U_a\right)}$
- Luikov number $Lu=\frac{a_m}{a_q}$
- Kossovitch number $K_0=\frac{\lambda C_m\left(U_0-U_a\right)}{C_q\left(T_a-T_0\right)}$
- Posnov number $Pn=\frac{\delta\left(T_a-T_0\right)}{C_q\left(U_0-U_a\right)}$
- Mass transfer Biot number $Bi_m=\frac{h_m d}{k_m}$
- Heat transfer Biot number $Bi_q=\frac{h_q d}{k}$

Then equations 48-51 can be written in dimensionless form as [40]:

$$\frac{\partial\vartheta}{\partial F_0}=\vec{\nabla}.\left[\left(1+\eta K_0 Lu\,Pn\right)\vec{\nabla}\vartheta-\eta K_0\,Lu\vec{\nabla}\Phi\right] \tag{52}$$

$$\frac{\partial\Phi}{\partial F_0}=\vec{\nabla}.\left[Lu\,\vec{\nabla}\Phi-Lu\,Pn\vec{\nabla}\vartheta\right] \tag{53}$$

In view of this consideration, the material balance for heat and moisture at the surface can be written as:

$$\frac{\partial \vartheta}{\partial n} = Bi_q(1-\vartheta) - (1-\eta)K_0\, Lu\, Bi_m(1-\Phi) \tag{54}$$

$$\frac{\partial \Phi}{\partial n} = Pn\frac{\partial \vartheta}{\partial n} + Bi_m(1-\Phi) \tag{55}$$

The objective is to determine the average dimensionless temperature and moisture content fields $\Phi(X,Y,Z,\tau)$, $\vartheta(X,Y,Z,\tau)$ respectively in wood sample. The numerical calculation was performed with FEMLAB software [41]. This is designed to simulate systems of coupled non-linear and time dependent partial differential equations (PDE) in one-, two-, or three-dimensions. The FEMLAB software operates in the MATLAB environment [42]. The simulation domain consisted of approximately 1466 nodes and 6580 elements.

The main conclusions of their research are as follows:

- The Luikov number affects both the heat and mass transfer. Luikov number is characterized the inertial properties of the transfer field of mass in comparison with the inertial properties of temperature field. Average temperature and moisture increase as a result of heating (note when the dimensionless moisture content increases, means that the dimensional moisture potential of the sample wood decreases). The heat supplied at the surfaces is used in the change of phase of the moisture from the liquid to the vapor at the surface. The moisture migrates toward the exchange surfaces by means of capillary, under the influence of a moisture content gradient. The temperature and moisture content converge to $\Phi = 1$, $\vartheta = 1$, which correspond to the fixed ambient conditions.
- The Kossovitch number affects strongly the heat transfer and has a slight effect on moisture content. The Kossovitch number represents the relation between the heat consumed on moisture evaporation and on heating of the moist body.
- The Posnov number affects only the transfer of moisture potential. Pn denotes the ratio of the thermal- moisture diffusivity to moisture diffusivity.
- The mass and heat transfer Biot numbers affect only mass transfer and heat transfer, respectively.

2.2. Computational Modeling of Heat and Mass Transfer During High Temperature Drying

Younsi. R. [43] conducted a three-dimensional mathematical model for high temperature thermal treatment of wood. Three moisture phases were accounted for: free (liquid) water, bound water and water vapor, which is the advantage of their model. The three-dimensional Navier–Stokes equations had been solved for the flow field, which would enable to study the conjugate drying (heat treatment) process for any three-dimensional shape. The problem considers a sample of wood exposed to convective heating in an inert atmosphere. A comparison with experimental results was performed under different operating conditions. The comparison of the numerical and experimental results implies that the proposed numerical algorithm can be used as a useful tool in designing high-temperature wood treatment processes. In the formulation of the mathematical mode, the main transport mechanisms have been identified and numerically expressed. The governing equations were solved numerically with the commercial package, FEMLAB [41]. The equations were written in partial differential form in line with program definitions, and initial and boundary conditions were determined. The mesh convergence was verified with refined mesh sizes. Time step of 1000 s, 1866 nodes and 8080 elements mesh size were considered to be appropriate. Comparison of results for temperature and moisture content profiles calculated by the model with those obtained from the experimental investigation has shown good agreement. There is some difference between the predicted and the measured results for moisture content at the end of the process. These might be due to chemical reactions taking place in the wood during the thermal treatment. Therefore, incorporating these effects into the model might decrease the discrepancies observed. The results show that the lower the higher gas velocity is, the higher the temperature is during treatment. The time necessary for moisture removal is greater if the gas velocity is lower, when all other treatment conditions are kept the same.

2.2.1. Governing Equations for Solid Wood

The equations describing three-dimensional heat and moisture transfer in wood during heat treatment are given as follows [44, 45].

- Mass balance

$$-\rho_d \frac{\partial(MC)}{\partial t} = \nabla\left(D_M \nabla MC + D_T \nabla T\right) \tag{56}$$

$$D_M = \underbrace{\frac{1.2146\times10^{-4} m_V \zeta(T)^{0.75}}{R} p_{SV} \frac{\partial\Psi}{\partial(MC)}}_{VAPOR\ \ TERM} + \underbrace{8.314\frac{D_b}{m_V}\frac{T}{\Psi}\frac{\partial\Psi}{\partial(MC)}}_{BOUND\ \ TERM} + \underbrace{\frac{\alpha.\beta.K_l \rho_l \left(MC_{\max} - MC_{FSP}\right)^{\beta}}{\mu_l \left(MC - MC_{FSP}\right)^{(1+\beta)}}}_{FREE\ \ WATER\ \ TERM} \tag{57}$$

$$D_T = \underbrace{\frac{1.2146\times10^{-4} m_V \zeta(T)^{0.75}}{R}\left(\frac{\partial p_{SV}}{\partial T}\Psi + p_{SV}\frac{\partial\Psi}{\partial T}\right)}_{VAPOR\ \ TERM} -$$
$$\underbrace{\frac{D_b}{m_v}.\left\{\left[187 + 35.1\ln\left(\frac{T}{298.15}\right) - 8.314\ln\left(\frac{p_V}{101325}\right)\right] - 8.314\frac{T}{P_V}\left(\frac{\partial p_{SV}}{\partial T}\Psi + p_{SV}\frac{\partial\Psi}{\partial T}\right)\right.}_{BOUND\ \ TERM} \tag{58}$$

- Energy balance

$$\left(\rho_d C_q\right)_{eff} \frac{\partial T}{\partial t} = \nabla\left(k_M \nabla(MC) + k_{eff}\nabla T\right) \tag{59}$$

$$\left(\rho_d C_q\right)_{eff} = \left(\rho_d C_q\right)_d + \left(\rho_d C_q\right)_V + \left(\rho_d C_q\right)_b + \left(\rho_d C_q\right)_f \tag{60}$$

$$k_M = \underbrace{h'_V \frac{1.2146\times10^{-4} m_V \zeta(T)^{0.75}}{R} p_{SV} \frac{\partial \Psi}{\partial (MC)}}_{VAPOR\,TERM} + \underbrace{h'_b 8.314 \frac{D_b}{m_V} \frac{T}{\psi} \frac{\partial \Psi}{\partial (MC)}}_{BOUND\,TERM} + \underbrace{h'_f \frac{\alpha.\beta.K_l \rho_l \left(MC_{\max} - MC_{FSP}\right)^{\beta}}{\mu_l \left(MC - MC_{FSP}\right)^{(1+\beta)}}}_{FREE\,WATER\,TERM} \tag{61}$$

$$k_{eff} = k + \underbrace{h'_V \frac{1.2146\times10^{-4} m_V \zeta(T)^{0.75}}{R}\left(\frac{\partial p_{SV}}{\partial T}\Psi + p_{SV}\frac{\partial \Psi}{\partial T}\right)}_{VAPOR\ \ TERM} - \tag{62}$$

$$\underbrace{h'_b \frac{D_b}{m_V}.\left\{\left[187 + 35.1\ln\left(\frac{T}{298.15}\right) - 8.314\ln\left(\frac{p_V}{101325}\right)\right] - 8.314\frac{T}{p_V}\left(\frac{\partial p_{SV}}{\partial T}\Psi + p_{SV}\frac{\partial \Psi}{\partial T}\right)\right\}}_{BOUND\ \ TERM}$$

2.2.2. Governing Equation for the Flow Field

The three-dimensional Navier–Stokes, energy and concentration equations are considered as follows [44, 45].

- Continuity

$$\nabla . v = 0 \tag{63}$$

- Momentum

$$\frac{\partial v}{\partial t} + (v.\nabla)v = -\frac{1}{\rho_f}\nabla p + v\nabla^2 v + g\beta\left(T - T_g\right) + g\beta'\left(C - C_g\right) \tag{64}$$

- Energy equation

$$\frac{\partial T}{\partial t} + (v.\nabla)T = \frac{k_f}{\rho_f C_{pf}}\nabla^2 T \tag{65}$$

- Concentration equation

$$\frac{\partial C}{\partial t} + (v.\nabla)C = D\nabla^2 C \tag{66}$$

2.3. Effect of High Temperature Drying on Heat and Mass Transfer Coefficients

Haque, N. described the analysis of heat and mass transfer coefficients for a single board of Pinus radiata timber over a range of high temperature and superheated steam drying conditions[46]. Heat and mass transfer coefficient were calculated and then the differences between measured and predicted values were discussed. Results showed that there was no significant relationship between the heat transfer coefficient and moisture content. The calculated heat transfer coefficients were significantly higher than the values derived from theoretical correlations. The mass transfer coefficients determined were significantly higher and more variable than the values derived from commonly used dimensionless groups and correlations used in chemical engineering applications. Possible reasons for this include measurement errors in harsh drying conditions, non-uniform flow, porous wood structure with pit aspiration feature, the presence of a dry shell, and the roughness of board surfaces compared with other processes.

The general form of the transfer equation used in drying models is:

Flux (rate of moisture transfer) = coefficient * driving force

Here, the coefficient is the heat or mass transfer coefficient. The driving force for heat transfer is the temperature difference between the assumed boundary layer and the bulk drying medium inside the kiln. The driving force for the mass transfer coefficient is either the vapor pressure difference or concentration difference between the boundary layer and the bulk.

2.3.1. Heat Transfer Coefficients

From energy balance analyzes it is evident that if a wet board is surrounded by hot kiln air, the heat will be transferred to the wood, which in turn will heat up and evaporate any moisture present in the wood. If only one face is exposed, and both edges, ends, and the opposite face are insulated from their surroundings, then:

$$Q = h.(T_{\infty} - T_{surf}).A..\Delta t = \Delta H_{wv}.N_V.A.\Delta t + C_{PS}.V.\rho_{surf}.\Delta T - Q_e \tag{67}$$

here N_V is the flux in mol $m^{-2}s^{-1}$. Although the bottom face and all edges of the sample were heat insulated, it was recognized that a small amount of heat could still be transferred from the insulation material. The amount of heat was estimated by:

$$Q_e = \Delta t.\frac{k_q}{a}.(T_{out} - T_{in}).\sum_{i=1}^{5} A_i \tag{68}$$

where k_q is the thermal conductivity of the insulation material, a is the thickness of the insulation material (half-thickness of board), T_{out} and T_{in} are the outside and inside temperatures of the insulation wall, and A_i is the area of the bottom surface and the four insulated edges.

The external heat transfer coefficient can be determined by:

$$h = \left[\frac{\Delta H_{wv}.N_V.A + C_{PS}.V.\rho_S.\Delta T / \Delta t - k_q / a.(T_{out} - T_{in}).\sum_{i=1}^{5} A_i}{(T_\infty - T_{surf}).A} \right] \quad (69)$$

The latent heat of water vaporization (ΔH_{wv}) decreases with temperature and can be calculated by [47]:

$$\Delta H_{wv} = 3.29 \times 10^5 - 2806.6.T \quad (70)$$

here, T is wood temperature (K).

The specific heat of wood includes that of bone-dry wood and that of water; thus:

$$C_{PS} = 4184.(0.324 + (MC)) \quad (71)$$

The wood drying model developed uses a heat transfer coefficient that is a strong function of velocity but weakly correlated (negatively) with dry-bulb temperature [48]. In general, for chemical engineering applications, heat transfer coefficients are estimated from the correlations and dimensionless groups such as Reynolds, Nusselt, and Prandtl numbers and using commonly found correlations between them for parallel flow over a flat plate [49]. This flow condition resembles the drying of timber boards.

The correlations are as follows and are valid for Reynolds numbers up to 10^5 and Prandtl numbers up to 0.7

$$Nu = 0.664.\mathrm{Re}^{0.5}.\mathrm{Pr}^{0.3} \quad (72)$$

$$\mathrm{Re} = \frac{v_{air} L}{\mu} \quad (73)$$

$$\mathrm{Pr} = \frac{C_{p\ air}\mu_{air}}{K_{air}} \quad (74)$$

$$h = \frac{Nu \times K_{air}}{d} \quad (75)$$

Past studies have also shown that drying curves predicted by the model based on these commonly found correlations do not agree very well in the case of drying wood [50]. Possible reasons for this include measurement errors in harsh drying conditions, non uniform flow, porous wood structure with pit aspiration feature, the presence of a dry shell, and the roughness of board surfaces compared with other processes.

2.3.2. Mass Transfer Coefficient

The mass transfer coefficient, based on the absolute humidity difference was calculated using the following equation:

$$k_{ma} = \frac{N_V}{\omega_S - \omega_g} \tag{76}$$

ω is the absolute humidity and can be expressed as:

$$\omega = \frac{\ell}{1-\varpi} \tag{77}$$

where ℓ is the density ratio or the molar mass ratio of water vapor relative to dry air, ϖ_G is the mole fraction of moisture vapor in the bulk air ($\varpi_g = p_g / P$); and y_s is the mole fraction adjacent to the surface ($\varpi_S = p_S / P$). p_S and p_G are the partial vapor pressure of the surface and bulk air, respectively, and P is the total pressure of the system. Absolute humidity was calculated using an equation that requires an estimate of partial pressure [51]. This was estimated using an equation proposed by Kayihan [52, 53] for high-temperature conditions up to the critical point of water (374°C) and that was also used elsewhere[54]:

$$p_S = \frac{10^{f(T)}.P}{760} \tag{78}$$

$$f(T) = 16.373 - \frac{2818.6}{T} - 1.6908\log T - 5.7546\times10^{-3}T + 4.0073\times10^{-6}T^2 \tag{79}$$

The mass transfer coefficient, based on the partial vapor pressure difference was calculated using the following equation:

$$k_{mp} = \frac{N_V}{p_S - p_g} \tag{80}$$

3. Mass Transport Model under Isothermal Conditions

Eriksson, J. et al. [55] analyzed various transport- based models and described a model to the drying of Swedish softwoods. Conservation equations for each of the moisture phases were considered separately and the transport laws of the different phases were employed.

Finally, simulations of the drying of a Norway spruce sample under isothermal conditions are carried out using the FEM numerical technique.

3.1. Mass Conservation

Mass conservation equations for the three phases of moisture in local form are summarized in equations 81– 83:

- Water vapor:

$$\frac{\partial}{\partial t}(\phi_g \rho_V) = -div(\rho_V v_V) + \dot{m}_{WV} + \dot{m}_{bV} \tag{81}$$

- Bound water:

$$\frac{\partial}{\partial t}(\phi_s \rho_b) = -div(\rho_b v_b) + \dot{m}_{bV} + \dot{m}_{wb} \tag{82}$$

- Free water:

$$\frac{\partial}{\partial t}(\phi_w \rho_w) = -div(\rho_w v_w) - \dot{m}_{wv} - \dot{m}_{wb} \tag{83}$$

where the velocity of the transported quantity is denoted by v_i, ρ_i is the density, and $\dot{m}_{ij}$ denotes the transition from phases i and j. From here on, the subscripts w, b, v, and s refer, respectively, to free water, bound water, water vapor, and the solid skeleton of wood. Denoting the total volume by V and the volume of the phase i by Vi, the volumetric fraction of this phase is:

$$\phi_i = \frac{V_i}{V} \tag{84}$$

with the geometrical constraint:

$$\phi_g + \phi_s + \phi_w = 1 \tag{85}$$

The total water flux consists of three parts as:

$$j = j_V + j_b + j_W = \rho_V v_V + \rho_b v_b + \rho_W v_W \tag{86}$$

3.2. Liquid Water Transport

The free water flow in the wood can then be described as given by Pang et al. [56]:

$$j_W = \frac{\rho_W K_W K_{rW}}{\mu_W} \nabla p_c \tag{87}$$

where K_W and K_{rW} are the absolute and relative permeability, respectively, while the viscosity of water is denoted by μ_W.

The capillary pressure is modeled by a function dependent only on the degree of saturation as:

$$p_c = p_g - p_l = A.S'^{-B} \tag{88}$$

where A and B are constants.

The saturation S' is defined as the amount of free water divided by the maximum amount of free water and can be expressed as a function of MC as:

$$S' = \begin{cases} \dfrac{MC - MC_{fsp}}{MC_{\max} - MC_{fsp}}, & MC \geq MC_{fsp} \\ 0, \quad MC < MC_{fsp} \end{cases} \tag{89}$$

where the maximum moisture content at which the entire structure is filled with water (full saturation) is given by:

$$MC_{\max} = \frac{\phi_{VS} \rho_W}{\rho_{wood}} + MC_{fsp} \tag{90}$$

The dry wood void fraction is given by:

$$\phi_{VS} = \frac{\rho_{CW} - \rho_{wood}}{\rho_{CW}} \tag{91}$$

where ρ_{CW} is the cell wall density and ρ_{wood} is the wood basic density.

3.3. Water Vapor Transport

The total vapor flow can be obtained by extending the one-dimensional expression in Nijdam et al. [57] to:

$$j_V = -\left(\frac{\rho_V K_V K_{rV}}{\mu_V} + \frac{p_g}{RT} D_V^{eff} \phi_g\right) \nabla p_V \tag{92}$$

where μ_V is the viscosity of water vapor, K_V and K_{rV} are the absolute and relative permeability, respectively.

The gaseous void fraction of the wood is:

$$\phi_g = \begin{cases} \phi_{VS}\left(\dfrac{MC_{\max} - MC}{MC_{\max} - MC_{fsp}}\right), & MC \geq MC_{fsp} \\ \phi_{VS}, \quad MC < MC_{fsp} \end{cases} \tag{93}$$

The effective diffusivity D_V^{eff} should include an attenuation factor due to the tortuous route of the void space as compared with mere diffusion through air. According to Perre´ and Turner [58], the gaseous diffusion has the form:

$$D_V^{eff} = \delta' K_{rV} D_V \tag{94}$$

where D_V is the binary bulk diffusivity of the gas mixture and the attenuation factors are collected in a diagonal matrix δ'. To compute the vapor density, the ideal gas law is used:

$$p_V = \frac{n_V}{V\phi_g} RT = \frac{\rho_V}{\phi_g} \overline{R} T \tag{95}$$

where the gas constant R is 8.314 $(J/mol.K)$, n_V is the number of moles, and T is the temperature (K).

The vapor pressure can be calculated through:

$$p_V = \psi p_S \tag{96}$$

where p_S is the vapor partial pressure at saturation, dependent only on temperature, and ψ is the relative humidity obtained from an inverse relation of a sorption isotherm curve.

The inverse relation is given by:

$$\psi = 1 - e^{B.X^D} \tag{97}$$

where $B = -\dfrac{b_2}{T}\left(1 - \dfrac{T}{T_c}\right)^{b_1}$ and $D = b_3 T^{b_4}$. (98)

All temperatures are given in Kelvin and the critical temperature of water is denoted by T_C. Values of coefficients b_1, b_2, b_3 and b_4 are given in Avramidis [59].

3.4. Bound Water Diffusion

The bound water flux is commonly described by Fick's law as:

$$j_b = -\rho_b D_b \nabla MC_b \tag{99}$$

where in 2D and 3D simulations D_b is a matrix containing the bound water diffusion coefficients, MC_b is bound water content and ρ_b is the bound water density.

4. Experimental Work

Two types of wood samples (namely; Guilan spruce and pine) were selected for drying investigation. A programmable domestic microwave oven (Deawoo, KOC-1B4K), with a maximum power output of 1000W at 2450MHZ was used. To establish the impact of each type of heat transfer mode on drying characteristics of wood and compare it with numerical solutions samples were dried in four methods: convection drying (150°C), microwave drying (270 W), infrared drying (100% power) and combination of microwave and convection drying. The dryer was run without the sample placed in, for about 30 min to set the desired drying conditions before each drying experiment. All wood samples were dried to a moisture content of approximately 30%. At 1 minute intervals throughout the drying process (until wood reached to 30% of its moisture content) the specimens were removed, weighted and then replaced to measure their moisture content.

4.1. Analytical Approaches

The evaluation of drying rate is very useful in understanding the mechanism of moisture movement within wood as well as the transport of moisture from the wood to the surrounding air. A drying rate curve is derived by drawing tangents from the drying curve (moisture content versus time) at various points along the length and plotting the slopes of the tangents against the corresponding moisture contents. Different forms of equations namely, the Newton, the Henderson and Pabis, the Logarithmic and the Wang and Singh models (table 1) have been used to characterize the drying of wood. The regression was performed in Labfit computer program.

Table 1. Drying models fitted to experimental data

Mode Model equations	Model name References
$\overline{E} = \exp(-at)$	Newton [60]
$\overline{E} = a\exp(-bt)$	Henderson and Pabis [61]
$\overline{E} = a\exp(-bt) + c$	Logarithmic [62]
$\overline{E} = 1 + at + bt^2$	Wang and Singh [63]

Two criteria were adopted to evaluate the goodness of fit of each model, the Correlation Coefficient (r) and the Standard Error (S'').

The standard error of the estimate is defined as follows:

$$S'' = \sqrt{\frac{\sum_{i=1}^{N}(\overline{E}_{\exp,i} - \overline{E}_{\text{Pred},i})^2}{n' - n}} \quad (100)$$

where $\overline{E}_{\exp,i}$ is the experimental moisture ratio at observation i; $\overline{E}_{\text{Pred},i}$ is the predicted moisture ratio at observation i; n' is the number of observation; n is the number of constants.

To explain the meaning of correlation coefficient, we must define some terms used as follow:

$$S_1'' = \sum_{i=1}^{N}(\overline{n} - \overline{E}_{\exp,i})^2 \quad (101)$$

where, the average of the data points ($\overline{n}$) is simply given by:

$$\overline{n} = \frac{1}{n'}\sum_{i=1}^{N}\overline{E}_{\exp,i} \quad (102)$$

We also define the deviation from the fitting curve as:

$$S_2'' = \sum_{i=1}^{N}(\overline{E}_{\exp,i} - \overline{E}_{pred,i})^2 \quad (103)$$

In view of the above, the improvement (or error reduction) due to describing the data in terms of a regression model can be quantified by subtracting the two quantities. Because the magnitude of the quantity is dependent on the scale of the data, this difference is normalized to yield:

$$r = \sqrt{\frac{S_1'' - S_2''}{S_1''}} \quad (104)$$

where, r is defined as the correlation coefficient. As the regression model better describes the data, the correlation coefficient will approach unity. For a perfect fit, the standard error of the estimate will approach $S''=0$ and the correlation coefficient will approach $r=1$.

5. Summary of Results

The model equations, method of drying, type of samples, model constants, standard error and correlation coefficient of these models used for moisture ratio change with time are presented in table 2. In this table 1 represents Combined Pine,2 Combined Guilan Spruce,3 Convection Pine,4 Convection Guilan Spruce ,5 Infrared Pine,6 Infrared Guilan Spruce,7 Microwave Pine and 8 Microwave Guilan Spruce respectively. All the models gave consistently good correlation coefficient I values in the range of 0.912-0.999. This indicates that all models could satisfactorily describe the air-drying of wood samples. Among the drying models, the Logarithmic model obtained the highest r values and the lowest S'' in the different drying methods of the study. It is clear that the r and S'' values of this model were changed between 0.996 and 0.999, 0.0021 and 0.0285, respectively. Thus, this model may be assumed to present the drying behavior of wood samples.

Table 2. Estimated values of coefficients and statistical analysis for the drying models

Models	Constants	r	S''
Newton	1) a = 0.012311	0.9135	0.135
	2) a = 0.0103481	0.942	0.1149
	3) a = 0.000572	0.971	0.0709
	4) a = 0.0004903	0.975	0.067
	5) a = 0.0007559	0.9219	0.1305
	6) a = 0.0008133	0.9573	0.0897
	7) a = 0.004203	0.912	0.1409
	8) a = 0.004173	0.94705	0.1101
Henderson and Pabis	1) a = 1.3376 b = 0.0173625	0.9708	0.0839
	2) a = 1.30317 b = 0.0134948	0.9784	0.0728
	3) a = 1.16522 b = 0.0006773	0.992	0.033
	4) a = 1.15555 b = 0.0005689	0.987	0.0479
	5) a = 1.2928 b = 0.00100667	0.9651	0.091
	6) a = 1.2219 b=0.00101251	0.9831	0.05846
	7) a = 1.30713 b=0.00566565	0.959	0.0998
	8) a = 1.28789 b = 0.00533431	0.978	0.071

Table 2. (Continued).

Models	Constants	r	S''
Logarithmic	1) a = 6.0309 b = 0.0018449 c = -4.8706	0.9964	0.031
	2) a = 1.793 b = 0.0058224 c = -0.62868	0.9971	0.0273
	3) a = 1.7103 b = 0.00028705 c = -0.64877	0.999	0.0021
	4) a = 1.4464 b = 0.00030291 c = -0.38344	0.999	0.011
	5) a = 10.024 b = 0.000058979 c = -8.907	0.998	0.0215
	6) a = 1.9971 b = 0.00035778 c =-0.90351	0.999	0.00693
	7) a = 72.712 b = 0.000044397 c = -71.585	0.996	0.0285
	8) a = 1.5642 b = 0.0028875 c = -0.38822	0.999	0.0114
Wang and Singh	1) a = -0.005561 b = -0.00003064	0.9865	0.0571
	2) a = -0.006556 b = 0.000005096	0.987	0.0544
	3) a = -0.000395 b = 0.0000000282	0.998	0.017
	4) a = -0.0003498 b = 0.00000002782	0.998	0.0171
	5) a = -0.0003682 b=0.00000007161	0.9925	0.0424
	6) a = -0.0005274 b = 0.000000028	0.9961	0.0281
	7) a= -0.001879 b = -0.000003003	0.9906	0.0485
	8) a = -0.002838 b = 0.00000156	0.9845	0.061

References

[1] Simpson, W.T.; Liu, J.Y. Dependence of the water vapor diffusion coefficient of aspen on moisture content. *Wood Science and Technology,* 1991, 26, pp. 9–21.

[2] Simpson, W.T. Determination and use of moisture diffusion coefficient to characterize drying of northern red oak. *Wood Science and Technology,* 1993, 27, pp. 409–420.

[3] Jia, D. and T. Afzal, M.T. Modeling of moisture diffusion in microwave drying of hardwood, moisture diffusion in microwave drying of hardwood, *Drying Technology,* 2007, 25, pp. 449–454.

[4] Koumoutsakos, A.; Avramidis, S.; Hatzikiriakos, S.G. Radio frequency vacuum drying of wood. I. Mathematical model. *Drying Technology,* 2001, 19, pp. 65–84.

[5] Resch, H.; Lofdahl, C.A.; Simth, F.J.; Erb,A. Moisture leveling in veneer by microwave and hot air. *Forest Products Journal*, 1970, 20 (10),pp. 50–58.

[6] Antti, A.L.; Perre, P. A Microwave applicator for on line wood drying: temperature and moisture distribution in wood. *Wood Science and Technology,* 1999, 33, pp.123–138.

[7] Jia, D.; Afzal, M.T. Moisture diffusion study in microwave drying of red maple (Acer rubrum L.). *Proceedings of the Asia-Pacific Drying Conference,* 2005, 2, pp. 685–694.

[8] Siau, J.F. Wood: Influence of moisture on physical properties; VPI: Blackburg, VA, 1995.

[9] Sherwood, T.K., Application of theoretical diffusion equations to the drying of solids. *Transactions of the American Institute of Chemical Engineers,* 1931, 27, pp. 190–200.

[10] Puyate, Y.T. Lawrence, C.J. Sherwood's models for the falling-rate period: A missing link at moderate drying intensity, *Chemical Engineering Science,* 2006, 61, pp.7177 – 7183.

[11] Marinos-Kouris, D., Maroulis, Z.B., Transport properties in drying of solids. In: Mujumdar, A.S. (Ed.)*, Handbook of Industrial Drying,* Marcel Dekker, New York, 1995, vol. 1, pp. 115.

[12] Sherwood, T.K., The drying of solids I. *Industrial and Engineering Chemistry,* 1929, 21, pp.12–14

[13] Gilliland, E.R., Sherwood, T.K., The drying of solids VI. *Industrial and Engineering Chemistry,* 1933, 25, pp. 1134–1136.

[14] Nijdam, J.J., Langrish, T.A.G., Keey, R.B., A high-temperature drying model for softwood timber, *Chemical Engineering Science,* 2000, 55, pp. 3585–3598.

[15] Puyate, Y.T., Diffusion in fine tubes and pores. Ph.D. Thesis, University of London, 1999.

[16] Suzuki, M., Maeda, S., On the mechanism of drying of granular beds—mass transfer from discontinuous source. *Journal of Chemical Engineering of Japan,* 1968, 1, pp. 26–31.

[17] Hougen, C.A., McCauley, H.J. and Marshall, W.R., Jr., Limitations of diffusion equations in drying, *Trans. AIChE,* 1940, 36, pp. 183-210.

[18] Meroney, R.N. The state of moisture transport rate calculations in wood drying, *Wood Fiber l,* 1969, l, pp. 64-74.

[19] Simpson, W.T. Determination and use of moisture diffusion coefficient to characterize drying of northern red oak, *Wood Science and Technology ,* 1993, 27(6), pp. 409-420.

[20] Skaar, C. Analysis of methods for determining the coefficient of moisture diffusion in wood, *Journal of the Forest Products Research Society,* 1954, 4(6), pp. 403-410.

[21] Van Arsdel, W.B. Approximate diffusion calculations for the falling rate phase of drying, *Trans. AIChE,* 1947, 43, pp. 13-24.

[22] Avramidis, S. and Siau, J.F. An investigation of the external and internal resistance to moisture diffusion in wood, *Wood Sciencs and Technology,* 1987, 21 (3), pp.249-256.

[23] Mounji, H., Bouzon, I. and Vergnaud, J.M. Modeling the process of absorption and desorption of water in two dimensions (Transverse) in a square wood beam, *Wood Science and Technology,* 1991, 26(1), pp. 23-37.
[24] Soderstrom, 0. Salin, J.G. On determination of surface emission factors in wood drying, *Holzforschung*, 1993, 47(5), pp. 391-397.
[25] Drain, A,, Taverdet, J.L. and Vergnaud, J.M, Modeling the kinetics of moisture adsorption by Wood, *Wood Science and Technology,* 1988, 22(1) pp. 11-20.
[26] Crank, J. *The mathematics of diffusion*, Clarendon Press, Oxford, 1975, Chap. 4, 2nd ed.
[27] Wadso, L., Surface mass transfer coefficients for wood, *Drying Technology,* 1993, 1l(6) pp. 1227-1249.
[28] Liu, J.Y. and Simpson, W.T. Mathematical relationship between surface emission and diffusion coefficients, *Drying Technology,* 1996, 14, pp. 677-699.
[29] Rosen, H.N. Functional relations and approximation techniques for characterizing wood drying curves, *Wood Science*, 1982, 15(1), pp. 49-55.
[30] Simpson, W.T. and Liu, J.Y. Dependence of the water vapor diffusion coefficient of aspen wood on moisture content, *Wood Science and Technology,* 1991, 26(1), pp. 9-22.
[31] Biggerstaff , T. Drying diffusion coefficient in wood as affected by temperature, *Forest products journal,* 1965, 15(3), pp.127-133.
[32] McNamara, W.S. and Hart, C.A. An analysis of interval and average Diffusion coefficients for unsteady-state movement of moisture in wood, *Wood Science,* 1971, 4(l),pp. 37-45.
[33] Liu, J. Y. And Simpson, W. T. Solutions of diffusion equation with constant diffusion and surface emission coefficients, *Drying Technology,* 1997, 15(10), pp. 2459-2477.
[34] Liu, J.Y. Mathematical relationship between desorption and sorption solutions, *Drying Technology*, 1993, l (5) pp. 961-975.
[35] Liu, J.Y. Diffusion coefficient of porous solid obtained from isothermal sorption tests, Research Paper FPL-UP-533, U.S. Department of Agriculture, *Forest Service,* Forest Products Laboratory, Madison, WI. , 1994.
[36] Younsi , R. Kocaefe, D. Kocaefe, Y. Three-dimensional simulation of heat and moisture transfer in wood, *Applied Thermal Engineering,* 2006, 26, pp. 1274 –1285.
[37] Luikov, A.V. Systems of differential equations of heat and mass transfer in capillary-porous bodies, *International Journal of Heat and Mass Transfer,* 1975, 8, pp. 1–14.
[38] Liu, J.Y. Cheng, S. Solution of Luikov equations of heat and mass transfer in capillary porous bodies, *International Journal of Heat and Mass Transfer,* 1991, 34 (7) pp. 1747–1754.
[39] Scheerlinck, N. Verboven, P. Stigter, J.D. Baerdemaeker, J.F.V. Impe, J.D. Nicolai , B.M. A variance propagation algorithm for stochastic heat and mass transfer problems in food processes, *International Journal for Numerical Methods in Engineering,* 2001, 51, pp. 961–983.
[40] Dantas, L.B. Orlande, H.R.B. Cotta, R.M. An inverse problem of parameter estimation for heat and mass transfer in capillary porous media, *International Journal of Heat and Mass Transfer,* 2003, 46, pp.1587–1598.
[41] Comsol, A.B. Femlab Version 2.0, *Reference Manual,* 2000.
[42] The Mathworks Inc., *MatLab Manual, Version* 5.3, 1998.

[43] Younsi, R. Kocaefe, D. Poncsak, S. Kocaefe, Y. Computational modeling of heat and mass transfer during the high-temperature heat treatment of wood, *Applied Thermal Engineering,* 2007, 27, pp. 1424–1431.

[44] Murugesan, K. Suresh, H.N. Seetharamu,. K.N. Narayana, P.A.A Sundararajan, T. A theoretical model of brick drying as a conjugate problem, *International Journal of Heat and Mass Transfer,* 2001, 44 (21) pp. 4075–4086.

[45] Oliveira, L.S. Haghighi, K. Conjugate heat and mass transfer in convective drying of porous media, *Numerical Heat Transfer,* 1998, 34, pp. 105–117.

[46] Haque, M. N. Analysis of heat and mass transfer during high-temperature drying of pinus radiate, *Drying Technology,* 2007,25:2, pp. 379–389.

[47] Pang, S. External heat transfer in moist air and superheated steam for softwood drying. *Chinese Journal of Chemical Engineering,* 2004, 12, 762–766.

[48] Pang, S.; Haslett, A.N. The application of mathematical models to the commercial high-temperature drying of softwood lumber, *Drying Technology,* 1995, 13, pp.1635–1674.

[49] Incropera, F.P.; DeWitt, D.P. *Fundamentals of heat & mass transfer*; John Wiley & Sons: New York, 2002.

[50] Tremblay, C.; Cloutier, A.; Fortin, Y. Experimental determination of the convective heat and mass transfer coefficients for wood drying. *Wood Science and Technology,* 2000, 34, pp. 253–276.

[51] Siau, J.F. *Transport processes in wood*; Springer-Verlag: New York, 1984.

[52] Kayihan, F. Simultaneous heat and mass transfer with local three phase equilibrium in wood drying. In: *Proceedings of 3rd International Drying Symposium*, Drying Research Limited, Wolverhampton, Birmingham, UK, 1982, pp. 123–134.

[53] Kayihan, F. Low and high temperature psychrometric relationship for the air–water vapor system and the equilibrium moisture contents for Wood; Research Report No. 045-4402-05; Science Unit, Wood Technology R&D, Weyerhaeuser Technology Centre: Tacoma, Washington, 1983.

[54] Pang, S. Some considerations in simulation of superheated steam drying of softwood lumber. *Drying Technology*, 1997, 15, pp.651–670.

[55] Eriksson, J., Johansson, H. and Danvind, J. , A mass transport model for drying wood under isothermal conditions, *Drying Technology*, 2007, 25, pp. 433–439.

[56] Pang, S.; Keey, R.B.; Langrish, T.A.G. Modeling the temperature profiles within boards during the high-temperature drying of Pinus radiata timber: The influence of airflow reversals. *International Journal of Heat and Mass Transfer,* 1994, 38 (2), pp.189–205.

[57] Nijdam, J.J.; Langrish, T.A.G.; Keey, R.B. A high temperature drying model for softwood timber. *Chemical Engineering Science,* 2000, 55, pp. 3585–3598.

[58] Perre´, P.; Turner, I.W. A 3-D version of TransPore: A comprehensive heat and mass transfer computational model for simulating the drying of porous media. *International Journal of Heat and Mass Transfer*, 1999, 42 (24), pp. 4501–4521.

[59] Avramidis, S. Evaluation of ''three-variable'' models for the prediction of equilibrium moisture content in wood. *Wood Science and Technology*, 1989, 23, pp. 251–258.

[60] Ayensu, A. Dehydration of food crops using solar dryer with convective heat flow. *Solar Energy,* 1997, 59, pp.121 –126.

[61] Doymaz, I. Convective air drying characteristics of thin layer carrots, *Journal of Food Engineering,* 2004, 61, pp.359 –364.

[62] Temple, S.J., Boxtel, A.J.B. Van, Thin layer drying of black tea, *Journal of Agriculture and Engineering Research*, 1999, 74, pp.167 –176.

[63] Ozdemir,M. Devres,Y.O. The thin layer drying characteristics of hazelnuts during roasting. *Journal of Food Engineering*, 1999, 42, pp. 225 –233.

In: Progress in Chemistry and Biochemistry
Editors: E. M. Pearce and G. E. Zaikov
ISBN: 978-1-60692-344-3

Chapter 20

STABLE RADICALS AS POSSIBLE MEDIATORS OF THE "LIVING" RADICAL POLYMERIZATION

M.V. Motyakin*, A.M. Wasserman*, P.E. Stott** and G.E. Zaikov***

*Institute of Chemical Physics, Russian Academy of Science,
ul. Kosygina 4, Moscow 119991, Russia, e-mail: spinchem@chph.ras.ru
**Crompton Corporation 199 Benson Road Middlebury, Connecticut 06749, USA
***Institute of Biochemical Physics, Russian Academy of Science,
ul. Kosygina 4, Moscow 119991, Russia

ABSTRACT

The stable radicals derived from different compounds were detected in the process of styrene autopolymerization. The nitroxide radicals are produced from nitrosocompound, hindered hydroxylamine, nitrophenols and nitroanisoles. The phenoxyl radicals are formed from quinine methides, and naphtoxyl radicals are generated from 2-nitro-1-naphtol. The radicals are identified, the kinetics of their formation and follow-up evolution are studied. These radicals can participate in the process of living radical polymerization as the mediators and can effect significantly on kinetics of polymerization and structure of the resulting polymer.

1. INTRODUCTION

"Living" free radical polymerization controlled by a stable radical ranks among the techniques for producing polymers with well-defined microstructure (see, for example, [1-11). Besides, the "living" free radical polymerization does not require high purity monomers and solvents that make this technique very attractive for industrial application. The basic concept consists of the reversible trapping of the propagating polymeric chain with a counter radical. The trapping reaction (reversible inhibition) makes the classical termination almost negligible, thus leading to narrow molecular weight distribution of the polymer. Either stable nitroxide radicals or species generating nitroxide radicals in situ are normally employed for the reversible trapping of the growing polymeric chains. Nitrons, nitrosocompounds, and alkoxyamines are the examples of such species [4,5,10,11]. There are few examples of employing of those chemical compounds which are neither stable nitroxide radicals nor the species which generate nitroxide radicals in situ. Therefore a searching for new compounds capable of effective participation in "living" radical polymerization is one of the topical problems.

In this connection we performed the EPR investigation of stable radicals derived from molecules of nitrosocompound, nitrophenols, nitroanisols, nitronaphtol, hindered hydroxylamine, and quinone methides in the process of styrene thermal autopolymerization. We have identified the radicals and studied kinetics of their formation and transformation.

2. EXPERIMENTAL

Styrene was purified by distillation under vacuum (20 torr) at 44-45° C. (n_D^{20}=1.5462). *m*-Xylene was purified by distillation with the following selection of fraction boiling at 139-139.5° C. m-Dichlorobenzene was purified by distillation under vacuum (14 torr) at 67°C.

Nitrosocompound, N,N-diethylhydroxylamine, quinone methydes have been obtained from Crompton corporation. 2,4-Dinitrophenol, 2,4-dinitroanysole and 2-nitroanysole have been obtained from Acros. These compounds were used without additional purification. 4-Nitrophenol (Chemapol) and 3-nitrophenol (Acros) were recrystallized from benzene, 2-nitrophenol (Acros) was recrystallized from ethanol with following sublimation in vacuum. Purity of the used compounds was controlled by thin-layer chromatography.

The compounds were weighted and brought to vial with inert solvent or styrene solution. The solution was stirred. After dissolving of compound ~0.2 ml of the solution was placed in a glass tube of 3 mm diameter. The solutions were studied in the presence of oxygen and in the absence of oxygen. The deoxygenated solutions were prepared by bubbling argon through the solution or by repeating freeze-thaw cycles. The experimental results did not depend on the method of deoxygenating. The glass tubes were sealed.

ESR spectra were recorded on the RADIOPAN ESR-spectrometer equipped with a flowing air thermostat-temperature controller. For determination of the stable radical concentration several nitroxyl radical concentration standards were prepared by dissolving known amounts of nitroxyl radical (4-hydroxy-2,2,6,6,-tetramethyl-pyperine-N-oxyl) in m-xylene. The ESR spectra were doubly integrated, and the value of the integral was plotted as a function of concentration. The radical concentration was read from the calibration curve. The glass tubes for all solutions were similar.

The Bruker programs were used for the theoretical analyze of ESR spectra and determination of hyperfine constants.

3. Results and Discussion

3.1. N-(1,4-Dimethylpentyl)-4-Nitroso-Aniline

Nitroso-compounds are widely used for the trapping of carbon-centered radicals. The trapping reaction results in formation of nitroxyl radical. This nitroxyl radical can take part in the process of living radical polymerization [4,5].

ESR spectrum of radicals in deoxygenated and oxygen-containing styrene solutions of N-(1,4-dimethylpentyl)-4-nitroso-aniline (structure is given in scheme 1) has been detected at 110° and 130° C. The ESR spectrum (figure 1a) consists of three basic spectral components and additional splitting. Three lines of the spectrum are due to interaction of uncoupling electron with nitrogen nuclei (a_N=11.5 G), additional splitting is assigned to interaction of uncoupling electron with protons of aromatic ring (a_H=2.5 G). The spectrum is very well known and is typical for aromatic nitroxyl radical [13-15].

Aromatic nitroxyl radicals are derived from reaction of N-(1,4-dimethylpentyl)-4-nitroso-aniline with active radicals R$^\bullet$, involved in polymerization. Scheme 1 shows the mechanism of nitroxyl radical formation.

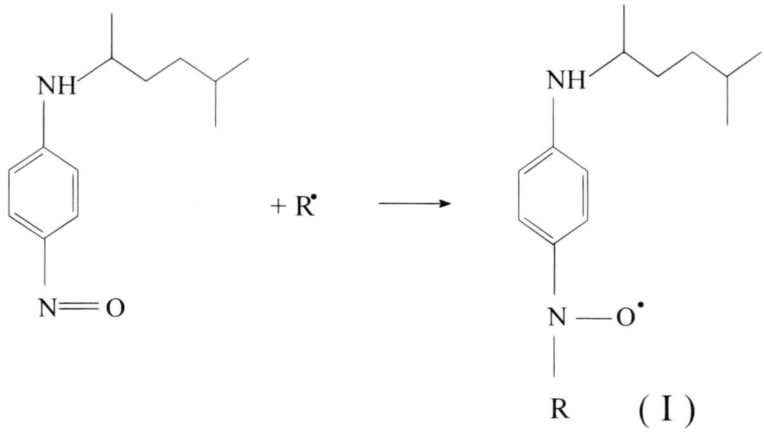

(I)

N-(1,4-dimethylpentyl)-4-nitroso-aniline

Scheme 1.

Kinetics of the nitroxyl radical formation generated during styrene polymerization in the presence of N-(1,4-dimethylpentyl)-4-nitroso-aniline is shown on figure 1b. It is important to notice that nitroxyl radicals appear already during the first minutes of thermal treatment. The radical concentration at temperature of 130° C is higher than at temperature of 110° C. Maximum nitroxyl radical concentration at 130° C is approximately 10^2 times less than the initial concentration of N-(1,4-dimethylpentyl)-4-nitroso-aniline in the styrene solution. At the same time concentration of the nitroxyl radicals in oxygen-containing samples is approximately two times less than concentration of the radicals in deoxygenated samples. It seems that nitroxyl radicals are terminated by the side reactions in the presence of oxygen.

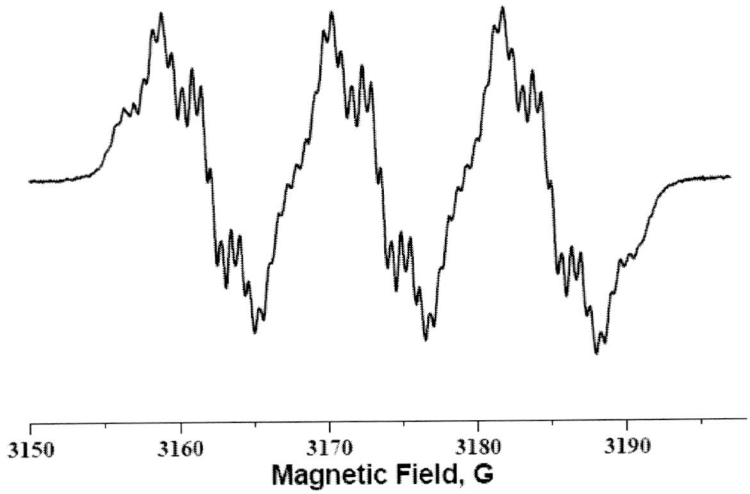

a

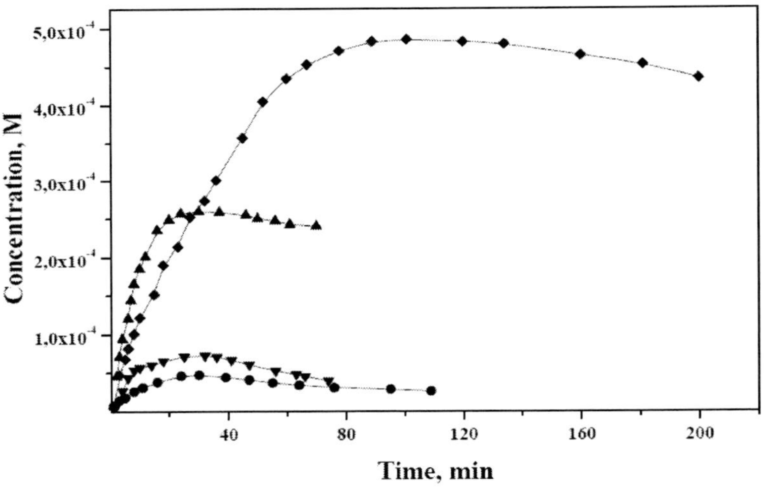

b

Figure 1. a. The ESR spectrum of nitroxyl radical (I) in styrene solution at 110° C. The same spectrum is observed at 130° C. b. Kinetics of nitroxyl radical (I) formation in styrene solution of N-(1,4-dimethylpentyl)-4-nitroso-aniline: (●) in the presence of oxygen at 110° C nirosocompound concentration is 4.5×10^{-3} M; (▲) in the presence of oxygen at 130° C, nirosocompound concentration is 3.19×10^{-2} M; (▼) in the presence of oxygen at 110° C, nirosocompound concentration is 3.19×10^{-2} M; (♦) in the absence of oxygen at 130° C, nirosocompound concentration is 2.42×10^{-2} M.

Nitroxyl radical concentration increases during the first minutes of observation and then varies only slightly. The stationary radical concentration can be explained by the reaction of reversible inhibition (Scheme 2).

We emphasize that a shape of the ESR signal does not change during thermal treatment. It means that only one type of radicals having the structure (I) has been detected.

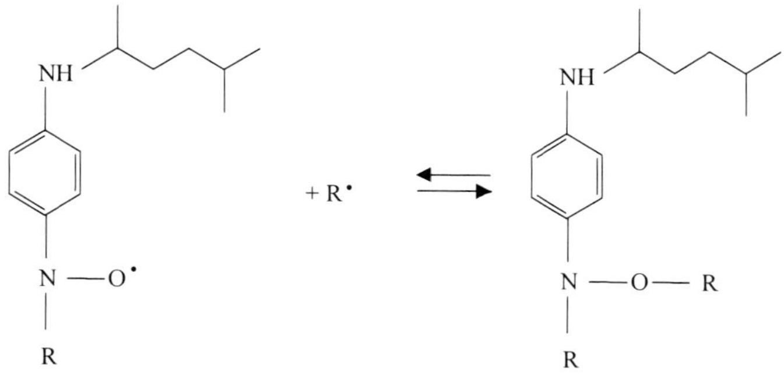

Scheme 2.

It should be noted that the reaction given at Scheme 2 is able to provide the "living chain" styrene polymerization.

3.2. N,N-Diethylhydroxylamine (DEHA)

ESR spectrum of radicals formed in DEHA solution in styrene at 130° C is presented on figure 2a. The similar spectrum was detected in DEHA solution in m-xylene at 130° C. The spectrum can be positively assigned to nitroxyl radical $(CH_3CH_2)NO^{\cdot}$. Analyze of spectrum was carried out by "Simphonia" program (Bruker). Theoretical spectra were calculated. Good agreement was observed with experimental spectra. Parameters used in simulation were following: hyperfine constant of interaction with nitrogen a_N=15.2 G, hyperfine constant of interaction with four protons of CH_2 group a_H=10.3 G, line width ΔH=1.4 G.

Consider possible mechanisms of nitroxyl radical formation in DEHA solutions.

Nitroxyl radicals in DEHA solution in m-xylene can be formed as a result of oxidation of DEHA by oxygen remained in the solution after deoxygenating. Hindered hydroxylamines react readily with oxygen [16]. Result of the reaction is formation of nitroxyl radical (scheme 3).

$$(CH_3CH_2)_2NOH \xrightarrow{O_2} (CH_3CH_2)_2NO^{\bullet}$$

Scheme 3.

Nitroxyl radicals in DEHA solution in styrene can be formed by two reactions. First reaction is an oxidation of DEHA by oxygen according to scheme 3. Second reaction is an interaction of DEHA with alkyl styrene radicals (scheme 4).

$$(CH_3CH_2)_2NOH + R^{\bullet} \longrightarrow (CH_3CH_2)_2NO^{\bullet} + RH$$

Scheme 4.

Kinetics of formation of radicals $(CH_3CH_2)NO^{\cdot}$ in DEHA solutions in m-xylene and in styrene at 130° C are presented on figure 2b.

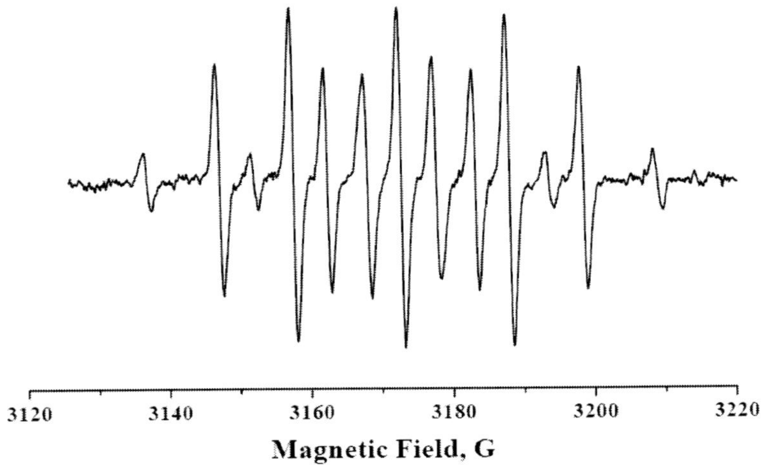

a

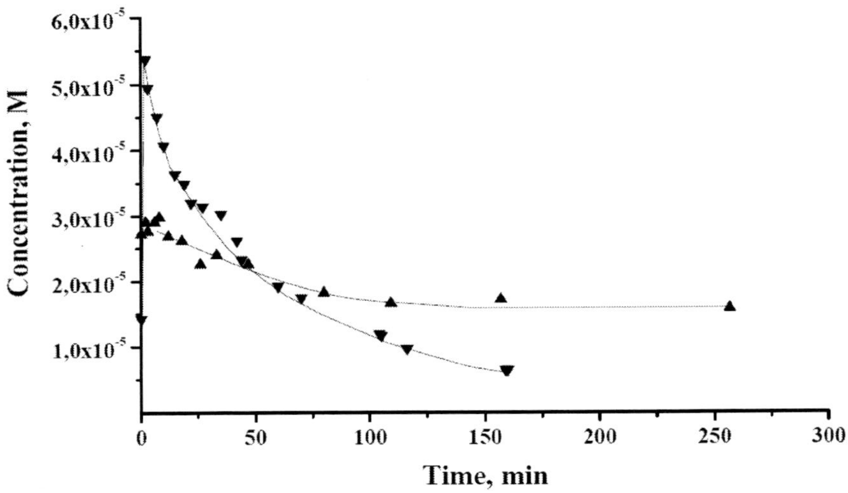

b

Figure 2. a.The ESR spectrum of radical $(CH_3CH_2)_2NO\bullet$ in DEHA solution in styrene (1.7x10^{-1}M) at 130° C. b. Variation of nitroxyl radical $(CH_3CH_2)_2NO\bullet$ concentration in DEHA solutions(1.7x10^{-1}M) in styrene(▲)and in m-xylene(▼) at 130° C.

Within first minutes of heating radicals are detected in DEHA solution in m-xylene. Concentration of the radicals gradually decreases during heating. Radical formation is apparently induced by oxidation of DEHA (scheme 3). Radical termination is apparently due to reaction of radicals with molecules of solvent or due to disproportional reaction leading to formation of diamagnetic products.

Concentration of radical in DEHA solution in styrene increases dramatically for first minutes. An increase of radical concentration is apparently induced by reaction of DEHA with alkyl radicals of styrene (scheme 4). With time the radical concentration decreases. A decrease of radical concentration is due to the reactons much as in DEHA solution in m-xylene. In addition, termination of nitroxyl radicals in styrene can be caused by reaction with alkyl radicals of styrene (scheme 5).

$$(CH_3CH_2)_2NO^{\cdot} + R^{\cdot} \rightleftharpoons (CH_3CH_2)_2NOR$$

Scheme 5.

The product $(CH_3CH_2)NOR$ is unstable and can participate in process of living radical polymerization.

3.3. Nitrocompounds

Aromatic nitrocompounds are widely used for inhibition and regulation of radical polymerization. Depending on structure nitrocompounds can be both strong and weak inhibitors [17]. Besides, stable radicals formed in process of radical polymerization in the presence of aromatic nitrocompounds can be involved in living free radical polymerization [4,5,18]. Reactions of nitro compound with active radicals involved in polymerization process can be schematically written as [19-22]:

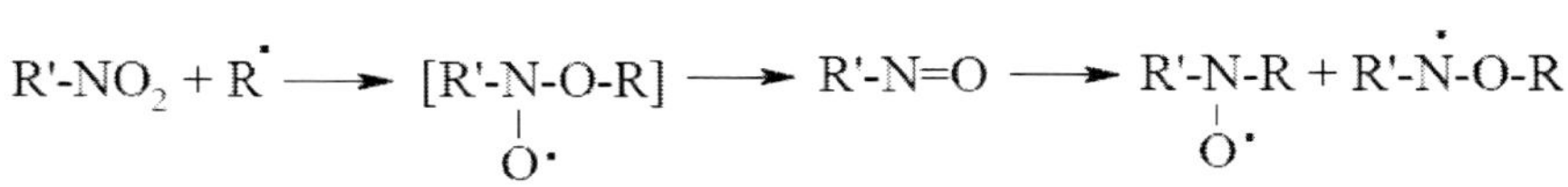

Scheme 6.

According the scheme 6 nitroxyl and aminyl radicals are formed. Nitroxyl radicals are more stable than aminyl radicals. Therefore, nitroxyl radicals accumulate in reaction system and can participate in free radical polymerization. To effect on polymerization it is necessary to understand mechanism of reaction in the presence of nitrocompounds and to clarify the structure of stable radicals formed in the result of nitrocompound transformation.

3.3.1. Nitrophenols

3.3.1.1. 2,4-Dinitrophenol

2,4- Dinitrophenol in M-Xylene

First of all it was necessary to clarify the possibility of stable radical formation during the heating of 2,4-dinitrophenol in inert solvent. The *m*-xylene was used as inert solvent. Solutions of 2,4-dinitrophenol ($1.36x10^{-1}$M) in *m*-xylene with oxygen and without oxygen were heated at 130° C for 3 hours. No radicals were detected by ESR spectroscopy during the heating. After the heating the solutions were stored at room temperature. The solutions gradually changed in color (yellow-green color appeared). Besides, ESR signal appeared in the deoxygenated solution stored for 24 days (figure 3a). The signal is weak (concentration is about $8x10^{-7}$ M). The signal can be presented as triplet of triplets and can be assigned to stable nitroxide radical with isotropic coupling constants of a_N=9.6 G and a_H=3.2 G. It is important to notice that aromatic nitrocompounds are the true oxidants [23]. ESR spectra of stable nitrogen-containing radicals are usually detected under heating of nitroaromatic compounds. Concentration of the radicals increases with increasing number of nitrogroups in molecule of aromatic compound [24,25]. According to [24], one of the possible ways of radical formation in the nitrophenol-containing systems is the disproportional reaction:

$$2NO_2C_6H_4OH \rightarrow NO_2C_6H_4O^{\bullet} + HN^{\bullet}O_2C_6H_4OH$$

Scheme 7.

Using schemes 6 and 7, the possible mechanism of nitroxide radical formation in 2,4-dinitrophenol solution of m-xylene can be written as:

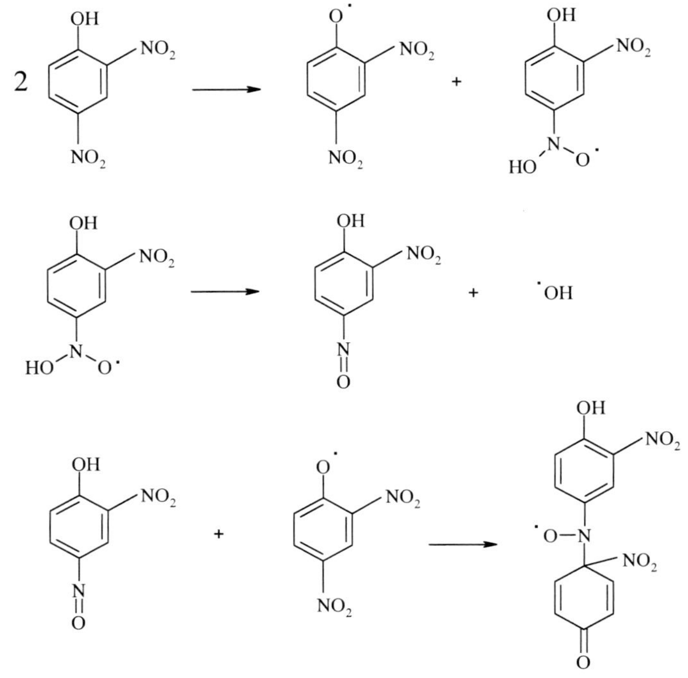

(II)

Scheme 8.

The spectrum presented on figure 3a is likely to be assigned to radical II. The value of a_N=9.6 G is caused by interaction of uncoupling electron with nitrogen atom, the value of a_N=3.2 G is caused by interaction with aromatic ring protons. Unfortunately, at present time it is impossible to make positive conclusion about mechanism of formation and structure of nitroxyl radicals formed in 2,4-dinitrophenol solution of m-xylene stored for a long time.

2,4-Dinitrophenol in Styrene

ESR spectrum (figure 3b) of stable radicals was detected in 2,4-dinitrophenol solution in styrene kept at room temperature in the presence of oxygen for 24 hours. The spectrum can be assigned to nitroxide radicals with hyperfine constants of a_N=10.0 G and a_H=3.2 G. Notice that spectra on figure 3a and figure 3b are distinctively different. It is reasonable to assume that formation of radicals, whose spectrum is presented on figure 3b, is related to two processes: firstly, to reactions presented on scheme 8, and, secondly, to the styrene oxidation by 2,4-dinitrophenol leading to formation of active radicals that finally transform to nitroxide radicals according to scheme 6. An extended analysis of formation and of structur of active radicals formed in 2,4-dinitrophenol solutions of styrene at room temperature needs some additional investigation.

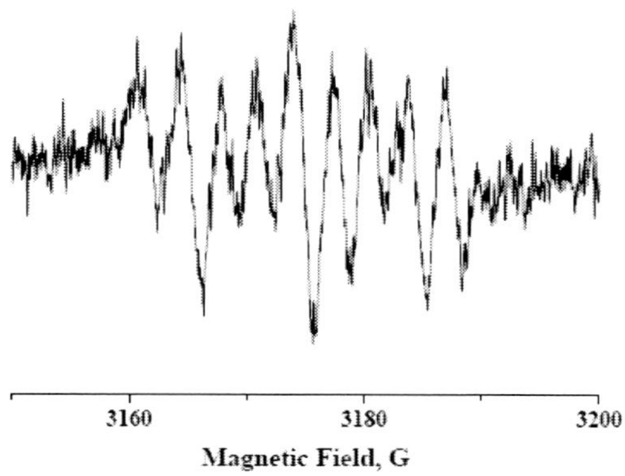

a

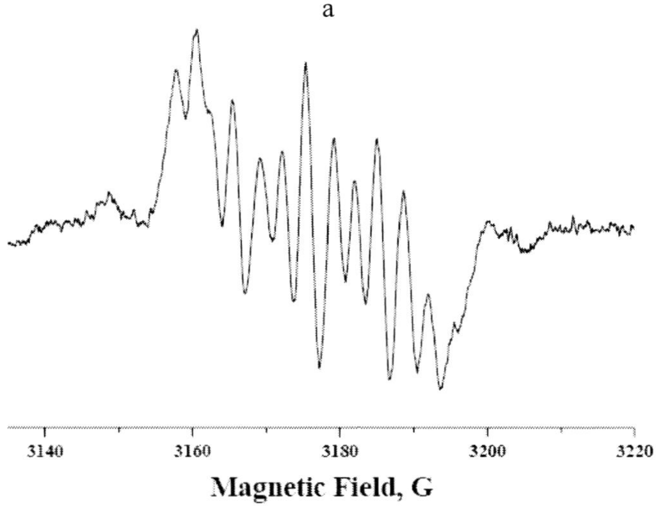

b

Figure 3. a. The ESR spectrum of radical observed in 2,4-dinitrophenol solution in m-xylene ($1.36x10^{-1}$M) after heating (180 min) at 130° C. The spectrum was recorded at room temperature after 24 days. b. The ESR spectrum of radicals observed in 2,4-dinitrophenol solution ($9.1x10^{-2}$M) in styrene in the presence of oxygen at room temperature. The spectrum was recorded after storage of sample for one day.

The transformation of ESR spectra of radicals formed in styrene solution of 2,4-dinitrophenol at 130° C in the presence of oxygen is on figure 4.

ESR spectra of the sample heated for a short time differ distinctively from ESR spectra of the sample heated for a long time. The general length of spectrum and distances between picks change during the heating. It is not inconceivable that the spectrum observed at initial stage of heating can be a superposition of spectra of nitroxide radicals III and IIIa, as well as, of alkoxy nitroxide radical IV and aminyl radical V formed in accordance with the scheme 6.

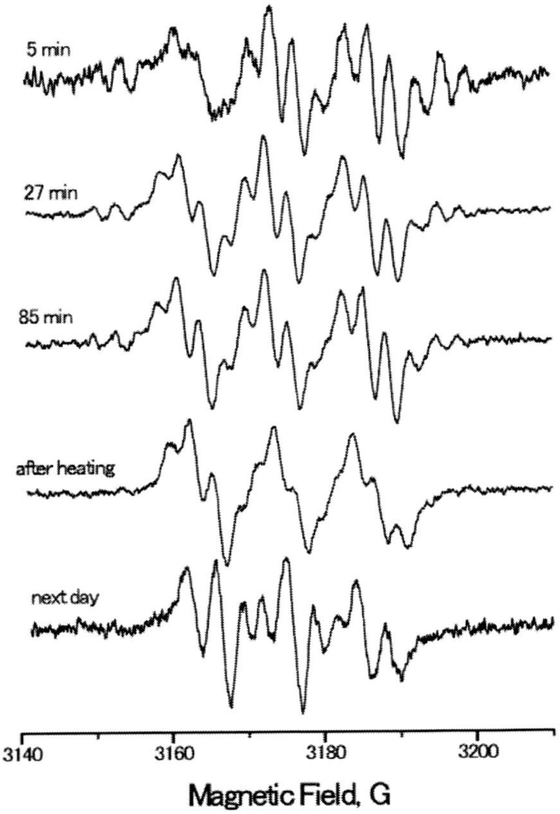

Figure 4. The ESR spectra of radicals formed in 2,4-dinitrophenol solution; ($9.1x10^{-2}M$) in styrene with oxygen at 130° C.

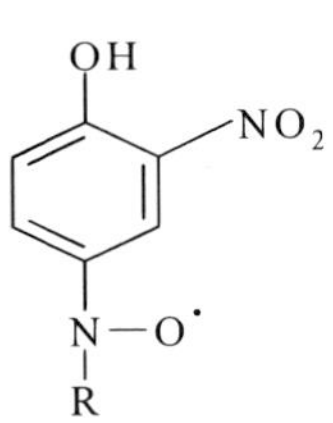

III

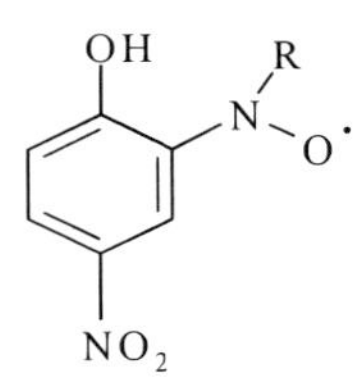

IIIa

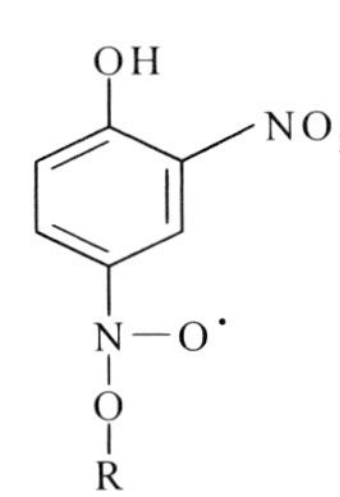

IV

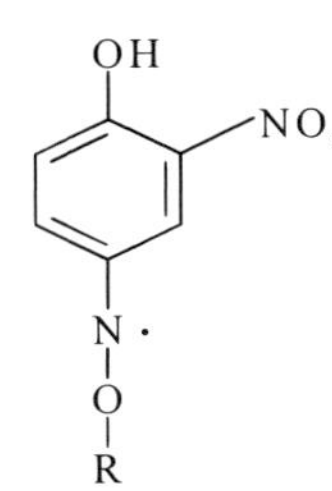

V

Radicals IV and V are much less stable than radicals III and IIIa, and are terminated during the heating and after heating [20,26]. Radicals III and IIIa accumulate at different stages of styrene polymerization reaction. That is why we detect nitroxide radical with hyperfine constants of a_N=11,3 G and a_H=2,8 G after 27 minute of heating whereas after heating we observe nitroxide radical with hyperfine constants of a_N= 9,4 G and a_H=3.8 G.

Kinetics of the nitroxyl formation at 130° C is presented in figure 5. Notice that nitroxyl radical concentration in styrene solution increases with the increasing of 2,4-dinitrophenol concentration in the solution. However, the maximum nitroxyl radical concentration is almost 10^4 times less than initial 2,4-dinitrophenol concentration in the solution. Nitroxyl radical concentration in deoxygenated solutions differs slightly from the concentration in oxygen-containing solutions. Nitroxyl radical concentration increases during the first minutes of thermal treatment and then decreases slowly. It is possible to assume that nitroxyl radicals derived from nitrocompounds can take part in the reversible inhibition.

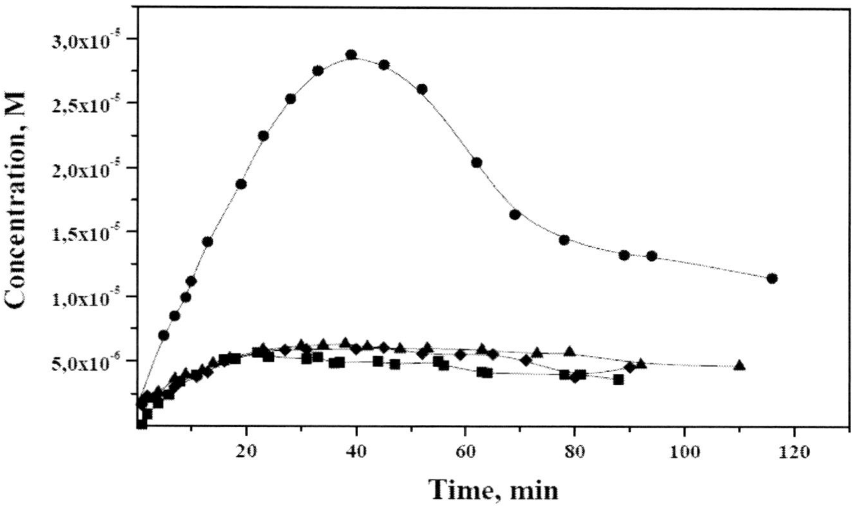

Figure 5. Kinetics of nitroxyl radical formation in styrene solution at 130° C: (●) in the presence of oxygen, 2,4-dinitrophenol concentration is 1.38x10^{-1} M; (▲) in the presence of oxygen, 2,4-dinitrophenol concentration is 4.12x10^{-2} M; (■) in the absence of oxygen, 2,4-dinitrophenol concentration is 4.34x10^{-2} M; (♦) in the absence of oxygen, 2,4-dinitrophenol concentration is 4.57x10^{-2} M.

3.3.1.2. 4-Nitrophenol in Styrene

ESR spectra of radicals formed during heating of 4-nitrophenol solution in styrene in the presence and in the absence of oxygen are presented on the figure 6A. Figure 6B presents kinetics of radical formation. All spectra are assigned to nitroxide radicals formed in accordance with the scheme 6.

It should be paid attention on three moments.

Firstly, The ESR spectra in the presence of oxygen do not differ from ESR spectra in the absence of oxygen. Nitroxide radicals formed during polymerization have hyperfine constants of a_N=11,7 G and a_H=3,0 G.

Secondly, there are no changes in ESR spectrum during heating.

Finally, the concentration of radicals generated in styrene solution of 4-nitrophenol is lower than in styrene solution of 2,4-dinirophenol (compare results presented on figure 5 and figure 6). The difference may be caused by different reactivity of nitro groups in the two compounds.

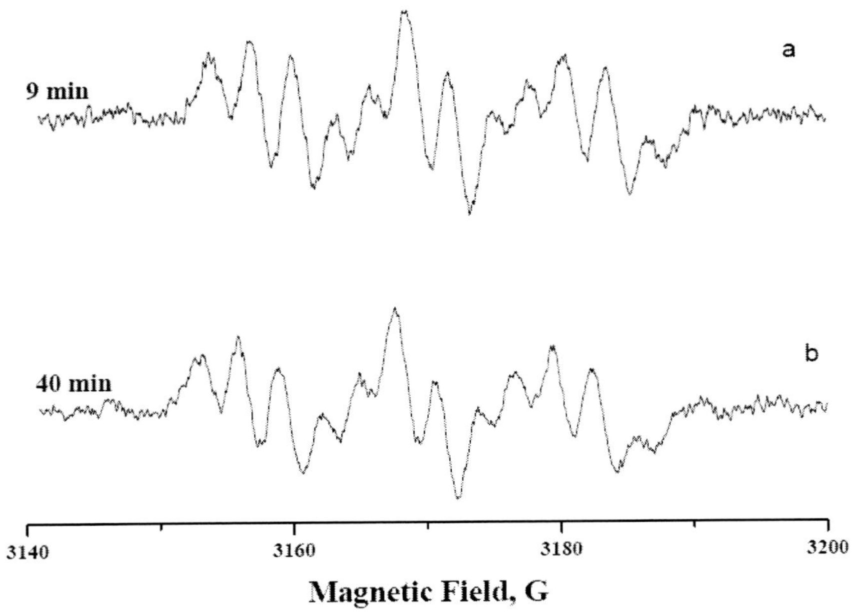

a

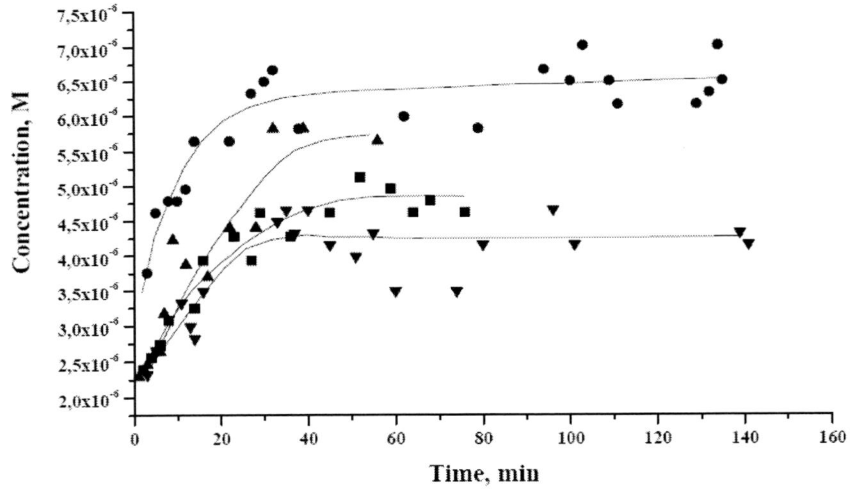

b

Figure 6.A.The ESR spectra of radicals observed in solutions of 4-nitrophenol in styrene (1.73x10^{-1}M ,a-with argon, b-with oxygen) at 130°C . B. Kinetics of radical formation in solutions of 4-nitrophenol in styrene at 130°C. Concentrations of 4-nitrophenol are 1.73x10^{-1}M (with oxygen(●) and with argon (▼)) and 1.83x10^{-1}M(with oxygen(▲) and with argon (■)).

3.3.1.3. 2- Nitrophenol in Styrene

The ESR spectra of radicals formed in styrene solution of 2-nitrophenol at 130° C on the air and in argon atmosphere are shown on figure 7 and figure 8. The kinetics of radical formation is shown on figure 9.

There are no practically difference between spectra and kinetics in the presence and the absence of oxygen. The radical concentration is about the same as in 4-nitrophenol solutions and is not changed after first ten minutes of heating. But there is transformation of ESR spectra. The general length of spectrum and distances between picks change during the heating. It seems that the spectra observed are a superposition of spectra of various types of stable radicals. Nitroxide radical spectrum with hyperfine constants of a_N=10,1 G and a_H=3,5 G is obtained after heating.

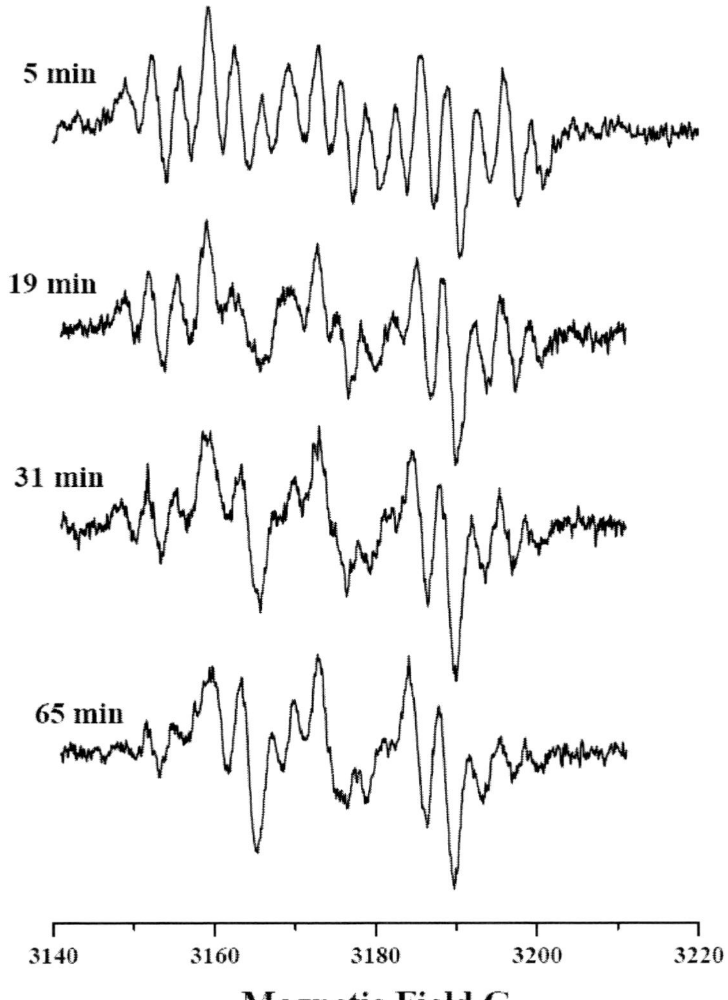

Figure 7. The ESR spectra of radicals observed in 2-nitrophenol solution; ($2.44x10^{-1}$M) in styrene with argon at 130° C.

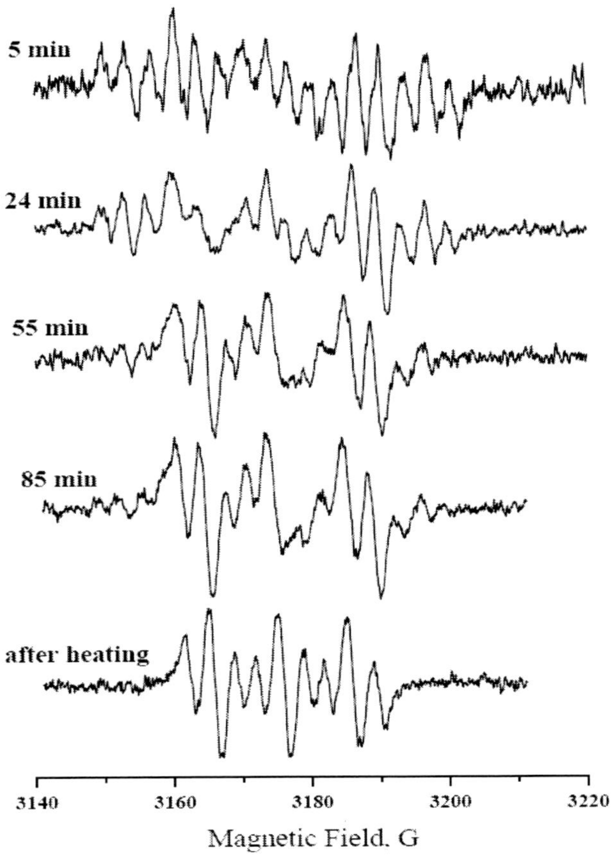

Figure 8. The ESR spectra of radicals observed in 2-nitrophenol solution; ($2.44x10^{-1}$M) in styrene with oxygen at 130°C.

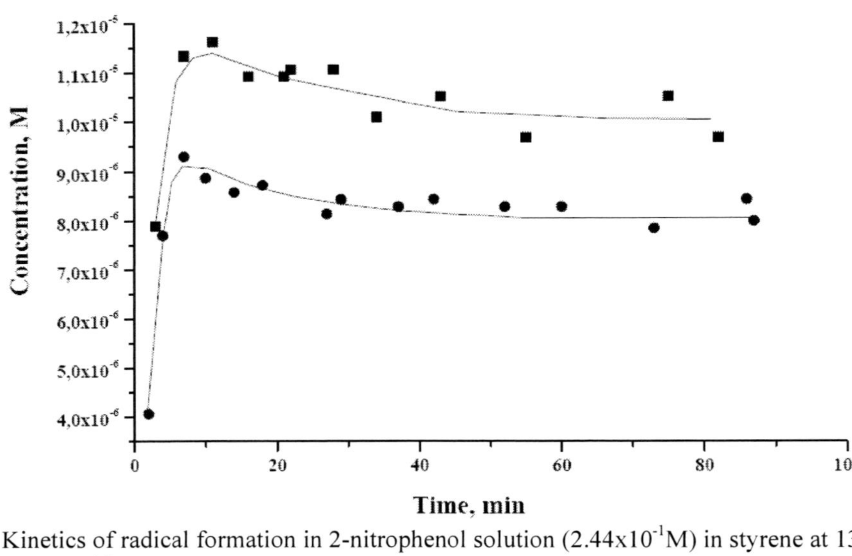

Figure 9. Kinetics of radical formation in 2-nitrophenol solution ($2.44x10^{-1}$M) in styrene at 130°C.●- with oxygen, ■- with argon.

3.3.1.4. 3-Nitrophenol in Styrene

ESR spectra of radicals formed during heating of styrene solution of 3-nitrophenol in the presence of oxygen are presented on the figures 10a.

The ESR spectra registered in the presence of oxygen do not differ from ESR spectra registered in the presence of argon. The ESR spectra are assigned to nitroxide radicals with hyperfine constants of a_N=11,3 G and a_H=3,0 G. There is no change in spectra of samples during heating at 130° C.

The kinetics of radical formation is shown on figure 10b.

It should be paid attention on two peculiarities of radical kinetics.

Firstly, the radical concentration in solution with oxygen is higher than the radical concentration in solution without oxygen. Perhaps, in this case oxygen initiates process of radical formation.

Secondly, concentration of the radicals formed in fresh made samples is less than in samples stored for one-two days at room temperature or at fridge temperature. Besides, kinetics of radical formation in fresh made samples differs from kinetics of radical formation in stored samples.

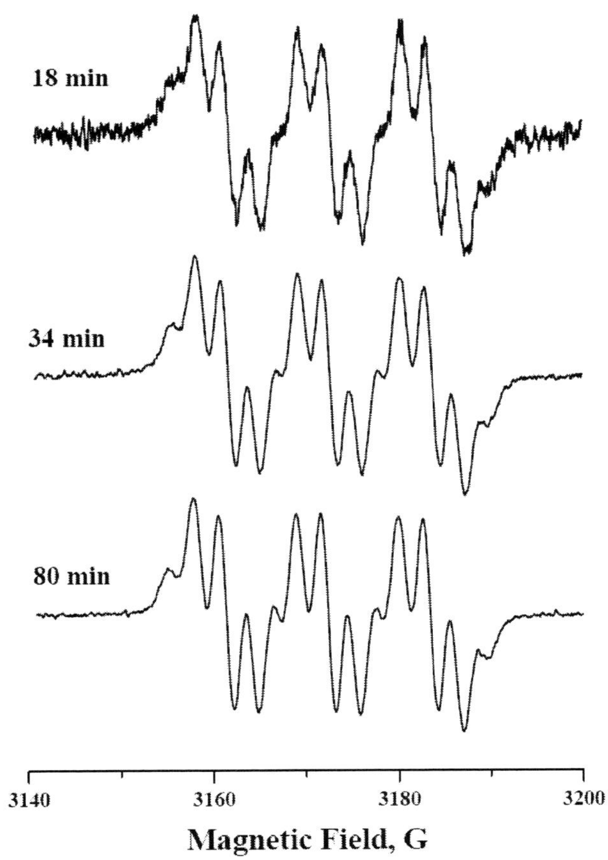

a

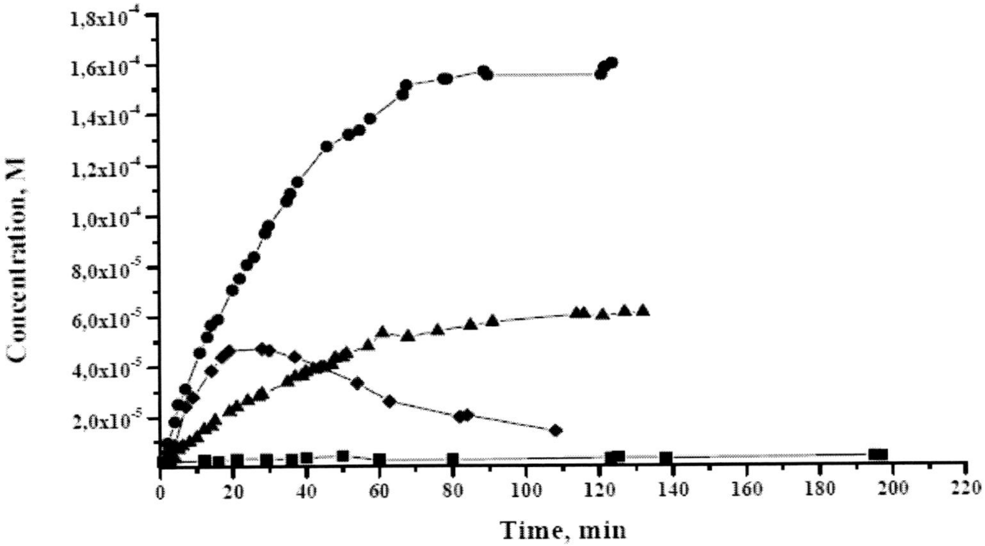

b

Figure 10.a. The ESR spectra of radicals observed in 3-nitrophenol solution; ($1.32x10^{-1}$M) in styrene at 130°C in the presence of oxygen. b. Kinetics of radical formation in 3-nitrophenol solution in styrene at 130°C.● – with oxygen, $1.81x10^{-1}$M, stored in fridge for 2 days; ■ –with argon, $1.73x10^{-1}$M; ♦- with oxygen, $1.73x10^{-1}$M; ▲ –with oxygen, $1.32x10^{-1}$M, stored at room temperature for 1 day.

3.3.2. Nitroanisoles

3.3.2.1. 2,4-Dinitroanisole

ESR spectra of radicals formed in styrene solutions of 2,4-dinitroanisole are given on figure 11a. The ESR spectra registered in the presence of oxygen do not differ from ESR spectra registered in the absence of oxygen. The ESR spectra obtained are assigned to nitroxide radicals with hyperfine constants of a_N=10,8 G and a_H=3,0 G. It is important to notice that the signal of ESR is detected not only in the samples heated at 130° C but in samples kept at room temperature as well.

Kinetics of radical formation and termination are shown on figure 11b. Oxygen does not affect on the radical concentration.

Comparison with 2,4-dinitrophenol results (figure 5) shows that 2,4-dinitrophenol produces more radicals than 2,4dinitroanisole with the same sample concentration.

It is necessary to pay special attention on the results of investigation of solution ($1.07x10^{-1}$M) stored for 6 days at room temperature. It was found a large ($5x10^{-5}$M) concentration of nitroxide radicals after storage (before heating). Besides, concentration of the radicals increases dramatically during heating of the sample.

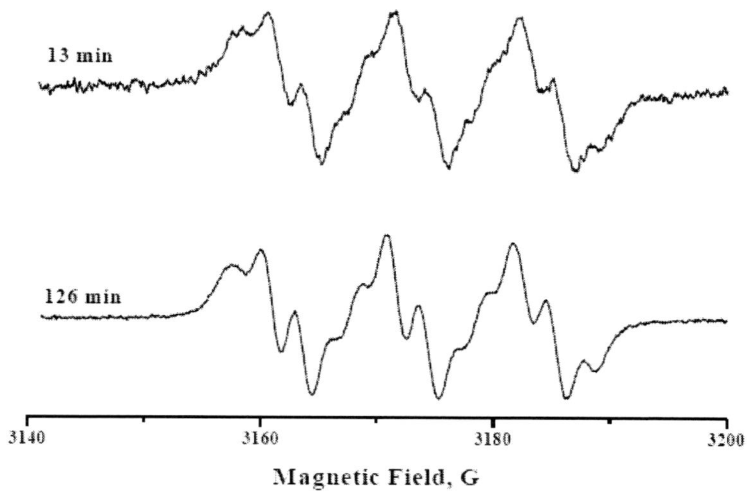

a

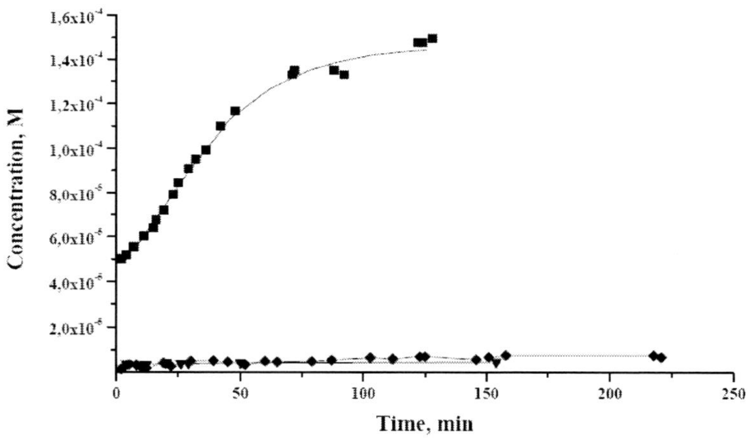

b

Figure 11.a. The ESR spectra of radicals observed in 2,4–dinitroanisole solution ($1.07x10^{-1}$M) in styrene in the presence of oxygen. b. Kinetics of radical formation in 2,4-dinitroanisol solutions in styrene at 130°C.`■ -$1.07x10^{-1}$M , with oxygen, stored at room temperature for 6 days; -$1.32x10^{-1}$M, with oxygen (♦) and with argon(▼).

3.3.2.2. 2-Nitroanisole in Styrene

The ESR spectra of radicals formed in 2-nitroanisole solution in styrene at 130° C are shown on figure 12a. The presence of oxygen does not affect on the ESR spectrum. The spectrum is poor-resolved triplet. The ESR spectra are assigned to nitroxide radicals with hyperfine constants of a_N=12,7 G and a_H=2,6 G. There is no change of spectra during heating at 130° C. Kinetics of radical formation is given on figure 12b. The presence of oxygen does not affect significantly on the concentration of radicals. In addition concentration of radicals in 2-nitroanisole solutions differs slightly from the concentration of radicals in 2-nitrophenol solutions (figure 9). It is important to pay attention on the following peculiarity: solution stored for one day at room temperature gives more radicals during the heating than fresh made solution.

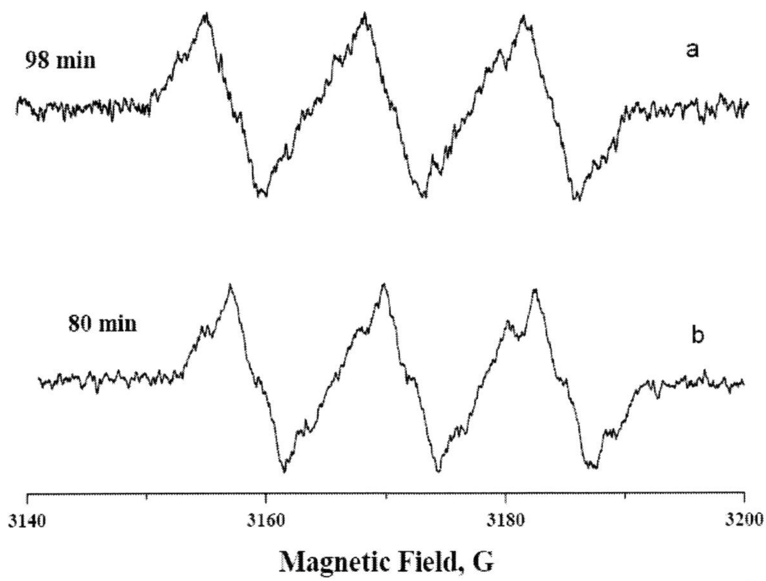

Figure 12.a. The ESR spectra of radicals observed in 2-nitroanisole solutions; (1.83×10^{-1}M) in styrene at 130° C. a- with oxygen, b- with argon. b. Kinetics of radical formation in 2-nitroanisole solutions in styrene (1.83×10^{-1}M) at 130°C., ● – with argon, ▲- with oxygen; ▼- with oxygen, stored at room temperature for 1 day; 2.74×10^{-1}M, ■ – with argon, ♦ - with oxygen.

Thus, the formation of stable nitroxide radicals in 2,4-dinitrophenol solution in xylene in the absence of oxygen, and in 2,4-dinitrphenol solution in styrene and 2,4-dinitroanisole solution in styrene in the presence of oxygen during their storage is observed. Nitroxide radical formation is observed during the heating of 2,4-dinitrophenol, 2-nitrophenol, 3-nitrophenol, 4-nitrophenol, 2,4-dinitroanisole, 2-nitroanisole solutions in styrene at 130° C either in the presence or in the absence of oxygen. Scheme 6 shows reactions of radical formation. In the cases of 2-nitrophenol and 2,4-dinitrophenol transformation of ESR spectra was observed during the heating. This transformation may be caused by a superposition of spectra of different stable radicals formed during the reaction. Concentration of radicals formed in 2-nitrophenol solution differs slightly from concentration of radicals formed in 2-nitroanisole solution. But fresh made 2,4-dinitrophenol solution in styrene produces more radicals than 2,4-dinitroanisole solution at the same conditions. Therefore, substitution OH group on OCH_3 affects on nitroxide radical concentration differently depending on phenol structure. Concentration of radicals formed during styrene polymerization in the presence of 3-nitrophenol, 2-nitroanisole, 2,4-dinitroanisole increases after preliminary storage of solutions at room or fridge temperature. Oxygen affects on concentration of radicals in 3-nitrophenol and 2,4-dinitroanisole solutions in styrene. The stable nitroxyl radicals from nitrophenols and nitroanisols can take part in living radical polymerization.

3.3.3. 2-Nitro-1-naphtol

The ESR spectrum (triplet with hyperfine constant of a_N = 27.1 G) is observed in solution of 2-nitro-1-naphtol (Scheme 9, compound VI) in m-xylene at 130° C in the presence of argon (figure 13a). The ESR signal can be assigned to iminoxyl radical (scheme 9, compound VII).

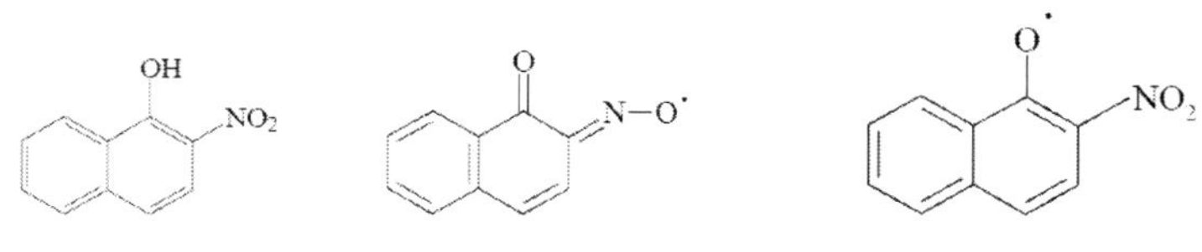

VI VII VIII

Scheme 9.

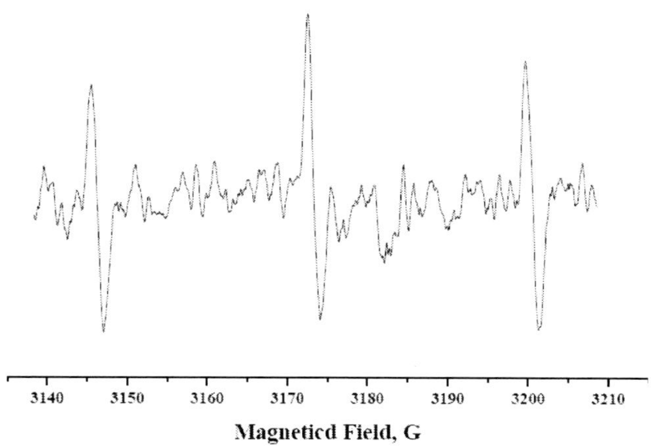

a

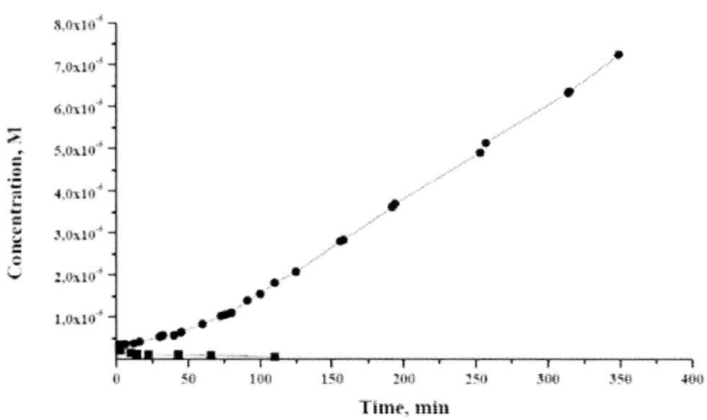

b

Figure 13.a. ESR spectrum of radicals observed in solution of 2-nitro-1-naphtol in m-xylene (2.11×10^{-1}M) with argon at 130° C for 8 minutes of heating. b. Kinetics of radical formation in solutions of 2-nitro-1-naphtol; (2.11×10^{-1}M) in styrene with argon (●) and in m-xylene with argon (■) at 130°C.

In solution nitronaphtols can be in tautomeric equilibrium with quinone shown on scheme 10.

VIa

Scheme 10.

Iminoxyl radical VII can be formed as a result of breaking of weak N-OH bond in compound VIa initiated by some impurity. It is important to note that concentration of iminoxyl radical is small and slightly decreases in time (figure 13b)

The ESR spectrum of radicals formed in solution of 2-nitro-1-naphtol in styrene at 130° C is shown on figure 14. The spectrum can be assigned to naphtoxyl radical (VII), in which interaction of uncoupling electron with 5 protons of aromatic rings (a_H=4.4 G) takes place. Formation of the naphtoxyl radical, for example, can be determined by hydrogen of OH-group abstraction reaction of 2-nitro-1-naphtol with active styrene radical. We note that the naphtoxyl radical is stable and the radical concentration increases all time (figure 13b).

So, naphtoxyl radicals form in solution of 2-nitro-1-naphtol in styrene at 130° C. It should like also to note the difference between 2-nitrophenol and 2-nitro-1-naphtol. Nitroxyl radicals were derived from 2-nitrophenol while naphtoxyl radicals were derived from 2-nitro-1-naphtol.

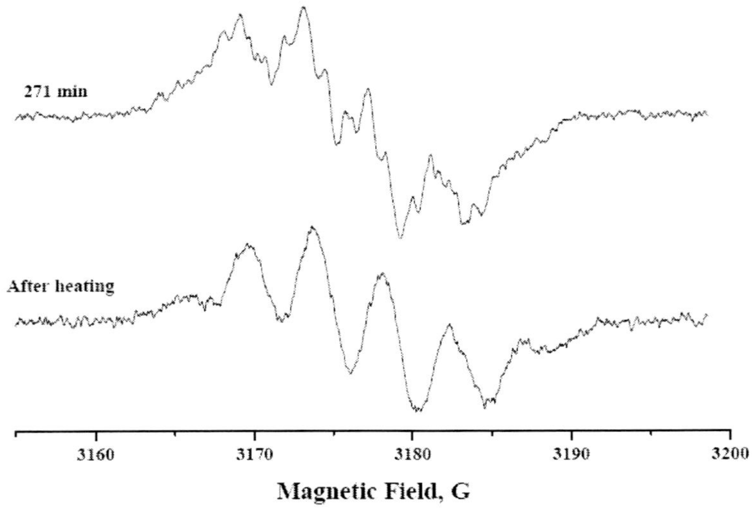

Figure 14. ESR spectra of radicals observed in solution 2-nitro-1-naphtol in styrene; ($2.11x10^{-1}$M) with argon at 130°C and at room temperature after heating.

3.4. Quinone Methides

Quinone methides belong to a class of unsaturated cyclic ketones[27,28]. Their high reactivity attracts great interest. The simplest quinone methides – unstable compounds. In the moment of formation they polymerize themselves or interact with the medium. High reactivity of quinone methides is determined mainly by low-lying excited level of the molecules[29]. Stability of quinone methides increases sufficiently by shielding of oxygen atom [29,30]. For instance, compound IX was obtained as individual compound. However, even at room temperature this compound undergoes transformations with formation of characteristic products of radical reactions[31,32]. ESR spectrum (triplet of triplets) was observed in solutions of compound IX in inert solvents. The spectrum was assigned to biradical X or phenoxyl radicals XI or XII [32-34].

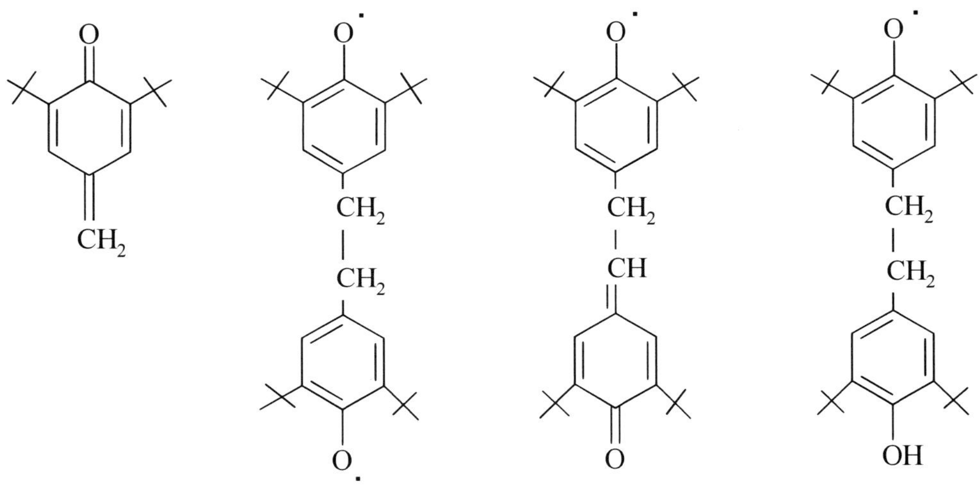

IX X XI XII

Scheme 11.

Quinone methides substituted at methylene group are more stable than compound IX. They form in processes of oxidation of hydrocarbons and polymers inhibited by hindered phenols [35,36]. Hindered quinone methides react with alkyl radicals and inhibit radical polymerization of styrene [37]. They inhibit liquid-phase oxidation of hydrocarbons [37] and oxidation of polymers [38]. Reactions of quinone methides with alkyl (R•) and peroxide ($RO•_2$) radicals can be presented by following schemes[37,38]:

In this work formation of stable radicals during styrene polymerization at 130° C in the presence of quinone methides Q (4-benzylidene-2,6-di-tert-butyl-cyclohexa-2,5-dienone), QE, (2,6-di-tert-butyl-ethylidenecyclohexa-2,5-dien-1-one) and TBSQ (3,5,3',5'-tetra-butylstilbenequinone, structures are given below) was studied.

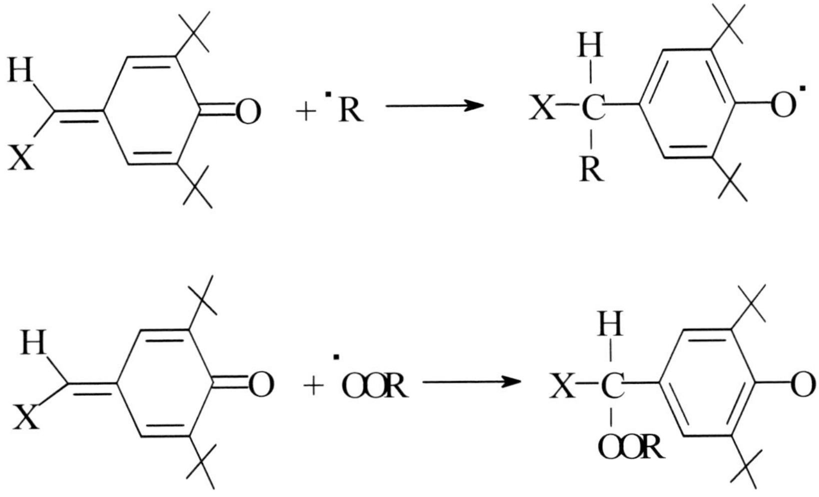

Scheme 12.

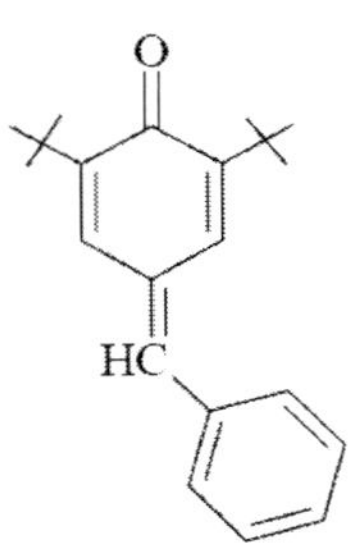

Q

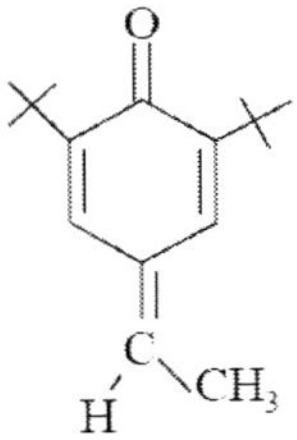

QE

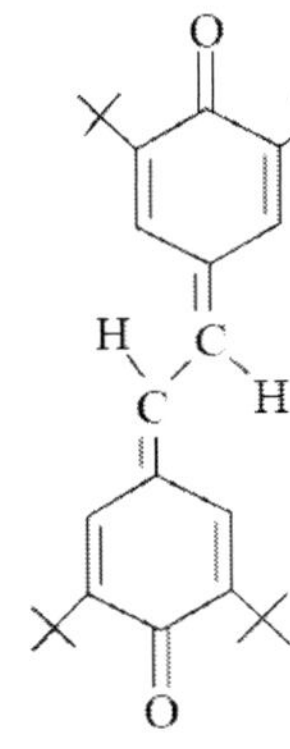

TBSQ

3.4.1. Quinone Methide "Q" (4-Benzylidene-2,6-Di-Tert-Butyl-Cyclohexa-2,5-Dienone)

ESR spectrum of radicals in deoxygenated and oxygen-containing styrene solutions of quinone methide "Q" has been detected at 130° C (figure 15a). The ESR spectrum obtained is a combination of ESR spectra of different radicals. During the first stage of polymerization process (up to 80 min) at 130° C ESR spectrum of phenoxyl radical (XIII) radicals dominates. This ESR spectrum corresponds to The reaction of phenoxyl radical (XIII) formation is presented in Scheme 13.

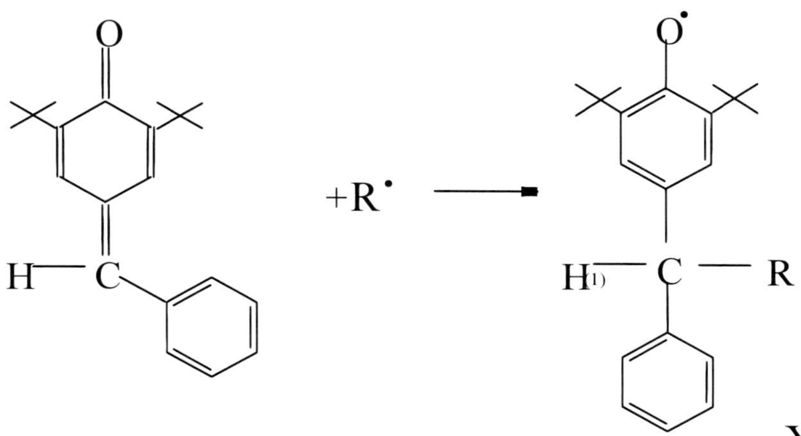

XIII

Scheme 13.

Note that $R^{\bullet}$ is active radical involved in process of styrene polymerization.

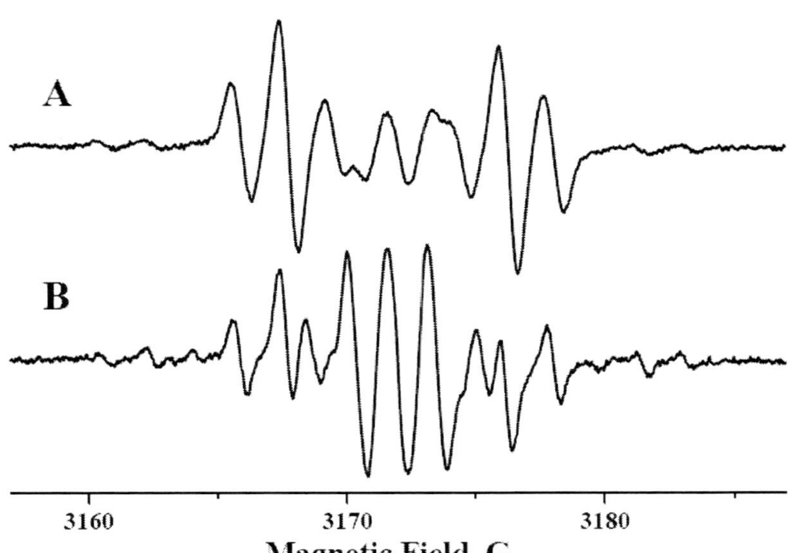

a

Hyperfine structure constants determined from ESR spectrum (figure 15a(A)) are following: $a_{H(1)}$=8.5 G, $a_{H(2)}$=1.8 G. The constant $a_{H(1)}$ is due to interaction of uncoupling electron with hydrogen atom H(1). The constant $a_{H(2)}$ is determined by interaction of uncoupling electron with hydrogen atoms of phenyl ring. The ESR spectra of phenoxyl radicals were studied extensively [13,14,39]. Values of the hyperfine structure constants allow us to determine the radical structure (XIII) uniquely.

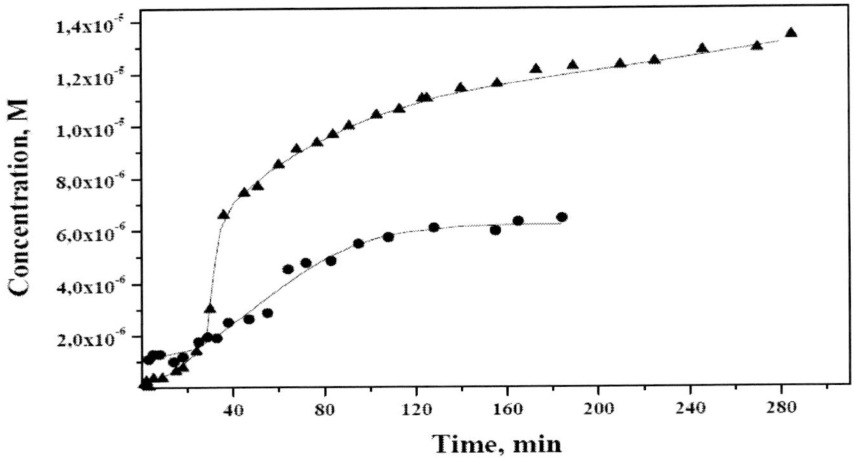

b

Figure 15.a. ESR spectra of radicals derived from quinine methide Q in styrene solution ($4.95x10^{-2}$M) at 130°C for 20min(A) and 150 min(B). b. Kinetics of radical formation in styrene solution of quinine methide Q at 130° C: (●) - $4.95x10^{-2}$ M , in the presence of oxygen; (▲) - $4.09x10^{-2}$ M, in the absence of oxygen.

Initial ESR spectrum transforms during the thermal treatment at 130° C. Signal intensity of radical (XIII) decreases, and at the same time signal intensity of secondary stable radical increases (figure 15a(B)). It means that radical (XIII) concentration decreases, and at the same time concentration of secondary phenoxyl radical increases. The ESR spectrum of this radical is triplet determined by interaction of uncoupling electron with two hydrogen atoms of phenyl ring (a_H=1.8 G). Secondary phenoxyl radical arises due to reaction of disproportionation[13] and reaction of active radical $R^{\bullet}$ with quinine XIV (Scheme 14). It is important to notice that the concentration of secondary phenoxyl stable radicals is high at later stage (300 min) of the process.

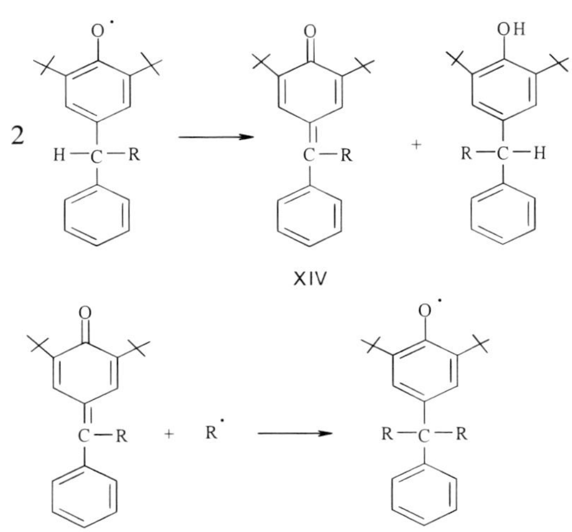

Scheme 14.

Kinetics of the radical formation at 130° C is presented on figure 15b. Notice that the stable radical concentration is approximately 10^3 times less than the concentration of quinone methide Q in styrene solution. The concentration of radicals in the presence of oxygen is significantly less than the radical concentration in the absence of oxygen. It can be explained by the fact that phenoxyl radicals are usually unstable in the presence of oxygen [13]. After 140 min of thermal treatment at 130° C radical concentration varies insignificantly. It seems that reversible inhibition takes place (Scheme 15).

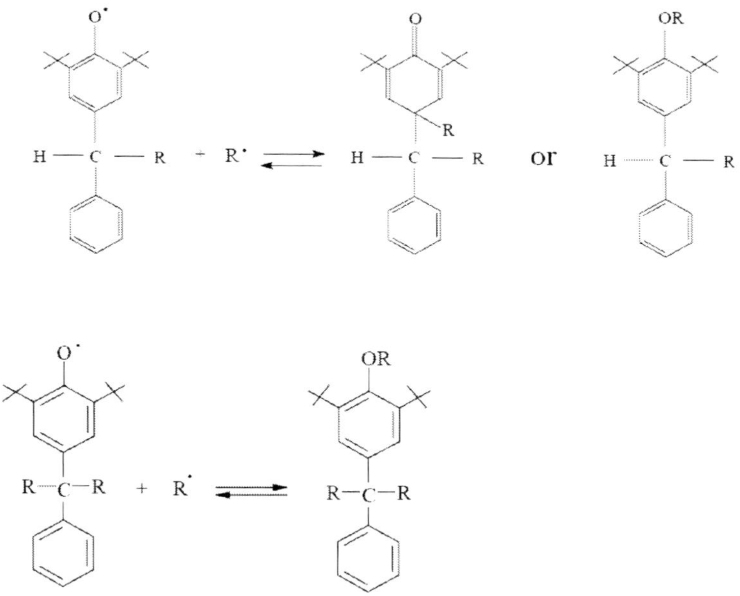

Scheme 15.

3.4.2. Quinone Methides QE (2,6-Di-Tert-Butyl-Ethylidenecyclohexa-2,5-Dien-1-One) and TBSQ (3,5,3',5'-Tetra-Butylstilbenequinone)

3.4.2.1. Quinone Methides QE and TBSQ in Inert Solvents

No ESR signal in solutions of QE in inert solvents (m-dichlorobenzene, m-xylene) was detected at room temperature. ESR spectra of radicals formed in these solutions with argon at 130° C are presented on figure 16a. ESR spectrum (triplet of triplets, $a_{H(1)}$=9.1 G, $a_{H(2)}$=1.8 G) was observed for the first minutes of heating. The spectrum can be assigned positively to phenoxyl radical XVa. Hyperfine constant, $a_{H(1)}$, is due to interaction of uncoupling electron with two protons of CH_2-group in *para*-position of phenoxyl radical. Hyperfine constant, $a_{H(2)}$, is determined by interaction of uncoupling electron with two *meta*-protons of aromatic ring. Similar spectra of radicals XVa are very well known and are described in literature in details [39]. The ESR spectrum changes during heating. It is due to formation and accumulation of stable radicals that are products of transformation of radical XVa.

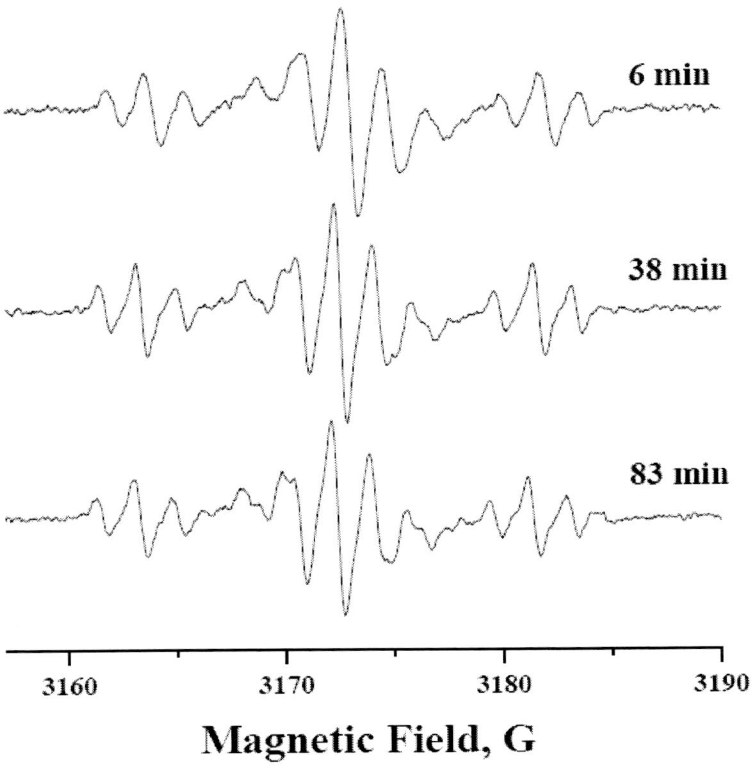

a

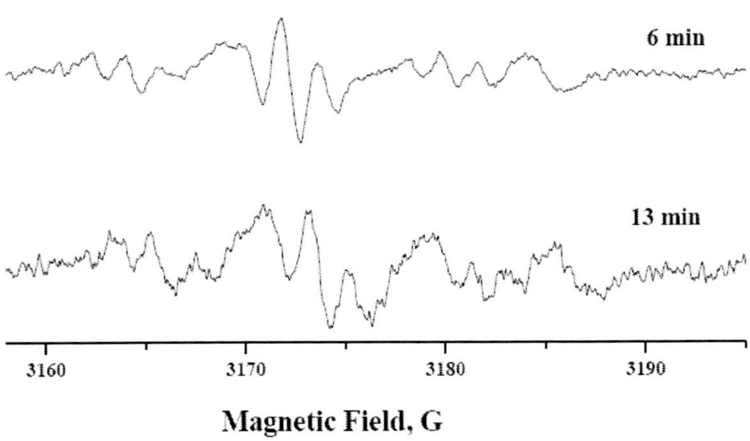

b

Figure 16.a. The ESR spectra of radicals formed in QE solution in m-xylene; (7.85x10^{-2}M) with argon at 130° C. b. The ESR spectra of radicals formed in TBSQ solution in m-xylene with argon (9.2x10^{-3}M) at 130° C.

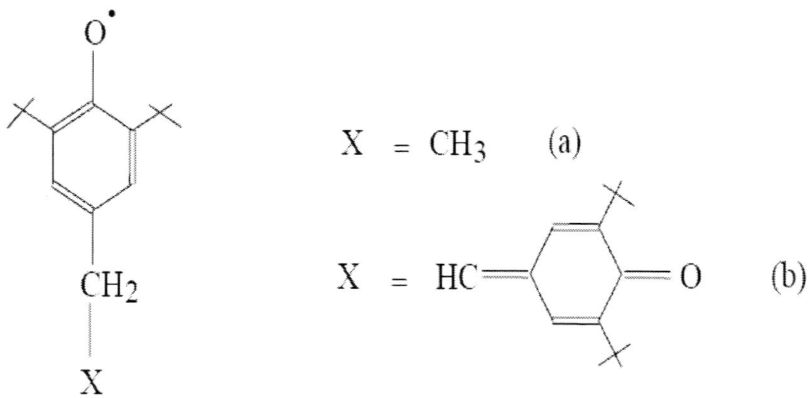

XV

Scheme 16.

ESR spectra of radicals formed in TBSQ solution in m-xylene with argon at 130° C are presented on figure 16b. The spectrum observed for first minutes of heating is similar the spectrum observed in solutions of QE. The spectrum can be assigned to radical XVb ($a_{H(1)}$=9.1 G, $a_{H(2)}$=1.8 G). During heating the shape of signal does not changes practically while the signal intensity decreases significantly. It seems likely that the radical termination proceeds without formation of new radical products.

Formation of stable phenoxyl radical in solutions of quinone methides QE and TBSQ in inert solvent can be understood in terms of involving of exited level of quinine methids[29]. Remind that molecule of quinone methide has low-lying exited level. The first step of reaction can be described by the formation of biradical XVI. (scheme 17).

Phenoxyl radical XV can be formed as the result of hydrogen abstraction reaction of carbon center of biradical XVI with solvent molecule (RH) or with other molecule of quinone methide (scheme 17). ESR spectrum of the phenohyl radical XV is detected for first minutes of heating at 130° C. With time this radical either changes slightly (solutions of QE) or terminates (solutions of TBSQ).

3.4.2.2. Quinone Methides QE and TBSQ in Styrene

No EPR signal in solutions of QE in styrene was detected at room temperature. ESR spectra of radicals formed in styrene at 130° C are presented on figure 17. The ESR spectrum (triplet of triplets, $a_{H(1)}$=9.1 G, $a_{H(2)}$=1.8 G) observed within first minutes can be positively assigned to phenoxyl radical XVa. Mechanism of radical formation was discussed above (scheme 17). With time the ESR spectrum changes. It seems that in addition to the reactions leading to formation of radical XVa reaction of QE with active alkyl radical (R•) of styrene (scheme 18) proceeds. This reaction leads to formation of phenoxyl radical XVII. Besides, isomerization and disproportional reactions of radicals XVa and XVII can take place also. As a result, the ESR spectrum observed is a superposition of spectra of different phenoxyl radicals formed in solution of QE in styrene at 130° C.

ESR spectra of radicals formed in TBSQ solution in styrene at 130° C are shown on figure 18. During first minutes of heating radicals XVb (triplet of triplets, $a_{H(1)}$=9.1 G, $a_{H(2)}$=1.8 G) form. Mechanism of radical formation was discussed above (scheme 17). With time concentration of radicals XVb decreases, and new radicals appear in solution. It seems, that in addition to the reactions, presented in Scheme 17, reaction of TBSQ with active alkyl radical (R•) of styrene (scheme 19) proceeds. Besides, further transformations of radicals XVb and XVIII leading to new phenoxyl radicals (for instance, in accordance with scheme 14) are also possible. As a result, superposition of ESR spectra of different phenoxyl radicals is observed.

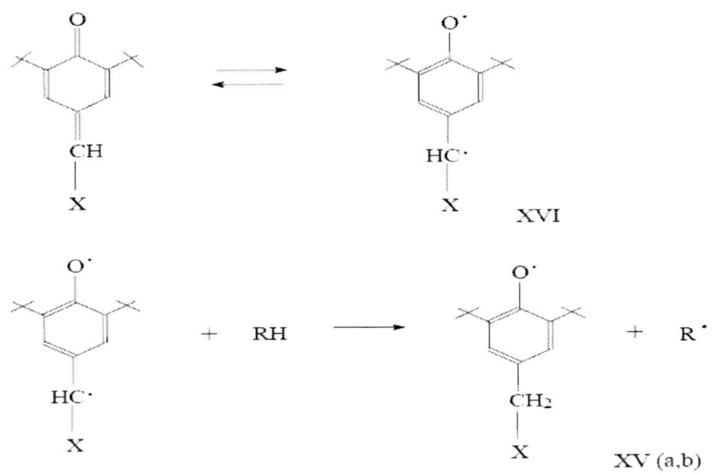

Scheme 17.

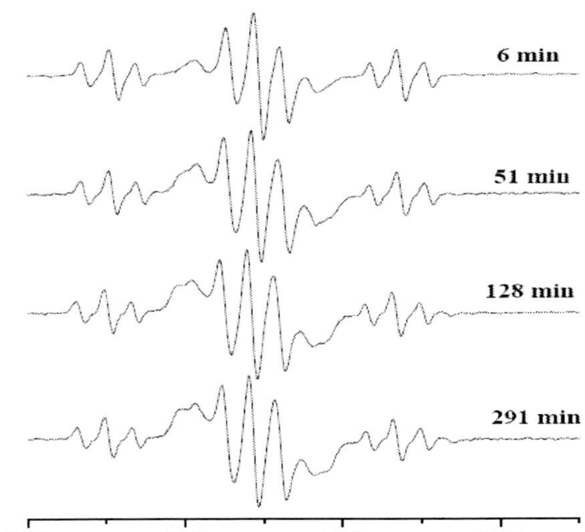

Figure 17. The ESR spectra of radicals formed in QE solution in styrene($8.4x10^{-2}$M) with argon at 130° C.

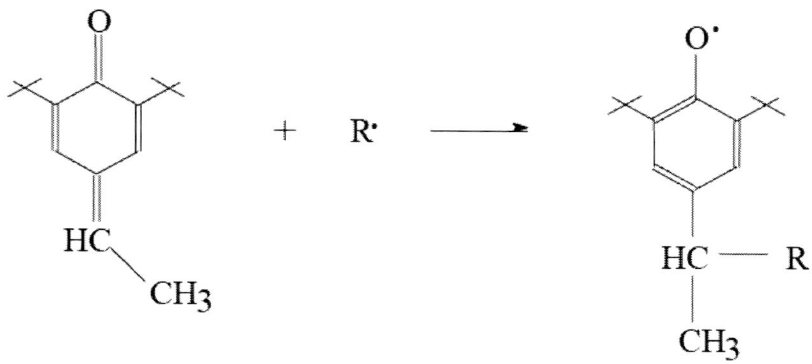

XVII

Scheme 18.

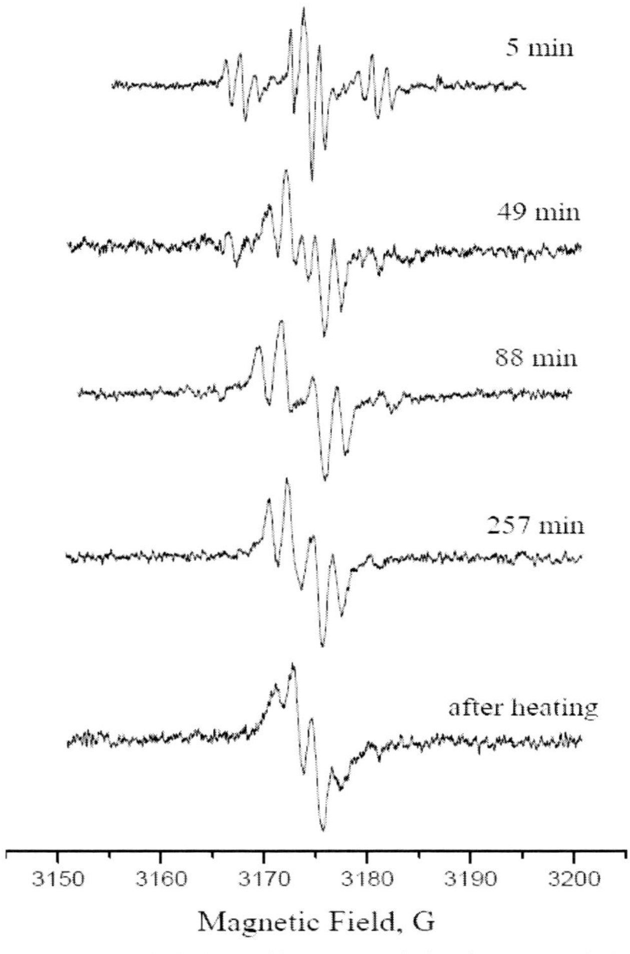

Figure 18. The ESR spectra of radicals formed in TBSQ solution in styrene; ($9.9x10^{-3}$M) with argon at 130°C.

Kinetics of formation of stable radicals in solutions of QE in m-xylene and in styrene at 130° C are shown on figure 19a.

The concentration of phenoxyl radicals in solution of QE in m-xylene increases during first 25 min and then changes insignificantly (at least during 250 minutes). Overall concentration of stable radicals in solution of QE in styrene increases during first 100 minutes and, then changes insignificantly. It is important to emphasize that stationary concentration of stable phenoxyl radicals in styrene solution remains invariable during a long time (at least during 300 minutes). Besides, the stationary concentration of stable radicals in QE solution in styrene is significantly higher than in QE solution in m-xylene.

Kinetics of formation of stable radicals in TBSQ solutions in m-xylene and in styrene at 130° C are shown on figure 19b. Phenoxyl radicals XVb form during first minutes in solution of TBSQ in m-xylene. The radicals disappear rapidly, and after 13 minutes the radical concentration is negligible. Phenoxyl radicals form during first minutes in the solution of TBSQ styrene. The concentration of the phenoxyl radicals in solution remains sufficiently high during the long time (300 minutes) and does not decrease.

So, stable phenoxyl radicals form in solutions of quinone methides QE and TBSQ in styrene at 130° C. The phenoxyl radicals can participate in reversible inhibition and in mediation of living radical polymerization.

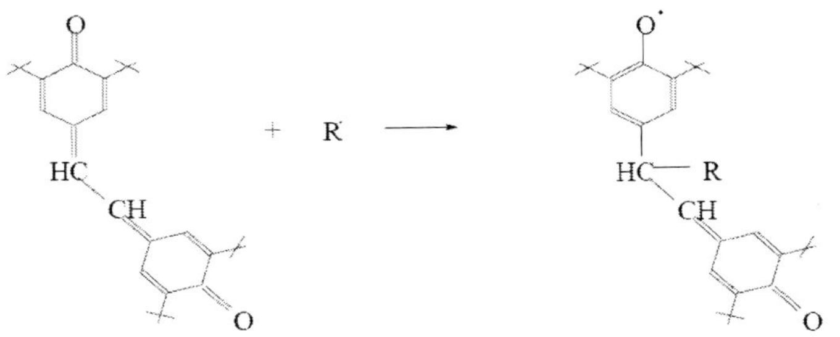

XVIII

Scheme 19.

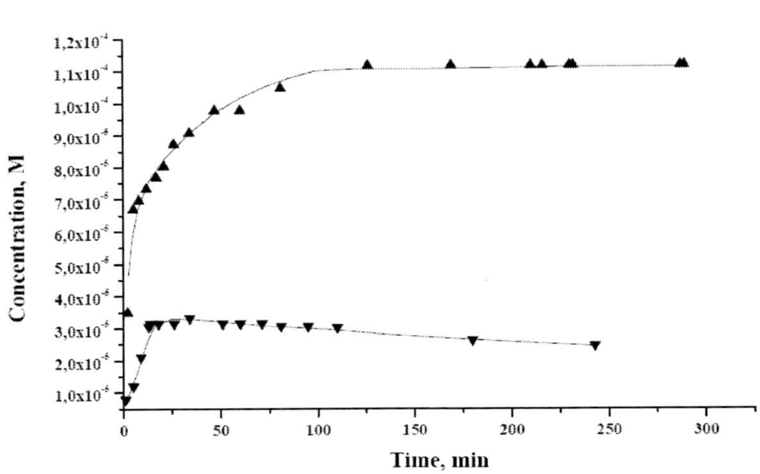

a

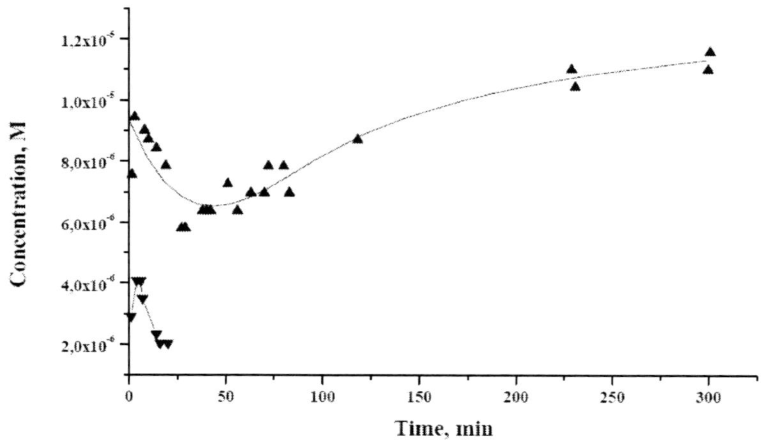

b

Figure 19.a. Kinetics of stable radical formation in QE solutions in styrene; ($8.4x10^{-2}$M) with argon (▲) and in m-xylene ($7.85x10^{-2}$M)with argon(▼) at 130°C. b. Kinetics of stable radical formation in TBSQ solutions in styrene; ($9.9x10^{-3}$M) with argon (▲) and in m-xylene ($9.2x10^{-3}$M) with argon(▼) at 130°C.

4. CONCLUSION

Results presented in this study show that stable radicals are derived from inhibitor molecules in process of styrene polymerization. The nitroxide radicals are produced from nitrosocompound, hindered hydroxylamine, nitrophenols and nitroanisoles. The phenoxyl radicals are produced from quinine methides and naphoxyl radicals are generated from 2-nitro-1-naphtol. These radicals can participate in the process of living radical polymerization as the mediators and can effect significantly on kinetics of polymerization and structure the resulting polymer.

To perform the function of reversible inhibitors, the stable radicals must fill several requirements [4].

First, they are bound to be stable at temperatures of the living polymerization being carried out.

Second, the concentration of stable radicals should fall in the range of 10^{-5} – 10^{-3} M at normal concentrations of the growing polymeric radicals. At lower concentrations, the living feature of polymerization is affected by strong contribution of quadratic termination of the growing polymeric radicals. At higher concentrations, the rate of polymerization is too slow.

Third, the concentration of the stable radicals-inhibitors may not vary essentially during the polymerization except for the initial time interval when the process attains the steady state regime. Essential reduction of the stable radicals concentration during polymerization causes a rise in fraction of "dead" chains. Generally, this happens due to side chemical reactions.

As is evident from our work, the majority of examined compounds fit satisfactorily the requirements formulated above. These compounds generate stable radicals in the necessary concentrations, which generally vary only slightly during polymerization. It must be emphasized that the listed requirements are necessary but not sufficient for realization of living radical polymerization. Nevertheless, there is reason to hope that the compounds represented in this paper offer the new contributors of stable radicals capable by means of reversible inhibition to control the growth of the polymeric chain in reactions of polymerization. Effectiveness of these compounds will be characterized in future researches.

ACKNOWLEDGEMENTS

The authors thank Prof. V.B. Golubev and Dr. M.Yu. Zaremski for helpful discussion.

REFERENCES

[1] *Advances in Controlled/Living Radical Polymerization*. Ed. K.Matyjaszewski. Oxford University Press, Oxford, 2003.

[2] K.A. Davis, K. Matyjaszewski.Statistical, Gradient, Block and Graft Copolymers by Controlled/Living Radical Polymerization. *Advances in Polymer Science.* V.159 , Springer Verlag, Berlin, 2002.

[3] C.J.Hawker, A.W.Bosman, E.Harth. *Chem.Rev.* 101(2001),3661.

[4] M.U. Zaremskii, V.B. Golubev. *Polymer Science.* 43C (2001), 81.

[5] G.V. Korolev, A.P. Marchenko. *Russian Chemical Reviews,* 69, (2000) , 409.

[6] D.F. Grishin, L.L. Semyonycheva, E.V. Kolyakina. *Mendeleev. Commun.* N6, p.250 (1999).

[7] `J-M.Catala, S.Jousset, J-P. Lamps. *Macromolecules.* 34(2001), 8654.

[8] M-O. Zink, A.Kramer, P.Nesvadba. *Macromolecules.* 33(2000), 8106.

[9] C.Detremleur, V.Sciannamea, C.Koulic et al. *Macromolecules*. 35(2002), 7214.

[10] V.B. Golubev, M.Yu.Zaremski, A.P.Orlova, A.V. Olenin. *Polymer Science,* 46A (2004), 295.

[11] M.Yu.Zaremski, A.P.Orlova, E.S. Garina,A.V. Olenin et al. *Polymer Science*, 45A (2003), 502.

[12] E.G. Jansen. *Account Chem. Res.* (1971) , 31.

[13] A.L. Buchachenko. Stable Radicals . Consultants Bureau, New York, 1965.

[14] A.L. Buchachenko, A.M. Wasserman. Stable Radicals. Chemical Structure, Reactivity and Applications. *Chemistry.* Moscow, 1973 (in Russian).

[15] G.R.Chalfont, M.I. Perkins, A. Horsfield, *J. Am. Chem. Soc.*, 90, (1968). 7141.

[16] E.G Rozantsev. Free iminoxyl radikals . M., *Chimija,* 1970 (in Russian).

[17] I.Gyongyhalmi, F.Foldes-Berezsnich, F. Tudesh, *Eur. Pol.* J. 31, (1995), 45.

[18] M.K. Georges, R.P.N. Veregin, P.M. Kazmaier, G.K. Hamer. *Polym. Mater. Sci. Eng.* .68, (1993), 6.

[19] W.Aherens, A.Berndt, *Tetraedron Lett.,* (1973),4281.

[20] S.Terabe, R.Koneka, *J. Chem. Soc., Perkin II,* (1973),369.`

[21] H.G. Aurich, W. Weiss. *Topics Curr. Chem.* 59, (1975), 65.

[22] E.G. Rozantsev, V.D. Sholle. Organic Chemistry of Free Radicals. Moscow, *Chemistry,* 1979 (in Russian).

[23] A.M.d'A. Rocha Gonsalves, J. M.T.B. Varejao, M. M.Pereira, *J. Hetecyclic Chem.,* 28, (1991), 640.
[24] E. G. Jansen, *J. Am. Chem. Soc.* V.87, N 15, p.3531 (1965).
[25] N.I. Boguslavskaya, G.F. Mart'yanova , O.E. Yakimchenko, Ya.S. Lebedev et al. Dokladi AN.SSSR v..220, №3, p.617 (1975).
[26] M.M. McMillian, R.O.C. Norman , *J. Chem. Soc.* (B), (1968),590.
[27] Volod'kin A.A., Ershov V.V. *Uspechi chimii ,* v. 57, №4 c. 595 (1988)
[28] E.R. Altwioker *Chem. Rev.,* v.67 N 5, p. 475(1967).
[29] V.V. Ershov., G.A Nikiforov., A.A. Volod'kin. Hindered Phenols. M., *Chimija,* 1972(in Russian)
[30] L.Filar., S.Winstein, *Tetrahedron Lett.* (1960),9.
[31] A.A Volod'kin., V.V Ershov., L.I Kudinova. Izvestija AN SSSR, ser. *Chimicheskaja,* (1978), 512.
[32] R.H Bauer., G.M Coppinger. *Tetrahedron,* (1963),1201
[33] G.M. Coppinger. *J. Amer. Chem. Soc.,*.86, (1964),4385.
[34] B.R Loy. *J. Org. Chem.* .31, (1966), 2386.
[35] L.R Mahoney., M.A Darooge., *J. Amer. Chem. Soc.,* .89, (1967), 5619.
[36] J Pospisil., S. Nespurek., H. Sweifel., *Polymer Degradation and Stability,* 54, (1996),7; 54, (1996), 15.
[37] N.V Zolotova. , M.B Tokareva., E.T. Denisov et al., *Kinetika I kataliz* ,20, (1979), 56 (in Russian).
[38] Zolotova N.V. , Galieva F.A., Tokareva M.B., Denisov E.T. et al. *Kinetika I kataliz*, 20, (1979),48 (in Russian).
[39] V.D. Pokhodenko. Phenoxyl radicals. Kiev. *Haukova dumka.* 1969 (in Russian).

In: Progress in Chemistry and Biochemistry
Editors: E. M. Pearce and G. E. Zaikov
ISBN: 978-1-60692-344-3

Chapter 21

USE OF THE EFFECT OF THERMAL EXPANSION OF POLYMER–GRAPHITE COMPOSITIONS FOR SYNTHESIZING LOW–DENSITY MATERIALS WITH CONDUCTION AND SORPTION PROPERTIES

To fond memories of my wife,
Valentina Dmitrievna, dedicated

V. N. Gorshenev
Emanuel Institute of Biochemical Physics,
Russian Academy of Sciences,
ul. Kosygina 4, Moscow, 119991 Russia

ABSTRACT

Results on the thermal expansion of polymer–graphite compositions prepared from plasticized poly(vinyl chloride) (PVC), SKTN-A low-molecular-weight organosilicon rubber, and polyethylene (PE) are presented. All the compositions tested contained graphite modified (MG) with a mixture of acids. The samples were heated in an oven to produce foamed low-density compositions. When heated, modified graphites are known to release gaseous products and experience thermal expansion in a thermoplastic composition to form a branched structure of current-conducting distributions of graphite particles. For all the polymer compositions, the temperature and concentration conditions for the nonconducting-to-conducting-state transition to occur were determined. The amounts of hydrocarbons sorbed and the oil-scavenging capacity of the thermally expanded polymer–graphite compositions were determined. A combined action of the sorbabilities of the polymer and thermally expanded graphite components makes it possible to localize oil spills on a water surface before appropriate mechanical means of oil collection are employed.

Keywords: Electroconducting polymer composites, modified graphites, sorbents.

INTRODUCTION

Foamed materials are widely used in solving numerous practical problems, such as the production of sound-absorbing and heat-insulation constructions, development of radiotechnological means of protection from electromagnetic radiation, manufacturing of sorption materials for localization of oil spills on a water surface. Traditional approaches to synthesizing foamed materials [1−4] are based on the introduction of gas-forming agents, which, when heated decompose to release gaseous products. In contrast to traditional gas-forming agents, we used modified graphites, which, when heated, release various gaseous products, largely CO, CO_2, and H_2O [5−8], and experience thermal expansion with the formation of a branched structure of current-conducting distributions of graphite particles.

In the present work, we report experimental data on the thermal expansion of modified graphites in composites, graphites that can be used for producing various low-density polymers and for solving other practical problems.

RESULTS AND DISCUSSION

It is known that the oxidation of graphite particles by sulfuric acid in the presence of nitric acid proceeds via the intercalation of sulfate anions into the interlayer space of the graphite crystal lattice. After washing and drying, a modified graphite (MG) with a loose packing density of ~300 g/l is obtained, which is used as a filler of polymers. Figure 1 shows SEM microimages of expanded MG particles (JSM-35 microscope, 300×). Initial graphite particles with a diameter within 200−600 μm appear as folded-structure plates 20 to 60 μm in thickness. As MG particles are heated, they bloat (frames *1* and *2* in figure 1), with the degree of expansion increasing with a temperature rise from 200 to 250°C. The thickness of particles bloated at 250°C was found to be ~500 μm. Note that graphite plates delaminate into thinner layers, with a thickness less than ~1 μm.

Depending on the nature of the polymer, the test composites were prepared by rolling, extrusion, pressing, and other methods at temperatures below 170°C. Flat samples, 1 to 2 mm in thickness, were studied using thermophysical, optical, electrophysical, and other methods [1−2].

Studies of the thermal expansion of MG−polymer composites by thermogravimetry (TG), thermomechanical analysis (TMA), and differential scanning calorimetry (DSC) demonstrated that this process occurs within 180−250°C. For some MG−polymer compositions the expansion of MG particles in the polymer matrix was of stepwise character [9].

Figure 2 shows a thermomechanical curve for the thermal expansion of a 6 wt % MG−PVC sample. At temperature above 180°C, the curve is indicative of a stepwise character of indentor movement, a behavior that may provide insights into the mechanism of the thermal expansion of MG. As the graphite particles expand, the distance between them decreases.

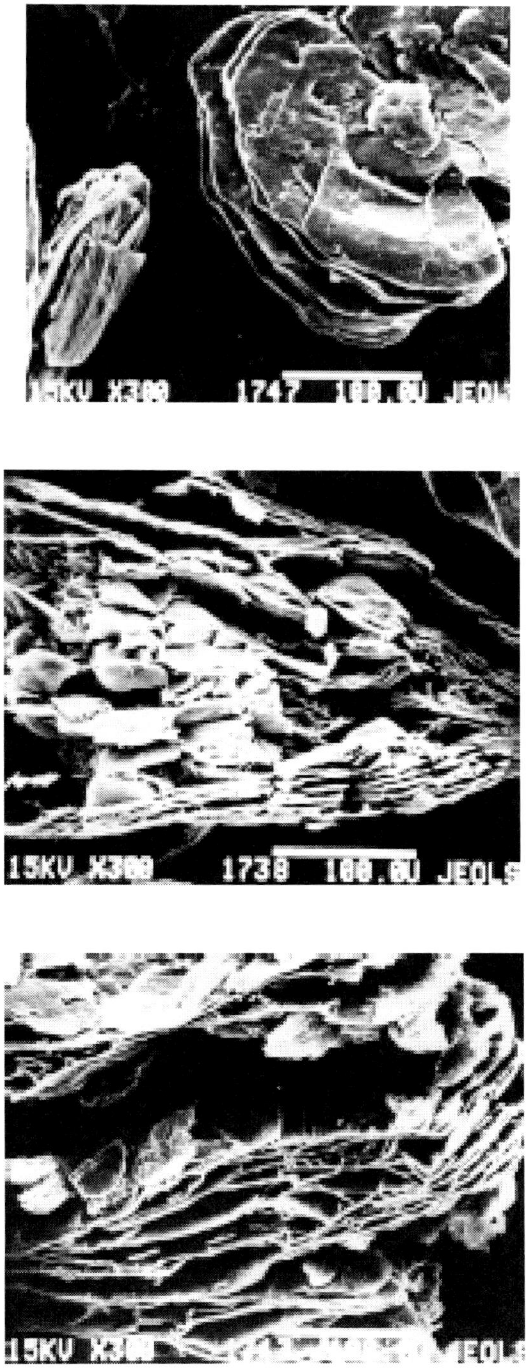

1

2

3

Figure 1. SEM microphotographs of expanded MG particles: (*1*) initial graphite and (*2, 3*) thermally expanded graphites at 200 and 250°C, respectively; JSM-35 microscope, magnification 300×.

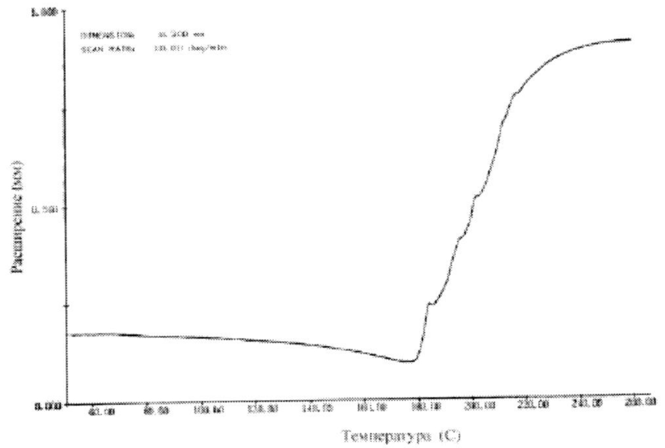

Figure 2. TMA curve for a 6 wt % MG–PVC plasticized sample.

The time evolution of the particle size and interparticle distance was monitored as follows. The sample of a composite material was attached to a microscope slide in the vertical plane, after which the support with the sample was placed into an oven. The microscope was preliminarily focused when the sample was still cold. The heating rate was set at 10–15 K/min with the help of an adjustable-ratio autotransformer. The temperature of the oven was measured with a thermocouple positioned near the sample surface and an ARRA-207 meter. Snapshots of the sample were taken at regular temperature intervals and processed with a PC computer. The snapshots show how the microstructure of the graphite particles changes with increasing temperature. Computer processing of snapshots yielded the mean particle diameter d_m and mean interparticle distance l_m. as functions of the temperature (figure 3). As can be seen, the mean diameter increases with the temperature severalfold (curve *1* in figure 3).

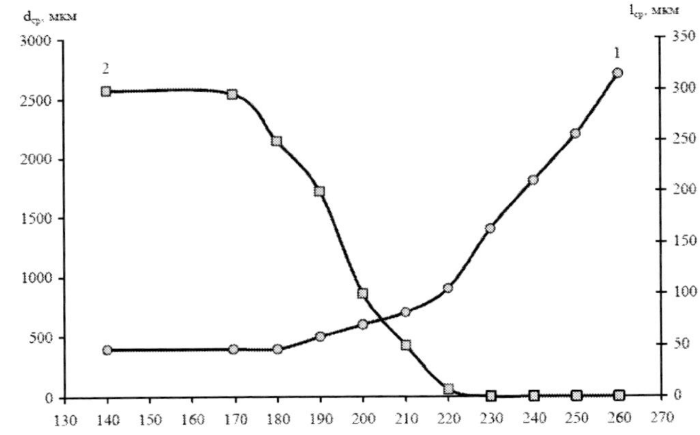

Figure 3. (*1*) Mean size of MG particles d_m and (*2*) mean interparticle distance l_m for a sample (Fig. 5) containing 6 wt % MG in plasticized PVC as a functions of the temperature.

The electric conduction arises as the particles approach each other closely enough to touch and form conducting chains throughout the sample (figure 4).

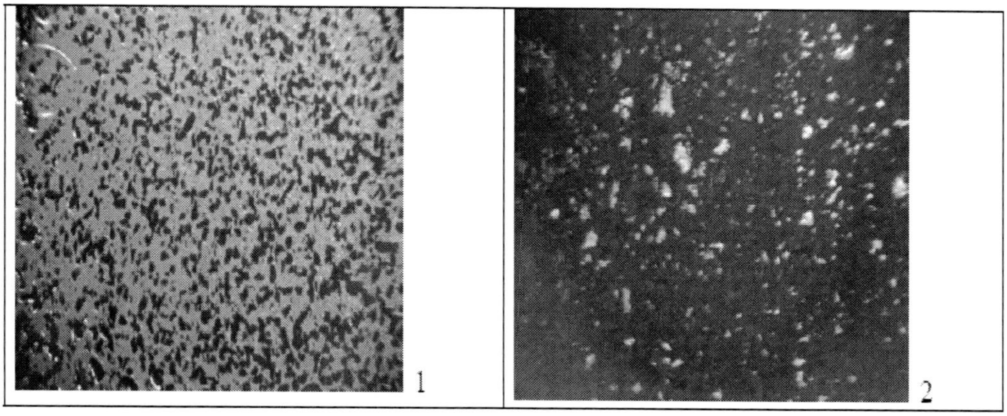

Figure 4. Photographs of the structure of the MG–PVC samples at (*1*) 140 and (*2*) 260°C.

For the sample prepared on the basis of the low-molecular-weight rubber cured with a cold-curing catalyst too, the mean particle diameter increases while the mean interparticle distance decreases. In contrast to the sample prepared on the basis of plasticized PVC, the expansion of graphite particles in cured CKTN is hampered (figure 5).

Observations of the expansion of MG particles in polymer composites demonstrated that, if a composite material is subjected to polymerization or curing, a network structure is formed that impedes the expansion of MG particles in the polymer matrix. In this case, elevated temperatures and large MG concentrations are needed to make the sample electroconductive. Studies of the thermal expansion of MG particles in polymer matrices demonstrated that, at moderate concentrations of MG in the polymer matrix after foaming, the sample acquires electroconductivity high enough for various practical applications [9, 10]. Thus, in heated thermoplastic MG–polymer compositions, along with various thermal processes, two additional processes occur: the bloating of graphite particle layers and the foaming of the composite due to gas release. Therefore, it is quite possible that the conductivity of the sample may diminish if the foaming of the polymer is faster than the expansion of the graphite particles, and, therefore, the volumetric density of current-conducting structures in the polymer becomes increasingly smaller.

Table 1 lists the experimentally measured characteristics of thermally expanded MG–PE samples. As can be seen, the originally (prior to thermal expansion) nonconducting samples become conducting upon foaming, despite the fact that their density is lower. As the concentration of MG in the PE matrix increases from 5 to 50 wt %, the degree of expansion increases from 2 to 28 (table 1).

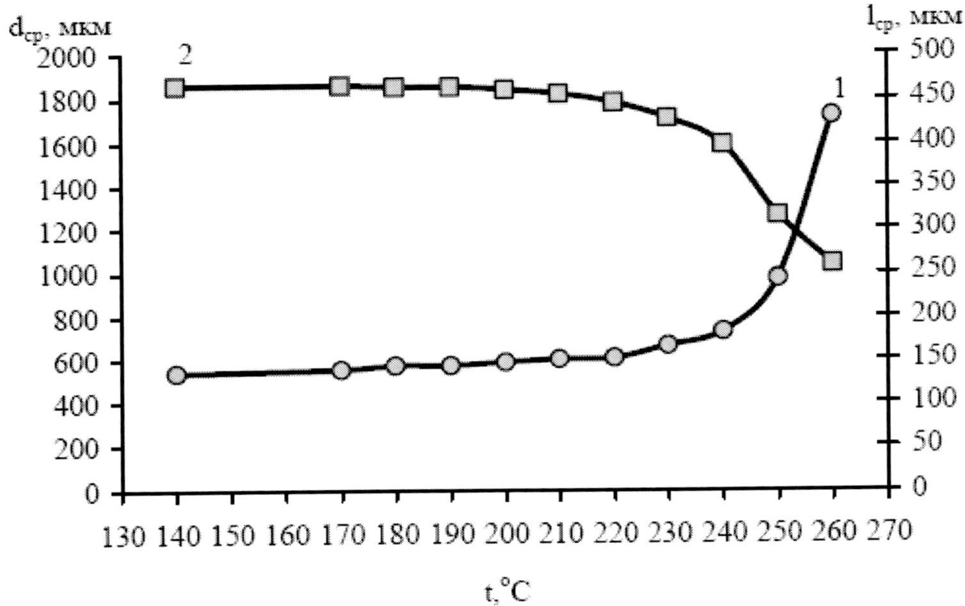

Figure 5. Temperature dependences of (*1*) mean size of MG particles d_m and (*2*) mean interparticle distance l_m for a sample containing 6 wt % MG in SKTN-A low-molecular-weight rubber.

Table 1. Degree of thermal expansion and conductivity of polyethylene-graphite compositions

No	Concentration of modified graphite in PE, wt %	Conductivity of compositions before thermal expansion, Ω^{-1} cm^{-1}	Conductivity of compositions before thermal expansion, Ω^{-1} cm^{-1}	Degree of expansion of the composites (ratio of volume before and after thermal expansion, V_2/V_1
1	5	–	4.3×10^{-4}	2.0
2	10	–	2.6×10^{-4}	4.0
3	15	–	4.4×10^{-4}	5.4
4	20	–	2.3×10^{-4}	7.0
5	25	–	8.1×10^{-4}	11.0
6	30	8.5×10^{-4}	7.7×10^{-4}	19.5
7	40	3.8×10^{-4}	–	25.0
8	50	1.6×10^{-4}	–	28.0

The electroconductive foamed composites exhibit substantial changes in the coefficients of transmission of electromagnetic radiation due to a better matching between the wave impedances of the composite and free space [4, 10]. Figure 6. displays the dependences of the transmission coefficients for the initial and foamed samples in the microwave range (1 cm).

We revealed that, depending on the technology of production of MG–polymer composites, it is possible to vary the degree of foaming, electric conductivity, and shape, for example, spherical or pyramidal items.

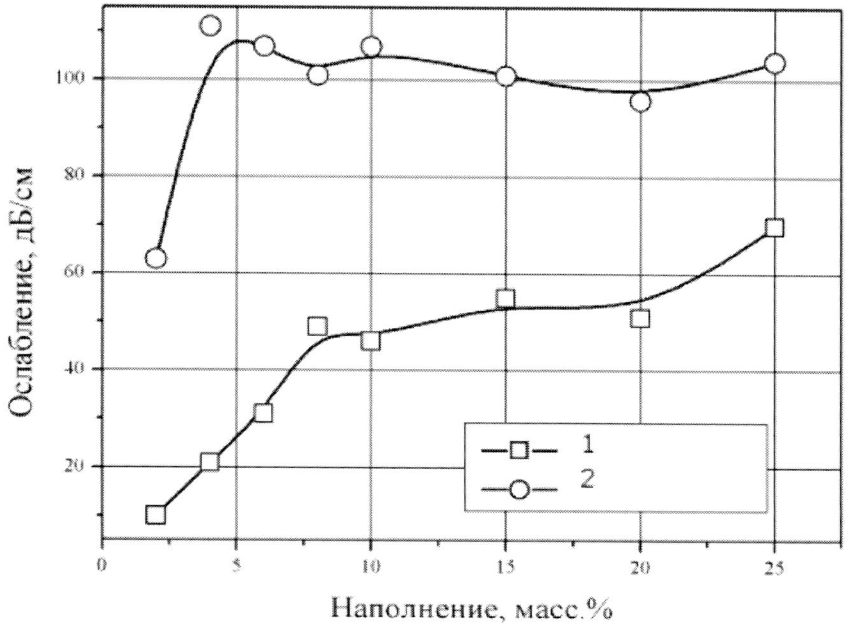

Figure 6. Dependence of the transmission coefficients for (1) initial and (2) foamed materials in the microwave range (1 cm).

The effect of thermal expansion of graphite–polymer compositions can be employed to decrease the combustibility of polymer composites. When subjected to the action of a flame, the composition expands, creating a layer that impedes flame propagation. The introduction of MG particles into polystyrene and into dyes for soaking foamed polyurethane made it possible to obtain low-combustibility materials.

A discussion of the results on the use of thermally expanded graphites for localization and collection of spills of oil and oil products can be found in [11, 12]. The development of economically efficient and high-performance methods for eliminating the consequences of spills of oil and oil products during production, transportation, and conversion is an urgent problem, in view of a large number of technogenic catastrophes and ensuing ecological disasters.

One of the main methods for collecting spills of oil and its products is based on the use of sorbents capable of localizing spills, a factor that simplifies subsequent removal of contaminations from the surface of soil or water. One of such sorbents is a thermally extended MG–LDPE composite material. LDPE has a low melting temperature, a factor that facilitates the preparation of the composite material. LDPE and MG-1 (10–50 wt %) were mixed in a Brabender-type mixer at 120–140°C. Granules of the composite material were prepared by mechanically crushing the mixture at room temperature and by extrusion. Granules were then heated to 230–270°C for 5 min. The sorption characteristics of composite materials were determined gravimetrically.

Foamed granules have a substantially larger size and significantly lower density compared to the initial composite material. In particular, the loose packing density of MG-based compositions (50 wt %) decreases from 250 g/l before expansion at 265±5°C to 24.6 g/l after.

Upon thermal expansion, the sorbability of the material with respect to aliphatic and aromatic hydrocarbons, diesel fuel, and oil increases dramatically (table 2). In this respect the synthesized materials are substantially superior to the known PE-based sorbents, the sorption capacity of which is only 4 g/g. At the same time, the sorbability of the aforementioned composite materials is noticeably inferior to that of thermally expanded MG [13]. The advantages of the developed composite materials compared to powdery sorbents concern the following aspects: (1) substantially lower density (floatable not only in water but also in hydrocarbons), a feature that facilitates their removal upon decontamination, (2) granular, nondusty form, and (3) the possibility of producing sorbent items of nearly arbitrary shape.

Along with PE, other thermoplasts and rubbers can be used as the sorbent matrix.

Thus, the low-density materials with sorption and electric conduction properties described in the present work can find applications in solving various fundamental and practical problems. Combining MG-type fillers with other foaming additives (for example, antipyrenes and modified lamellar compounds) may widen the scope of applications of low-density polymer-based composite materials.

Table 2. Coefficients of sorption of hydrocarbons and oil products by sorbents prepared from thermally expanded polyethylene–graphite compositions (50 wt % MG)

Hydrocarbon	Polyethylene–MG	
	Before thermal expansion	After thermal expansion
Heptane	0.82	9.8
Toluene	1.54	13.6
Diesel fuel	1.33	12.2

REFERENCES

[1] A. A. Berlin and F. A. Shutov, Reactive-Oligomers-Based Foamed Materials (*Khimiya,* Moscow, 1978) [in Russian].

[2] O. G. Trakanov, I. V. Shamov, and V. D. Al'pern, Filled Foamplastics (*Khimiya,* Moscow, 1989).

[3] V. S. Pirumov, A. G. Alekseev, and B. V. *Aizikovich Zarub. Radioelektr*. No. 6, 2 (1994).

[4] V. N. Gorshenev, S. B. Bibikov, and V. N. Spektor, *Synth. Met.* 86, 2255 (1997).

[5] W. Rudorff and U. Hofmann, *Z. Anorg. Allg. Chem.* 238 (1), 1 (1938).

[6] A. Metrot and J. E. Fisher, *Synth. Met.* 3 (3), 201 (1981).

[7] I. G. Chernysh, I. I. Karpov, G. P. Prikhod'ko, and V. M. Shai, Physicochemical Properties of Graphite and Its Compounds (*Naukova Dumka,* Kiev, 1978) [in Russian].

[8] G. I. Titelman, V. N. Gelman, Yu. V. Isaev, and Yu. N. Novikov, *Materials Science Forum.* 91-93, 213 (1992).

[9] V. N. Gorshenev and A. N. Shchegolikhin, *Khim. Fiz.* No. 1 (2008).

[10] V. N. Gorshenev, S. B. Bibikov, E. I. Kulikovskii, and Yu. N. Novikov, A Composite Material for Absorbing Electromagnetic Radiation and the Method of Its Preparation, RF Patent No. 2242487, *Byull. Izobret.,* No. 35 (2004).
[11] V. N. Gorshenev, A. A. Ovchinnikov, and Yu. N. Novikov, *Zh. Fiz. Khim.* 75 (6), 1058 (2001).
[12] V. N. Gorshenev and V. I. Filippov, *Zh. Fiz. Khim.* 76 (6), 1119 (2002).
[13] V. N. Gorshenev and V. B. Ivanov Thermally Expanded Polyethylene-Graphite Sorbents for Localization and Elimination of the Consequences of Spills of Oil Products, *Abstracts of Int. Conf. on Chemical Technology.* "CT'07" (Moscow, 2007), vol. 2, p. 278.

In: Progress in Chemistry and Biochemistry
Editors: E. M. Pearce and G. E. Zaikov

ISBN: 978-1-60692-344-3

Chapter 22

SPATIAL-ENERGY PARAMETER AS A MATERIALIZED ANALOG OF WAVE FUNCTION

G.A. Korablev* and G.E. Zaikov**

* Basic research-educational center of chemical physics and mesoscopy,
Udmurt research center, Ural Division, RAS, Izhevsk, E-mail: korablev@udm.net
**Institute of biochemical physics after N.M. Emanuel, RAS, Russia,
Moscow 119991, 4 Kosygina St., E-mail: chembio@sky.chph.ras.ru.

ABSTRACT

The notion of spatial-energy parameter (P-parameter), a complex characteristic of the most important atomic values, was introduced for the relative motion of two interacting material points based on the modified Lagrangian equation. The physical sense of P-parameter as an averaged energy of valence electrons in an atom responsible for inter-atom interactions and having a direct connection with electron density in an atom is demonstrated. Wave properties of P-parameter are found, and its wave equation formally analogous to the equation of Ψ-function is given.

Keywords: Lagrangian equations, wave functions, spatial-energy parameter, electron density, Hamiltonian function.

1. SPATIAL-ENERGY PARAMETER

When summarizing, analyzing and introducing criteria of isomorphism, mutual solubility, phase-formation the quasi-classical notions referring to atom composition (or to more complex structures) are used, for instance, atom radius, electronegativity [1], electron density or combined characteristics (atom radius - electronegativity). At the same time, the application of quantum-mechanical calculations for the same purposes is much less effective.

However the existing models often use only one parameter, though many properties can be interpreted only taking different atom characteristics into account. Therefore the use of the parameter of unilateral dependence does not appear to be quite universal, and criteria applied can succeed only for certain cases and classes of compounds.

A definite compensation of volume energy of interacting structures takes place during the interaction of differently charged heterogeneous systems leading to the decrease of resulting volume energy (for instance, during the hybridization of atom orbitals). But this is not a direct algebraic deduction of corresponding volume energies. The comparison of multiple regularities of physical and chemical processes allows assuming that in such and analogous cases a principle of adding inverse values of volume energies or kinetic parameters of interacting structures is fulfilled. Here are some examples:

1) During the ambipolar diffusion when mutual transition of differently charged particles takes place in the given medium (in plasma, in electrolyte), the diffusion coefficient is found using the following relation:

$$\frac{\eta}{Д} = \frac{1}{a_+} + \frac{1}{a_-}$$

where a_+ and a_- - mobilities of charges of both atoms, η – constant coefficient.

2) Total rate of topochemical reaction (υ) between a solid and gas is defined as follows:

$$\frac{1}{\upsilon} = \frac{1}{\upsilon_1} + \frac{1}{\upsilon_2}$$

where υ_1 – diffusion rate of a reagent, υ_2 – reaction rate between gaseous reagent and a solid.

3) Changes in the light velocity ($\Delta\upsilon$) during the transition from vacuum into the given medium is calculated based on the principle of algebraic deduction of inverse values of corresponding velocities:

$$\frac{1}{\Delta\upsilon} = \frac{1}{\upsilon} - \frac{1}{c},$$

where c – light velocity in vacuum.

4) Lagrangian equation for relative motion of the system of two interacting material points with masses m_1 and m_2 in coordinate x is a s follows:

$$m_r x'' = -\frac{\partial U}{\partial x}$$

where

$$\frac{1}{m_r} = \frac{1}{m_1} + \frac{1}{m_2} \qquad (1)\ (1a)$$

here U – mutual potential energy of material points; m_r – reduced mass. At the same time $x'' = a$ (characteristic of system acceleration).

For elementary parts of interactions Δx we can take:

$$\frac{\partial U}{\partial x} \approx \frac{\Delta U}{\Delta x}$$

Then:

$$m_r a\Delta x = -\Delta U;$$

$$\frac{1}{1/(a\Delta x)} \cdot \frac{1}{(1/m_1 + 1/m_2)} \approx -\Delta U$$

or:

$$\frac{1}{1/(m_1 a\Delta x) + 1/(m_2 a\Delta x)} \approx -\Delta U$$

Since in physical sense $m_i a\Delta x$ equals the potential energy of each material point ($-\Delta U_i$), then:

$$\frac{1}{\Delta U} \approx \frac{1}{\Delta U_1} + \frac{1}{\Delta U_2} \qquad (2)$$

Thus the resulting energy characteristic of the interaction system of two material points is found when adding two inverse values of initial energies of interacting subsystems.

Therefore assuming that the energy of valence orbitals of an atom (responsible for inter-atom interactions) can be calculated by the principle of adding two inverse values of some initial energy components, we postulate the introduction of P-parameter as an averaged energy characteristic of valence orbitals according to the following equations:

$$\frac{1}{q^2/r_i} + \frac{1}{W_i n_i} = \frac{1}{P_E} \qquad \frac{1}{P_0} = \frac{1}{q^2} + \frac{1}{(Wrn)_i} \qquad P_E = P_0/r_i \qquad (3),(4),(5)$$

where W_i – orbital energy of electrons; [2] r_i – orbital radius of i–orbital [3]; $q=Z^*/n^*$ [4,5], n_i – number of electrons of this orbital, Z^* and n^* - effective charge of the nucleus and effective main quantum number.

P_O will be called spatial-energy parameter (SEP), and P_E – effective P–parameter (effective SEP). Effective SEP has a physical sense of some averaged energy of valence electrons in atom and is measured in energy units, for instance, in electron-volts (eV).

The values of P_0-parameter are tabulated constants for electrons of the given orbital of the given atom.

For the dimensionality of SEP we can write down the following:

$$[P_0]=\left[q^2\right]=[E]\cdot[r]=[h]\cdot[\upsilon]=\frac{kgm^3}{c^2}=J\,m$$

where [h] and [υ] – dimensions of Planck's constant and velocity.

The introduction of P-parameter should be discussed as further development of quasi-classical notions using quantum-mechanical data on atom composition for obtaining criteria of energy conditions of phase-formation. At the same time, for similarly charged homogeneous systems (for example, orbitals in the given atom) the principle of algebraic adding of such parameters is preserved:

$$\sum P_E=\sum(P_0/r_i) \qquad \sum P_E=\frac{\sum P_0}{r} \qquad (6),(7)$$

(where r – main dimension characteristic if atom), or:

$$\sum P_0=P_0'+P_0''+P_0'''+\ldots \qquad r\sum P_E=\sum P_0 \qquad (8),(9)$$

Here all P-parameters are summed up by all valence orbitals of atom.

To calculate P_E-parameter at the given distance from the nucleus, either atom radius (R) or ion radius (r_I) can be used depending upon the bond type.

Applying the equation (8) to hydrogen atom, we can write down the following:

$$K\left(\frac{e}{n}\right)_1^2=K\left(\frac{e}{n}\right)_2^2+mc^2\lambda \qquad (10)$$

where: e – elementary charge, n_1 and n_2 – main quantum numbers, m – electron mass, c – electromagnetic wave velocity, λ – wave length, K – constant.

Using the known relations ν = C/λ and λ = h/mc (where h – Planck's constant, ν – wave frequency) from the formula (10), we can obtain the equation of spectral regularities in hydrogen atom, in which $2\pi^2e^2$ /hC = K.

2. Effective Energy of Valence Electrons in Atom Using SEP Method. Comparison with Statistic Model

Modified equation of Thomas-Fermi transformed into a simple form when introducing non-dimensional variables [6] looks as follows:

$$U = e\left(V_i - V_0 + \tau_0^2\right) \tag{11}$$

where: V_o – count potential; e – elementary charge; τ_0 – exchange and correlation corrections; V_i – interatomic potential at the distance r_i from nucleus, U – total energy of valence electrons.

The given value U was compared with the values of P_E-parameter for 21 elements that are partly given in table 1.

As it is shown in table 1 the values of U and P_E-parameters are practically the same (with deviations in most cases below 1-2%) *without any transition coefficients*. Multiple corrections introduced into the statistic model are compensated with the application of simple rules of adding inverse values of energy parameters, and SEP rather accurately presents the known solutions of Thomas-Fermi equation for inter-atom potential of atoms at the distance r_i from the nucleus. Namely, the following equality takes place:

$$U = P_E = e\,(V_i - V_0 + \tau^2_0) \tag{12}$$

Using a known relation [6] between electron density (β_i) and inter-atom potential (V_i), we have:

$$\beta^{2/3}_i \approx (3e/5)\cdot(V_i - V_0);\ \beta^{2/3}_i \approx Ae\cdot(V_i - V_0 + \tau^2_0) = [Ae\cdot r_i\cdot(V_i - V_0 + \tau^2_0)]/r_i \tag{13}$$

where A – constant.

According to formulas (12 and 13) we have the following relation:

$$\beta_i^{2/3} = A\ P_0/r_i \tag{14}$$

that establishes the connection between P_0-parameter and electron density in the atom at the distance r_i from the nucleus.

Since the value $e(V_i\text{-}V_0+\tau_0^2)$ in the model of Thomas-Fermi there is a function of charge density, P_0-parameter is an immediate characteristic of charge density in the atom.

Table 1. Comparison of the total energy of valence electrons in atom calculated in the statistic model of the atom Thomas-Fermi *(U)* and with the help of approximation

Atom	Valence electrons	r_i(Å)	X	φ(X)	*U*(eV)	W(eV)	n_i	q^2(eV Å)	P_E(eV)
1	2	3	4	5	6	7	8	9	10
Ar	$3P^4$	0,659	3,548	0,09- 0,084	35,36- 33,02	12	4	73,196	33,45
	$3S^2$	0,607	3,268	0,122- 0,105	47,81- 44,81	34.41- 29,0	2 2	96,257 96,157	47,99 42,47
	$2P^4$	0,146	0,785	0,47	834,25	246	4	706,3	817,12
Kr	$4P^4$	0,795	5,501	0,0495- 0,0459	32,28- 34,94	11	4	85,883	31,27
	$4S^2$	0,691	4,781	0,0629- 0,0588	47,19- 37,50	27,5 24	2 2	115,22 115,22	41,36 37,26
	$3d^{1o}$	0,2135	1,481	0.2976- 0,296	722,2 718,5	89	10	680,68	695,6
	$3P^4$	0,244	1,688	0,267- 0,265	567,2- 562,7	214	4	668,01	652,0
Xe	$5P^4$	0,986	7,8600	0,0311 0,0257	24,50 20,28	7	4	88,906	21,37
	$5P^2$	0,944	7,525	0,0214	17,62	11,1	2	88,906	17,97
	$5S^2$	0,837	6,672	0,0357 0,0402	33,21 37,28	18 23,4	2 2	116,43 116,43	28,60 35,01
V	$4S^2$	1,401 8,26	8,508* 8,26	0,0325 0,0345	7,680 8,151	7,5	2	22,33	7,730
Cr	$4S^2$	1,453	8,95 8,70	0,0295 0,0313	7,013 7,440	7	2	23,712	7,754
Mn	$4S^2$	1,278	7,76	0,0256м	10.89 7,214	6,6	2	25,12	7,895

Table 1 (Continued)

Atom	Valence electrons	r_i(Å)	X	φ(X)	*U*(eV)	W(eV)	n_i	q^2(eV Å)	Atom
1	2	3	4	5	6	7	8	9	1
Mg	3S2	1,279	5,90 5,90	0,055м 0,0183м	7,43 2,47	6 2	2 2	17,50 17,50	6,40 3,10
Sc	4S2	1,57	8,668	0,0315	6,07	6,6	2	19,34	6,37

Note: 1) Bond energy of electrons (Wi) are obtained: theoretically (by Hartri-Fock method), "+" – by ESCA method, the other – by the results of optical measurements; 2) * - energy of valence electrons (U) is calculated with Fermi-Amaldi correction.

3. Electron Density in Atom Calculated via P-Parameter

The correctness of the equality (14) was additionally checked using Clementi functions [7] to estimate electron density β_i at the distance r_i. The density was calculated for atoms from ^{3}Li to ^{35}Br for external s- and p- orbitals taking into consideration small contributions from electrons of inner shells of atom into β_i. For atoms of 3d-metals

β_i was calculated for $4S^2$-electrons taking into account the density of 3d-electrons, whose partial contribution into β_i is 30-40%. P-parameter was estimated either via the bond energy [8] or ionization energy [9]. The results of these calculations are given in figure 1(a,b,c,d) with firm lines. With p- and d- orbitals electrons on them were distributed based on Hunda rule. The calculations for equations (12-14) are given in figure 1(a,b,c,d) with dots. As it is seen there is a good correlation between the values of β_i, calculated via P_0 and obtained from atomic functions. Consequently it is possible to compare electron densities of orbitals of interacting atoms (at the given distance from the nucleus) via the values of P_0/r_i. The value, inverse to the electron density, characterizes the value of partial volume of the given spatial-energy parameter of the corresponding system, i.e. basically the energy intensity of the given structure ($1/P = V_E$). And the principle of inverse adding of P-parameters is a principle of algebraic addition of energy intensities of interacting subsystems:

$$\frac{1}{P_S} = \frac{1}{P_E^{'}} + \frac{1}{P_E^{''}} + \ldots$$

or

$$V_E = V_E^{'} + V_E^{''} + \ldots \qquad (15),(16)$$

where P_C – resulting P-parameter of structure, V_E – its energy intensity, $P_E^{'}$ and $P_E^{''}$ - P_E-parameters of subsystems, $V_E^{'}$ and $V_E^{''}$ – their energy intensities.

Or taking into consideration the number of homogeneous atoms N_1 and N_2 in subsystems we have:

$$\frac{1}{P_S} = \frac{1}{N_1 P_E^{'}} + \frac{1}{N_1 P_E^{''}} + \ldots \qquad (17)$$

Such an equation (17) is further used to calculate the resulting P_S-parameter of complex structures.

4. Dependence of P-Parameter on the Module of Maximum Values of Radial Part of Ψ-Function

The modules of maximum values of radial part of Ψ-function Д (n,ℓ,r) with SEP values. Here Д (n,ℓ,r) – radial wave function of (n,ℓ,r). For initial data we used wave functions, bond energies and inner orbital radii [9], orbital radii of external orbitals [3], value q по [4,5].

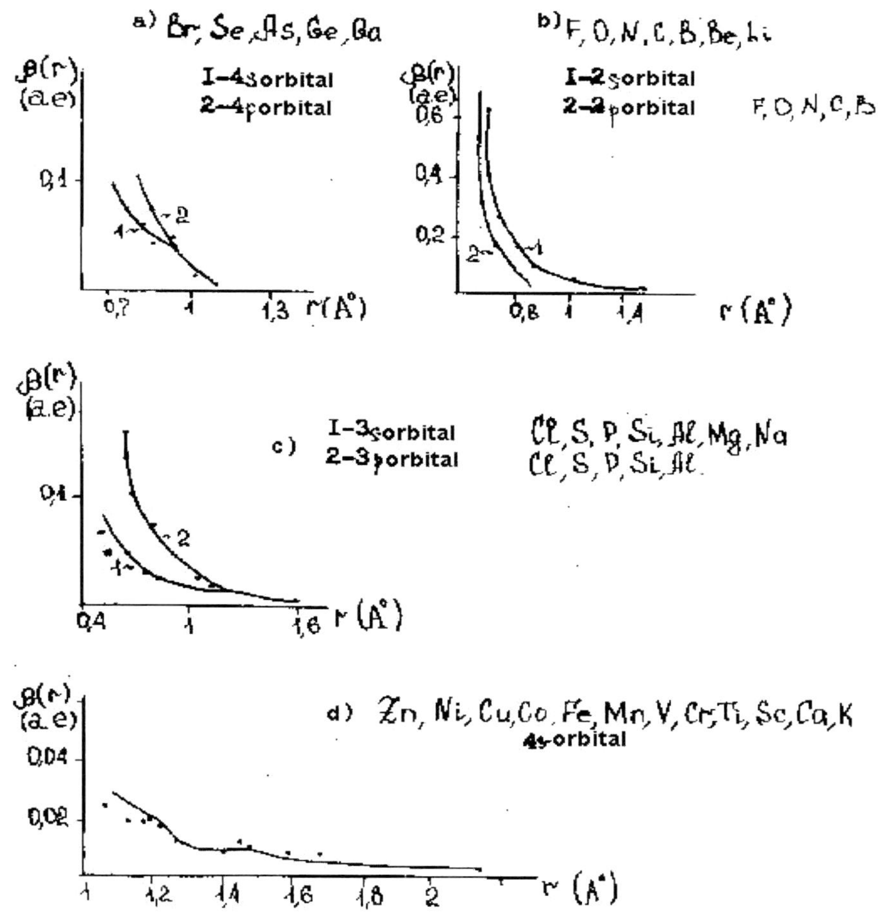

Figure 1. Electron density at the distance r_i calculated via Clementi functions (evan lines) and with the help of P-parameter (dots).

Such calculations revealed a linear dependence between the relation Po/Д (n,ℓ,r) = N_X and serial number of an element (Z) as shown in fig 2. From the figure we can see that the relation N_X/Z is a characteristic value for all elements with the same orbital quantum number (ℓ) of valence orbitals. Therefore the inflections in the figure occur mainly in the points corresponding to the changes in valence electron structure of elements (by orbital quantum number), e.g.: $B(2P^1)$, $Ne(2P^6)$, $Sc(3S^1)$,$Ga(4P^1)$, $Cu(3d^{1○}4S^1)$.

When analyzing the figure, we see that N_X = f(Z*), where Z^* - effective charge of the orbital.

N_X was compared with the values of $(Wr)^{1/2}$ based on the following equation:

$$\beta_0 Po / Д(n,\ell,r) = [Wr\,(n+\ell+1)]^{1/2} \qquad (18)$$

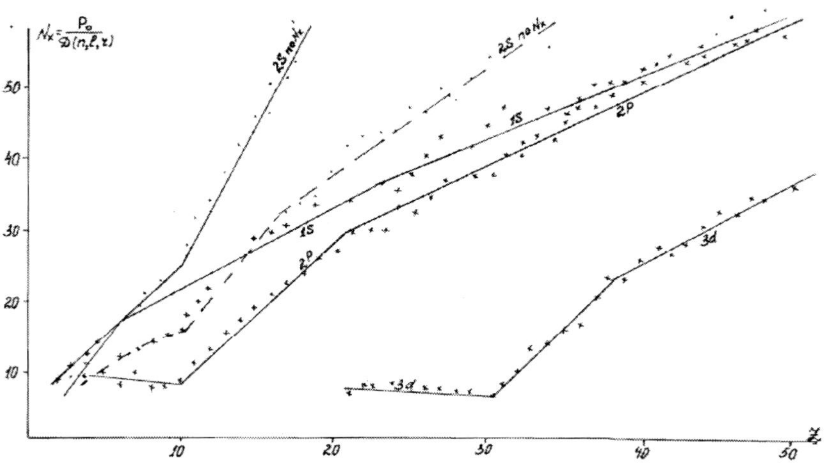

Figure 2. Dependence between P-parameter, wave function and element number.

Where β_0 – proportionality coefficient, and quantum numbers n and ℓ are introduced when considering the character of inflections of graphical dependence of N_X upon Z. The results of multiple calculations on initial data [2,7] show that coefficient β_0 for the most elements is a relatively constant value.

Thus for 64 elements coefficient β_0 is in the range of 0.59-0.66.

So we can consider that SEP has a direct connection with a maximum value of radial part of ψ-function and characterizes the distribution of electron charge density in the atom. But the redistribution of electron charge density has the main meaning when forming the chemical bond.

5. Wave Equation of P-Parameter

To describe spatial-energy properties of an atom two types of P-parameters are introduced with simple connection between them:

$$P_E = \frac{P_0}{R}$$

where R – dimensional characteristic of an atom.

Taking into consideration additional quantum characteristics of atom sublevels, this equation in the coordinate x can be written down as follows:

$$\Delta P_E \approx \frac{\Delta P_0}{\Delta x}$$

or

$$\partial P_E = \frac{\partial P_0}{\partial x}$$

where ΔP equals the difference between P_0-parameter of *i*-orbital and P_I–parameter of indication (parameter of main state at the given set of quantum numbers).

According to the established rule of adding P-parameters [10] of similarly charged or homogeneous systems for two orbitals in the given atom with different quantum characteristics and in accordance with the law of energy conservation we have:

$$\Delta P_E'' - \Delta P_E' = P_{E,\lambda}$$

where $P_{E,\lambda}$ – spatial-energy parameter of quantum transition.

Taking $\Delta\lambda=\Delta x$ as the dimensional characteristic of the interaction, we have:

$$\frac{\Delta P_0''}{\Delta\lambda} - \frac{\Delta P_0'}{\Delta\lambda} = \frac{P_0}{\Delta\lambda} \text{ or: } \frac{\Delta P_0'}{\Delta\lambda} - \frac{\Delta P_0''}{\Delta\lambda} = -\frac{P_0\lambda}{\Delta\lambda}$$

We will divide it termwise by $\Delta\lambda$:

$$\left(\frac{\Delta P_0'}{\Delta\lambda} - \frac{\Delta P_0''}{\Delta\lambda}\right)\Big/\Delta\lambda = -\frac{P_0}{\Delta\lambda^2},$$

where:

$$\left(\frac{\Delta P_0'}{\Delta\lambda} - \frac{\Delta P_0''}{\Delta\lambda}\right)\Big/\Delta\lambda \sim \frac{d^2 P_0}{d\lambda^2},$$

then:

$$\frac{d^2 P_0}{d\lambda^2} + \frac{P_0}{\Delta\lambda^2} \approx 0$$

Taking into consideration only those interactions when $2\pi\Delta x=\Delta\lambda$ (closed oscillator), we have the following equation:

$$\frac{d^2 P_0}{dx^2} + 4\pi^2 \frac{P_0}{\Delta\lambda^2} \approx 0 \qquad (19)$$

Since

$$\Delta\lambda = \frac{h}{mv},$$

then:

$$\frac{d^2 P_0}{dx^2} + 4\pi^2 \frac{P_0}{h^2} m^2 v^2 \approx 0$$

or

$$\frac{d^2 P_0}{dx^2} + \frac{8\pi^2 m}{h^2} P_0 E_k = 0$$

where

$$E_k = \frac{mV^2}{2}$$ - kinetic energy of an electron.

Schredinger equation for the stationary state in coordinate x:

$$\frac{d^2 \psi}{dx^2} + \frac{8\pi^2 m}{h^2} \psi E_k = 0 \qquad (20)$$

when comparing the equations (19 and 20) we see that P_0-parameter is numerically approximately equal to the value of Ψ-function,

$$P_0 \approx \Psi$$

and in general is proportional to it: $P_0 \sim \Psi$. Taking into consideration the practical capabilities of applying the methodology of P-parameter, we can consider that this parameters is a materialized analog of Ψ-function.

Since P_0-parameters as well as Ψ-function have wave properties, the principles of superposition have to be executed for them, this determines a linear character of equations of adding and changing P-parameters.

6. SPECTRAL CHARACTERISTICS OF SOME ATOMS AND THEIR CONNECTION WITH SPATIAL-ENERGY PARAMETER

We use some properties of wave function as applied to P-parameter. In particular, by the analogy with wave function we can assume that the square of P_E-parameter multiplied by the volume element (Δr^3) has to characterize the electron density in this element of the volume – (e^2).

or:

$$P_E^2(\Delta r)^3 = \frac{P_0^2}{\Delta r^2}\Delta r^3 = P^2_0\Delta r \sim e^2 = \frac{\hbar c}{137} \quad (21)$$

where c – velocity of speed; $\hbar = \dfrac{h}{2\pi}$. Here h – Planck's constant.

When estimating quantum transitions between energy levels w_2 and w_1, the substitution of Δr for the wavelength of radiation (λ) is possible:

$\Delta r = \lambda$

The calculations made for simple atomic quantum transitions ns^1-nP^1 in elements C, N, O, Al and Si [9] showed that the following relation is fulfilled:

$$(\Sigma P_0)^2\,\lambda = 48\,\frac{c\hbar^3}{2m_e}$$

or:

$$(\Sigma P_0)^2\,\lambda = 24\,\frac{c\hbar^3}{m_e}; \quad (22)$$

$\dfrac{2m_e}{\hbar^2}$ - coefficient of Schredinger equation, where: $\Sigma P_0 = P_0'\,(ns^1) + P_0''\,(np'')$

here n – main quantum number.

The average value of $(\Sigma P_0)^2\,\lambda$ equals 364118 eV^{2}Å^3 = $9.3184 \cdot 10^{-63}$ J^{2}m^3 with deviations in the range of ± 6% from calculated values (table 2).

Table 2. P-parameters of some atoms and their spectral characteristics of S-P transitions

Atom	Valence electrons	W(eV)	r_i (Å)	q_i^2 (eVÅ)	P_0 (eVÅ)	ΣP_0 (eVÅ)	Transition	λ (Å)	$\Sigma P_0 \lambda$ ($eV^2Å^3$)
C	$2P^1$	11.792	0.596	35.395	5.8680	14.889	$2S^1$-$2P^1$	1549	343386
	$2P^2$	11.792	0.596	35.395	10.061				
	$2S^1$	19.201	0.620	37.240	9.0209				
	$2S^2$	19.201	0.620	37.240	14.524				
N	$2P^1$	15.445	0.4875	52.912	6.5916	17.301	$2S^1$-$2P^1$	1238	370564
	$2P^2$	15.445	0.4875	52.912	11.710				
	$2P^3$	15.445	0.4875	52.912	15.830				
	$2S^1$	25.724	0.521	52.283	10.709				
	$2S^2$	25.724	0.521	52.883	17.833				
O	$2P^1$	17.195	0.4135	71.383	6.4663	$2S^1$+$2P^1$	$2S^1$-$2P^1$	1031	374557
	$2P^2$	17.195	0.4135	71.383	11.858				
	$2P^3$	17.195	0.4135	71.383	16.423				
	$2P^4$	17.195	0.4135	71.383	20.338				
	$2S^1$	33.859	0.450	72.620	12.594				
	$2S^2$	33.859	0.450	72.620	21.466				
Al	$3P^1$	5.7130	1.312	26.443	5.840	13.755	$3S^1$-$3P^1$	1854	350777
	$3S^1$	10.706	1.044	27.119	7.9149				
	$3S^2$	10.706	1.044	27.119	12.253				
Si	$3P^1$	8.0848	1.068	29.377	6.6732	16.5458	$3S^1$-$3P^1$	1393	381306
	$3P^2$	8.0848	1.068	29.377	10.876				
	$3S^1$	14.690	0.904	38.462	9.8716				
	$3S^2$	14.690	0.904	38.462	15.711				

Average value: 364118 ($eV^2Å^3$).

For complex multiplet transitions the equation (22) was modified:

$$(\Sigma P_0)^2 \lambda \frac{N_H}{N_K} = 24\frac{c\hbar^3}{m_e} \tag{22a}$$

where N_H and N_K – number of registered electrons on *nS* and *nP* sublevels at the beginning and end of quantum transition.

The analysis of all these calculations by the equations (22,22a) confirms that P-parameter has a direct connection with the wave function Ψ and can characterize energy values of atom spectral transitions, in particular – for the simplest transitions between sublevels with the same main quantum numbers (*n*).

7. Wave Properties of P-Parameter and Principles of Adding P-Parameters

Since P-parameter has wave properties (as analogous to Ψ'-function), in the process of structural interactions mainly the regularities for the interference of corresponding waves have to be carried out.

The minimum interference, oscillation weakening (in antiphase) takes place if the difference of wave motion (Δ) equals the odd number of semi-waves:

$$\Delta = (2n+1)\frac{\lambda}{2} = \lambda(n + \frac{1}{2}),$$

where n = 0, 1, 2, 3, … (23)

As applied to P-parameter, this rule means that the minimum of interaction occurs if P-parameters of interacting structures are also in "anti-phase" – there is an interaction of either differently charged systems or heterogeneous atoms (for example, during the formation of valence-active radicals CH, CH_2, CH_3, NO_2 …, etc).

The summation of P-parameters in this case follows the principle of adding inverse values of P-parameters – equations (3,4,17).

The difference in wave motion (Δ) for P-parameters can be estimated via their relative value ($\gamma = \frac{P_2}{P_1}$) or via the relative difference of P-parameters (coefficient α), which at minimum interactions give an odd number:

$$\gamma = \frac{P_2}{P_1} = (n + \frac{1}{2}) = \frac{3}{2}; \frac{5}{2} \ \ldots$$

At n=0 (main state) $\frac{P_2}{P_1}=\frac{1}{2}$ (24)

Let us point out that for stationary levels of one-dimensional harmonic oscillator the energy of these levels $\varepsilon=h\nu(n+\frac{1}{2})$, therefore in quantum oscillator, in contrast to the classical one, the least possible energy value does not equal zero.

In this model the minimum of interaction does not produce energy equal to zero corresponding to the principle of adding inverse values of P-parameters – equations (3,4,17). *Maximum of interference,* increase of oscillations (in phase) takes place if the difference of wave motion equals the even number of semi-waves:

$$\Delta=2n\frac{\lambda}{2}=\lambda n \text{ or } \Delta=\lambda(n+1).$$

As appropriate to P-parameters the maximum intensification of interaction in the phase corresponds to the interactions of similarly charged systems or systems homogeneous by their properties and functions (for example, between fragments or blocks of complex organic structures, such as CH_2 and NNO_2 in octogen).

Then:

$$\gamma=\frac{P_2}{P_1}=(n+1) \quad (25)$$

Analogously for "confluent" systems (with the same values of functions) of 2D harmonic oscillator the energy of stationary states is as follows:

$$\varepsilon=h\nu(n+1)$$

By this model the maximum of interactions corresponds to the principle of algebraic adding of P-parameters – equations (6-8). When n=0 (basic state) we have $P_2=P_1$, or: maximum of interactions of structures occurs if the have the same P-parameters. This concept was used [10] as the main condition of isomorphic replacements.

8. Functions of Lagrange-Hamilton and Analogous Comparisons

The Lagrange function is the difference between kinetic (T) and potential (U) energies of the system:

$$L = T - U \tag{26}$$

Hamilton function can be considered as the total of potential and kinetic energies, i.e. as the total mechanical energy of the system:

$$H = T + U \tag{27}$$

From equations (1) and (2) and in accordance with the law of energy preservation we have:

$$H + L = 2T$$

$$H - L = 2U \qquad (28)\ (29)$$

We try to evaluate the motion of isolated system of free atom as a relative motion of its two subsystems: nucleus and orbital.

An atom structure is formed from differently charged masses of nucleus and electrons. In this system the orbital energy of electrons (W_i) and effective energy of atom nucleus, that takes into consideration screening effects, form an atom structure.

In a free atom its electrons move in coulomb field of nuclear charge. The effective charge nucleus, characterizing the potential energy of such subsystem, with screening effects equals:

$$q^2 / r_i$$

where

$$q = Z^* / n^* \qquad (30)\ (31)$$

We can assume that orbital energy of electrons during their motion in coulomb field of nucleus charge is mainly determined by the value of kinetic motion of such motion.

Thus it is assumed that:

T ~ W

and

$$U \sim q^2 / r_i \qquad (32{,}33)$$

In such approach the total of values of W and q^2/r_i is analogous to Hamilton function (H):

$$W + q^2/r_i \sim \mathrm{H} \tag{34}$$

Since the value of P_E-parameter has to be proportional to L, then

$$P_E \sim L \tag{35}$$

Equation (28) is:

$$\left(W + \frac{q^2}{r_i}\right) + P_E \approx 2W \tag{36}$$

P_E-parameters of free atoms [10] were calculated based on the equations (3-5) using the values of bond energies of electrons as their orbital energies [2]. When calculating the values of effective P_E-parameter as dimensional characteristics of the atom (R), mainly the values of their radii by Belov-Bokiy or covalence radii (for non-metals) were applied.

Mutual overlapping of orbitals of valence electrons in the atom can be considered, in the first approximation, via the averaging of their total energy dividing it by the number of valence electrons recorded (N):

$$\left(\frac{q^2}{r_i} + W\right)\frac{1}{N} + P_E \approx 2W \tag{37}$$

Thus, $\left(\frac{q^2}{r_i} + W\right)\frac{1}{N}$ characterizes the averaged value of the total energy of i-orbital as converted to one valence electron and in this approach it is the analog of Hamilton function – H.

In free atoms of 1a and 2a groups of periodic System S-orbital is the only valence orbital the peculiarities of deformation of electron cloud of which were considered through the introduction of coefficient $K = n/n^*$, where n – main quantum number, n^* – effective main quantum number by the following equation:

$$\left(\frac{q^2}{r_i}+W\right)\frac{1}{KN}+P_E \approx 2W \tag{38}$$

This means that for subgroups 1a and 2a in small periods K=1 and then K=$\frac{4}{3{,}7};\frac{5}{4};\frac{6}{4{,}2}$ - for periods 4,5 and 6 of the System of these subgroups. For the rest of the cases K=1.

And one more peculiarity as applied only to group 1a of periodic System: $2r_i$ (orbital diameter of *i*-orbital) was used as dimensional characteristic in the first term of equation (38).

Taking into consideration the aforesaid observations by initial equations (37) and (38) for valence orbitals of practically all the elements, the calculations and comparisons of the values $\left(\frac{q^2}{r_i}+W\right)\frac{1}{KN}+P_E$ and $2W$ were carried out, some results are given in table 3. Besides, the equation (38) can be transformed as follows:

$$\left(\frac{q^2}{r_i}+W\right)\frac{1}{KN}-P_E \approx 2(W-P_E) \tag{39}$$

And we can consider the relation obtained in the same way as the equation (39). Actually, following the given analogy, W ~ T and P_E ~ L. Thus (W-P_E) ~ (T-L), and in the equation (26) T-L=*U*. Therefore (W-P_E) ~ *U*.

The analysis of all the data shows that the nearness of values of the investigated magnitudes compared is mostly within 5% and only sometimes – over 5% but never exceeds 10%.

All this proves that there is a certain analogy of the equations (28) and (38), as well as (29) and (39), therefore in the first approximation P_E-parameter can be the analog of Lagrange function and $\left(\frac{q^2}{ri}+W\right)\frac{1}{KN}$ - the analog of Hamilton function [11].

Table 3. Comparison of main energy atomic characteristics

Element	Valence electrons	W (eV)	r_i (Å)	q^2 (eVÅ)	P_0 (eVÅ)	R (Å)	$P_E=P_0/R$ (eV)	N	$\left(\frac{q^2}{ri}+W\right)\frac{1}{N}+P_E$	2W (eV)
1	2	3	4	5	6	7	8	9	10	11
Li	$2S^1$	5.3416	1.586·2	5.8892	3.475	1.55	2.2419	1	9.440	10.683
Be	$2S^2$	8.4157	1.040	13.159	7.512	1.13	6.6478	2	17.182	16.831
B	$2P^1$	8.4315	0.776	21.105	4.9945	0.91	5.4885	3	17.365	16.863
	$2S^2$	13.462	0.769	23.890	11.092	0.91	12.189	3	27.032	26.924
C	$2P^2$	11.792	0.596	35.395	10.061	0.86	11.699/2	4	23.645	23.584
	$2S^2$	19.201	0.620	37.240	14.524	0.77	18.862	4	38.824	38.402
N	$2P^3$	15.445	0.4875	52.912	15.830	0.71	22.296/3	5	32.228	30.890
	$2S^2$	25.724	0.521	53.283	17.833	0.71	19.788	5	50.775	51.448
O	$2P^1$	17.195	0.4135	71.383	6.4663	0.66	9.7979/4	6	34.087	34.390
	$2S^2$	33.859	0.450	72.620	21.466	0.66	32.524	6	65.064	67.718
F	$2P^1$	19.864	0.3595	93.625	6.6350	0.64	10.367/5	7	42.115	39.728
	$2S^2$	42.792	0.396	94.641	24.961	0.64	39.002	7	79.257	85.584
Na	$3S^1$	4.9552	1.713·2	10.058	4.6034	1.89	2.4357	1	10.327	9.9104
Mg	$3S^1$	6.8859	1.279	17.501	5.8588	1.60	3.6618	2	13.946	13.772
Al	$3P^1$	5.713	1.312	26.443	5.840	1.43	4.084	3	12.707	11.426
	$3S^2$	10.706	1.044	27.119	12.253	1.43	8.5685	3	20.796	21.412
Si	$3P^1$	8.0848	1.068	29.377	6.6732	1.17	5.7036	4	14.600	16.170
	$3S^2$	14.690	0.904	38.462	15.711	1.17	13.428	4	27.737	29.380
P	$3P^3$	10.659	0.9175	38.199	16.594	1.30	12.765/3	3	21.686	21.318
	$3S^1$	18.951	0.803	50.922	11.716	1.10	10.651	3	38.106	37.902

CONCLUSIONS

1. The notion of spatial-energy parameter (P-parameter) that numerically equals energy expression in Thomas-Fermi equation is introduced based on the simultaneous recording of the most important atomic characteristics.

2. It is shown that P-parameter is a direct characteristic of electron density in an atom. The dependence of P-parameter on maximum value of radial part of ψ-function is obtained.

3. Wave equation of P-parameter, which is formally analogous to the equation of ψ-function, is given.

4. Since P-parameter comprises main constants characterizing the state of valence electrons inside the atom, it helps to correlate and predict physical and chemical properties of compounds and processes of phase formation in simple and complex systems, by bond and hybridization energy of atom orbitals, as well [12-14].

REFERENCES

[1] Batsanov S.S., Zvyagina R.A. Overlapping integrals and problem of effective charges. – Novosibirsk: *Nauka,* 1966, - 386 p.

[2] Fischer C.F. Average-Energy of Configuration Hartree-Fock Results for the Atoms Helium to Radon.// *Atomic Data,*-1972,- № 4,- p. 301-399.

[3] Waber J.T., Cromer D.T. Orbital Radii of Atoms and Ions //*J. Chem. Phys.* —1965,-V 42,-№ 12,-p. 4116- 4123.

[4] Clementi E., Raimondi D.L. Atomic Screening constants from S.C.F. Functions, 1.// *J. Chem. Phys.*-1963,-v.38,-№11,-p.2686-2689.

[5] Clementi E., Raimondi D.L. Atomic Screening Constants from S.C.F. Functions, II.//*J. Chem. Phys.*-1967,-V.47, № 4,-p. 1300-1307.

[6] Gombash P. Statistic theory of atom and its applications. – M.: I.L.,1951, - 398 p.

[7] Clementi E. Tables of atomic functions // *J.B.M. S. Res. Develop. Suppl.,* 1965, V. 9, № 2, p.76.

[8] Zigban K., Nordling K., Falman A., et al. Electron spectroscopy. - M.: *Mir,* 1971,- 493 p.

[9] Allen K.U. Astrophysical magnitudes. - M.: *Mir,* 1977, 446 p.

[10] Korablev G.A. Spatial Energy Principles of Complex Structures Formation. Netherlands, Leiden, Brill Academic Publishers and VSP. 2005. 426 p. (Monograph).

[11] Korablev G.A., Kodolov V.I., Lipanov A.M. Analogous comparison of Lagrange and Hamilton functions with spatial-energy parameter. – *Chemical physics and mesoscopy,* URC RAS, 2004, №1, v.6, p. 5-18.

[12] Korablev G.A., Zaikov G.E. Energy of chemical bond and spatial-energy principles of hybridization of atom orbitals.//*J. of Applied Polymer Science.* USA, 2006, V.101, n.3, p.2101-2107.

[13] Korablev G.A., Zaikov G.E. Formation of Carbon Nanostructures and Spatial-Energy Stabilition Criterion//*Polymers and Composites*. (Polymer Year book Series, V.22), Nova Science Publishers, Jnc., New York, 2007, p.31-42.
[14] Korablev G.A., Zaikov G.E. P-Parameter as and Objective Characteristics of Electronegativity//*Reactions and Properties of Monomers Polymers,* Nova Science Publishers, Inc., New York, 2007, p.203-213.

In: Progress in Chemistry and Biochemistry
Editors: E. M. Pearce and G. E. Zaikov

ISBN: 978-1-60692-344-3

Chapter 23

ENERGY CRITERIONS OF PHOTOSYNTHESIS

G.A. Korablev and G.E. Zaikov
Basic research-educational center of chemical physics and mesoscopy,
Udmurt research center, Ural Division, RAS, Izhevsk, E-mail: korablev@udm.net
Institute of biochemical physics after N.M. Emanuel, RAS, Russia,
Moscow 119991, 4 Kosygina St., E-mail: chembio@sky.chph.ras.ru

ABSTRACT

The application of methodology of spatial-energy interactions (P-parameter) to main stages of photosynthesis is given. Their energy characteristics are calculated. The values obtained correspond to the reference and experimental data.

Keywords: Spatial-energy parameter, free radicals, structural interactions, photosynthesis.

INTRODUCTION

Photosynthesis – process of converting electromagnetic energy of the sun's rays into the energy of chemical bonds of vital organic substances … [1].

It is the only natural process through which the organic world obtains the reserve of free energy and which provides all bio-organisms with chemical energy. From the moment photosynthesis was discovered by D.M. Priestly (1771-1850), the researches passed several important stages. The first works connected with photosynthesis energy refer to 1850-1900 (R. Mayer, D.G. Stocks, J. Sax). The application of physiological concepts – in 1900-1950 (M.S. Tsvet, A.A. Richter, W. Arnold). Further development of bio-physicochemical aspects of synthesis till now resulted in its modern model and clarified the way of carbon during photosynthesis (M. Calvin), concept of two photosystems (R. Emerson), structure of reaction center (I. Deizinhoffer, H. Michel, R. Huber), etc.

The basis of photosynthesis – consecutive chain of redox reactions, during which electrons are transferred from donor-reducer to acceptor-oxidizer with the formation of reduced compounds (carbohydrates) and oxygen isolation.

It is known that excitation energy for complex organic molecules of chlorophyll type lasts for 10^{-8}-10^{-9} sec and can be stored only for insignificant fractions of a second. But during photosynthesis the energy of absorbed light quantum is stored for a long period (from several minutes to millions of years). The energy is stored here in molecular as chemical bonds rich with energy in complex organic structures. Therefore photosynthesis energy can be presented based on the analysis of changes in energies of chemical bonds of molecular structures in dynamics of all main types of photosynthesis.

This is the aim to use the methodology of spatial-energy interactions (P-parameter) in this paper.

SPATIAL-ENERGY PARAMETER (P-PARAMETER)

Structural and interatomic interactions are sure to have an electron nature. Thus the registration of the extent to which electrons fill the atom valence states is the basis of the method of valence bonds in chemistry and is numerically expressed through coulomb electrostatic interaction.

Also important are exchange-promotional structural interactions that determine isomorphism, solubility of components in solid, liquid and molecular media [2].

During the interactions of oppositely charged heterogeneous systems the volume energy of interacting structures is compensated to a certain extent thus leading to the decrease in the resultant volume energy.

The analysis of different physical and chemical processes allows assuming that in many cases the principle of adding reciprocals of volume energies or kinetic parameters of interacting structures is executed.

Some examples: ambipolar diffusion, total rate of topochemical reaction, change in the light velocity when transiting from vacuum into the given medium, resultant constant of chemical reaction rate (initial product – intermediary activated complex – final product).

Lagrange equation for relative movement of isolated system of two interacting material points with masses m_1 and m_2 in coordinate x can be presented as follows:

$$\frac{1}{\Delta U} \approx \frac{1}{\Delta U_1} + \frac{1}{\Delta U_2} \tag{1}$$

where ΔU_1 and ΔU_2 – potential energies of material points in elementary section of interactions, ΔU - resulting (mutual) potential energy of these interactions.

The atom system is formed from oppositely charged masses of nucleus and electrons. In this system energy characteristics of subsystems are the orbital energy of electrons (W_i) and effective energy of nucleus that takes into consideration the screening effects (by Clementi).

Therefore, assuming that the resultant interaction energy of the system orbital-nucleus (responsible for interatomic interactions) can be calculated based on the principle of adding reciprocals of some initial energy components, we substantiate the introduction of P-parameter [2] as an averaged energy characteristic of valence orbitals in accordance with the following equations:

$$\frac{1}{q^2/r_i} + \frac{1}{W_i n_i} = \frac{1}{P_E} \tag{2}$$

$$P_E = \frac{P_0}{r_i} \tag{3}$$

$$P_E = \frac{P_0}{R} \tag{3a}$$

$$\frac{1}{P_0} = \frac{1}{q^2} + \frac{1}{(Wrn)_i} \tag{4}$$

$$q = \frac{Z^*}{n^*} \tag{5}$$

Here: W_i – bond energy of electrons [3]; r_i – orbital radius of *i*–orbital [4]; n_i – number of electrons of the given orbital, Z^* and n^* – effective charge of nucleus and effective main quantum number [5], R – numerical characteristic of atom (bond).

The P_0 value will be called a spatial-energy parameter, and the P_E value – effective P–parameter. Effective P_E–parameter has a physical sense of some averaged energy of valence electrons in atom and is measured in energy units, e.g. in electron-volts (eV).

The values of P_0- and P_E-parameters of some elements calculated based on equations (2-5) are given in table 1.

Modifying the rules of adding reciprocals of energy characteristics of subsystems as applied to complex structures we can obtain [6] the equation for calculating P_C-parameters of complex structure:

$$\frac{1}{P_C} = \left(\frac{1}{NP_E}\right)_1 + \left(\frac{1}{NP_E}\right)_2 + \ldots \tag{6}$$

where N_1 and N_2 – number of homogeneous atoms in subsystems.

The calculation results of some complex structures based on equation (6) are given in table 2.

The calculations for 21 elements showed that the values of P_E-parameters are similar to corresponding values of total energy of valence electrons according to the statistic model of atom.

Simple dependence between P_E-parameter and electron density at the distance r_i can be obtained (according to the statistic model of atom):

$$\beta_i^{2/3} = A \cdot \frac{P_0}{r_i} = A P_E, \qquad (7)$$

where A-constant

When the solution is formed in the places of atom-components contact, the unified electron density has to be established. The dissolving process is accompanied by the redistribution of this density between valence areas of both particles and transition of some electrons from external spheres to the neighboring ones.

It is obvious that if electron densities in free atom-components of the solution at the distances of orbital radius r_i are similar, the transition processes between boundary atoms of particles are minimal thus favoring the solution formation.

Thus the task of evaluating the solubility in many cases comes to comparative evaluation of electron density of valence electrons in free atoms (on averaged orbitals) participating in the solution formation.

In this regard the maximum total solubility evaluated through the coefficient of structural interaction and isomorphism α are determined by the state of minimal value that represent relative difference of effective energies of external orbital:

$$\alpha = \frac{P'o/r_i' - P''o/r_i''}{(P'o/r_i' + P''o/r_i'')/2} 100\%; \qquad (8)$$

$$\alpha = \frac{P'_C - P''_C}{P'_C + P''_C} 200\% \qquad (9)$$

Multiple calculations and comparisons with the experiment allowed arranging the unified averaged figure-nomogram of degree of structural interaction and solubility (ρ) dependence upon coefficient α [2].

Table 1. P-parameters of atoms calculated via the bond energy of electrons

Atom	Valence electrons	W (eV)	r_i (Å)	q^2_0 (eVÅ)	P_0 (eVÅ)	R (Å)	P_0/R (eV)
1	2	3	4	5	6	7	8
H	$1S^1$	13.595	0.5295	14.394	4.7985	0.5295 0.46 0.28 R_I^-=1.36	9.0624 10.432 17.137 3.525

Atom	Valence electrons	W (eV)	ri (Å)	q2 0 (eVÅ)	P0 (eVÅ)	R (Å)	P0/R (eV)
1	2	3	4	5	6	7	8
C	$2P^1$	11.792	0.596	35.395	5.8680	0.77	7.6208
						0.69	8.5043
	$2P^2$	11.792	0.596	35.395	10.061	0.77	13.066
						0.69	14.581
	$2P^1_r$				4.4044		
	$2P^3_r$				13.213		
	$2S^1$	19.201	0.620	37.240	9.0209	0.77	11.715
	$2S^2$				14.524	0.77	18.862
	$2S^1+2P^3_r$				22.234	0.77	28.875
	$2S^1+2P^1_r$				13.425	0.77	17.435
	$2S^2+2P^2$				24.585	0.77	31.929
N	$2P^1$	15.445	0.4875	52.912	6.5916	0.70	9.4166
	$2P^2$				11.723	0.70	16.747
	$2P^3$				15.830	0.70	22.614
	$2P^4_r$				19.193	0.55	34.896
	$2P^5_r$				21.966	0.55	39.938
	$2S^1$	25.724	0.521	53.283	10.709	0.70	15.299
	$2S^2$	25.724	0.521	53.283	17.833	0.70	25.476
	$2S^2+2P^3$				33.663	0.70	48.09
O	$2P^1$	17.195	0.4135	71.383	4.663	0.66	9.7979
	$2P^1$					R_I=1.36	4.755
	$2P^1$					R_I=1.40	4.6188
	$2P^2$	17.195	0.4135	71.383	11.858	0.66	17.967
						0.59	20.048
						R_I=1.36	8.7191
						R_I=1.40	8.470
	$2P^4$	17.195	0.4135	71.383	20.338	0.66	30.815
						0.59	34.471
	$2S^1$	33.859	0.450	72.620	12.594	0.66	19.082
	$2S^2$	33.859	0.450	72.620	21.466	0.66	32.524
	$2S^2+2P^4$				41.804	0.66	63.339
						0.59	70.854
Ca	$4S^1$	5.3212	1.690	17.406	5.929	1.97	3.0096
	$4S^2$				8.8456	1.97	4.4902
	$4S^2$					R^{2+}=1.00	8.8456
	$4S^2$					R^{2+}=1.26	7.0203
S	$3P^1$	11.901	0.808	48.108	6.0143	1.04	7.7061
	$3P^2$	11.901	0.808	48.108	13.740	1.04	13.215
	$3P^4$	11.904	0.808	48.108	21.375	1.04	20.553
	$3S^1$	23.933	0.723	64.852	13.659	1.04	13.134
	$3S^2$	23.933	0.723	64.852	22.565		
	$3S^2+3P^4$				43.940	1.04	42.250
Se	$4P^1$	10.963	0.909	61.803	8.5811	1.17	7.3343
	$4P^2$				15.070	1.17	12.880
	$4P^2$				15.070	1.6	9.4188
	$4P^2$				15.070	1.14	13.219
	$4P^4$				24.213	1.17	20.710
	$4P^4$					1.6	15.133
	$4S^1$	22.787	0.775	85.678	14.642	1.17	12.515
	$4S^2$				25.010	1.17	21.376
	$4S^2+4P^4$				49.214	1.17	42.066
	$4S^2+4P^4$				49.214	1.6	30.759

Table 1. (Continued).

Atom	Valence electrons	W (eV)	ri (Å)	q2 0 (eVÅ)	P0 (eVÅ)	R (Å)	P0/R (eV)
1	2	3	4	5	6	7	8
P	$3P^1$	10.659	0.9175	38.199	7.7864	1.10	7.0785
	$3P^1$					R^{3-}=1.86	P_3=4.18
	$3P^3$	10.659	0.9175	38.199	16.594	1.10	62
	$3P^3$					R^{3-}=1.86	15.085
	$3S^2$	18.951	0.803	50.922	19.050	1.10	8.9215
	$3S^2+3P^3$				35.644	1.10	17.318
							32.403
Mg	$3S^1$	6.8859	1.279	17.501	5.8568	1.60	3.6618
	$3S^2$				8.7787	1.60	5.4867
						R^{2+}=1.02	8.6066
Mn	$4S^1$	6.7451	1.278	25.118	6.4180	1.30	4.9369
	$4S^2$				10.223	1.30	7.8638
	$3d^1$	17.384	0.3885	177.33	6.5058	1.30	5.0043
	$4S^1+3d^1$				12.924	1.30	9.9414
	$4S^2+3d^2$				22.774	1.30	17.518
	$4S^2+3d^5$				38.590	1.30	29.684
Na		4.9552	1.713	10.058	4.6034	1.89	2.4357
	$3S^1$					R^+_I=1.18	3.901
						R^+_I=0.98	4.6973
Cl		13.780	0.7235	59.849	8.5461	1.00	8.5461
	$3P^1$					R^-_I=1.81	4.7216
Fe	$4S^1$	7.0256	1.227	26.572	6.5089	1.26	4.8325
	$3d^1$	17.603	0.364	199.95	6.2084		
	$4S^1+3d^1$				12.717	1.26	10.093
	$4S^2+3d^1$				16.664	1.26	13.226
K	$4S^1$	4.0130	2.612	10.993	4.8490	2.36	2.0547
						R^+_I=1.45	3.344
	$4S^2$(*)				7.2115	2.36	3.0557
						R^+_I=1.45	4.9734

Table 2. Structural P_C-parameters calculated via the bond energy of electrons

Radicals, fragments of molecules	P_i' (eV)	P_i'' (eV)	P_C (eV)	Orbitals
OH	17.967	10.432	6.5999	O ($2P^2$)
	9.7979	9.0624	4.7080	O ($2P^1$)
	9.7979	10.432	5.0525	O ($2P^1$)
	17.967	17.138	8.7712	O ($2P^2$)
H_2O	2·9.0624	17.967	9.0226	O ($2P^2$)
	2·10.432	17.967	9.6537	O ($2P^2$)
	2·17.138	17.967	11.788	O ($2P^2$)
CH_2	28.875	2·17.138	15.674	C ($2S^12P^3_r$)
	31.929	2·17.138	16.531	C ($2S^22P^2$)
	28.875	2·9.0624	11.125	C ($2S^12P^3_r$)
CH_3	31.929	3·17.138	19.696	C ($2S^22P^2$)
	28.875	3·9.0624	14.003	C ($2S^12P^3_r$)

Radicals, fragments of molecules	P_i' (eV)	P_i'' (eV)	P_C (eV)	Orbitals
CH	28.875 31.929 31.929	17.138 9.0624 17.138	10.755 7.059 11.152	C $(2S^1 2P^3{}_r)$ C $(2S^2 2P^2)$ C $(2S^2 2P^2)$
H_3O	3·17.138	17.967	13.314	O $(2P^2)$
C_2H_5	2·31.929	5·17.138	36.590	C $(2S^2 2P^2)$
CH_2	31.929	2·9.0624	11.562	C $(2S^2 2P^2)$
CH_3	28.875	3·17.138	18.491	C $(2S^2 2P^3{}_r)$
CH_3	31.929	3·9.0624	14.684	C $(2S^2 2P^2)$
CH	28.875	10.432	7.6634	C $(2S^2 2P^3{}_r)$
CH	31.929	10.432	7.8630	C $(2S^2 2P^2)$
CO	31.929	20.048	12.315	C $(2S^2 2P^2)$
C=O	14.581	20.048	8.4416	C $(2P^2)$
C=O	17.435	20.048	9.3252	C $(2S^1 2P^1{}_r)$
C-O_2	28.875	2·20.048	16.786	C $(2S^1 2P^3{}_r)$
C-O_2	31.929	2·20.048	17.774	C $(2S^2 2P^2)$
CO-OH	12.315	8.7712	5.1226	C $(2S^2 2P^2)$
CH-OH	11.152	8.7712	4.9159	C $(2S^2 2P^2)$
CO-H	8.4416	9.0624	4.3705	C $(2P^2)$

The following spatial-energy principles defining the character of structural spatial-energy interactions were determined:

1. Complete (total-lot) isomorphic interaction takes place at relative difference of P-parameters of valence orbitals of interchanging atoms (within 4-6%).
2. P-parameter of the smallest value defines the orbital that is mainly responsible for isomorphism.
3. Qualitatively the isomorphism character is defined by geometrical similarity of orbital shapes responsible for isomorphism. At the same time, the more similar are the extensions, trajectories and inclination angles of such orbitals, the more perfect is isomorphism.

According to the degree of isomorphic similarity of interchanging structures they can be classified into three types (I, II, III) given for some cases in table 3.

PHOTOSYNTHESIS. INITIAL STAGE

Magnesium atom that is four-coordinated with nitrogen atoms is included into chlorophyll in the central cavity of the whole structure. The porphinated chlorophyll ring is located in aqueous medium. Each central Mg atom forming chelate compound has two bonds by donor-acceptor mechanism and two covalence bonds. Two molecules of bacterio-chlorophyll are located close to each other (about 3 Å) and form competent-structure – dimer chlorophyll. In the dynamics of structural permutations all four bonds of each Mg atom become equivalent [7]. All this allows assuming that total effective P_E-parameter of Mg will be approximately two times greater than from $2S^2$-orbital (5.4867x2=10.973 eV).

At the first stage of photosynthesis in the system of PS-2 dimensional characteristics of hydrogen atom can change in structured water molecules under the radiation with energy hν from boron radius (0.529 Å) to atomic ("metal") – 0.46 Å, this corresponds to the obtaining of P_E-parameter equal 10.432 eV by hydrogen that is similar to P_E-parameter of 2Mg. It should be pointed out that general change in the scale of photosynthesis potentials PS-2 approximately equals 1.5 eV, and the difference between the data of P-parameters of hydrogen atoms equals 1.37 eV.

The rest of hydrogen atoms with "boron" P_E-parameter equaled to 9.0624 eV have similar values with P_E-parameters of $2P^1$-orbitals of nitrogen atoms surrounding magnesium.

Other data are not less important: initial value of P_E-parameter of $2S^2$-orbital of magnesium atom gives from P_E-parameter (table 3) of radical (O-H) α=8.24 % and $\rho \approx$ 77-82 %. This ρ value can increase to even 100 % under the light action due to minor changes in dimensional characteristics of atoms-components. Absolute difference of these P-parameters equals 0.43 eV, thus corresponding to the changes in the scale of potentials during the synthesis of ATP.

Total spatial-energy action upon the bond H-O-H of magnesium and nitrogen atoms (table 3) results in the possibility of breaking this bond with the isolation of free hydrogen and oxygen atoms.

This initial process finishes with the participation of manganese-containing system connected with proteins of reaction center PS-2. Structural reconstruction can take place in manganese cluster (two-nucleus or four-nucleus) under the action of radiation [8, 9] from univalent state (4.9369 eV – this is similar to initial values of Mg P_E-parameter) to bivalent (9.9414 eV) and further – to quadrivalent state (17.518 eV).

All this provides enzymatic action of Mn upon the bond H-O-H, both upon oxygen and hydrogen atoms, and hydroxyl group in general. This is confirmed by the approximate equality of P_E-parameters of bi- and quadrivalent Mn with P_E-parameters of $2P^1$ and $2P^2$-orbitals of oxygen atom (table 3). Thus, all the above interactions and structural re-groupings inducted with light result in the formation of oxidized chlorophyll based on the following reaction [10]:

$$H_2O+2h\nu \rightarrow \frac{1}{2}O_2+2e^-+2H^+$$

with the isolation of two electrons and two protons. These electrons, broken off from the water, through the chain of "dark" reactions go further to PS-1 that utilizes them in the next photosynthesis stages to reduce $NADP^+$ to NADPN that is carried out also with the help of proton transfer system.

For double bond of $2P^1$-orbital the carbon atom has P_E-parameter – (8.5043 eV) – similar to P_E-parameter of hydrogen atom (table 1). Therefore one of the freed hydrogen atoms join the double bond C=C available in NADPN with the formation of single bond with carbon atom [9].

Table 3. Photosynthesis structural interactions

Atoms, molecules, radicals	1 component		2 component		α (%)	ρ (mol%)	SEI types
	Orbitals	P_E, P_C (eV)	Orbitals	P_E, P_C (eV)			
O-P	$2P^2$	8.470*	$3P^3$	8.9215*	5.19	100	I
O-P	$2P^1$	4.6188*	$3P^1$	4.1862*	9.83	60-65	I
Mg^{2+}-H	$3S^2$	8.6066	$1S^1$	9.0624	5.16	100	I
H_2O-CH_2	$1S^1$-$2P^2$	11.788	$2S^12P^3_r$-$1S^1$	11.125	5.79	100	II
C-O	$2S^1$-$2P^1$	17.435	$2P^2$	17.967	3.01	100	I
CO-OH	$2P^2$-$2P^2$	8.4416	$2P^2$-$1S^1$	8.7712	7.51	90-95	II
CO-H_2O	$2S^12P^1_r$-$2P^2$	9.3252	$1S^1$-$2P^2$	9.0226	2.21	100	II, III
CH_2-CO_2	$2S^22P^2$-$1S^1$	16.531	$2S^12P^3_r$-$2P^2$	16.785	1.53	100	III
2Mg-H	$3S^2(3S^2)$*	10.973	$1S^1$	10.432	5.05	100	I
Mn-H	$4S^13d^1$	9.9414	$1S^1$	10.432	4.82	100	I
Mn-O	$4S^13d^1$	9.9414	$2P^1$	9.7979	1.45	100	II
Mn-O	$4S^23d^2$	17.518	$2P^2$	17.967	2.53	100	II
N-H	$2P^1$	9.4166	$1S^1$	9.0624	3.83	100	II
Mn-OH	$4S^1$	4.3969	$2P^1$-$2S^1$	4.7080	4.75	100	II, III
Mg-(O-H)	$2S^2$	5.4867	$2P^2$-$1S^1$	5.0575	8.24	77-82	II
K^+-H^-	$4S^1$	3.344	$1S^1$	3.525	5.27	100	I
Fe-S	$4S^23d^1$	13.226	$2P^2$	13.215	0.08	100	II
Na^+-H^-	$3S^1$	3.901	$1S^1$	3.525	10.1	55-60	I

PHOSPHORYLATION

It is considered [7,11] that directed transition of protons serves as energy source during phosphorylation. Between the numbers of transported protons and electrons certain stoichiometric relations are revealed. Thus, in the course of electron transfer (along the whole transport system) ATP molecules are formed.

Apparently, ATP phosphorylation energy can also be estimated through the system of electron transfer.

In particular, electron transfer results in that phosphoric acid molecules present in ATP, NADP and NADPN contain oxygen atoms in the form of O^-. Spatial-energy interactions (including isomorphic) are objectively expressed both at similar and opposite electrostatic charge of atoms-components. Such interactions can also take place between two heterogeneous atoms, if only their P_E-parameters are roughly equal, and geometric shapes of orbitals are similar or alike.

The radiation energy hν in PS-1 promotes, apparently, the changes in dimensional characteristics of phosphorous and oxygen atoms from covalent to anion ones. Therefore, P_0-parameters of free phosphorus and oxygen atoms are distributed at the distance of their anion radii 1.86 Å and 1.40 Å, respectively. This similarity of values of their P_E-parameters: α=5.19 % for $2P^3$-orbitals of phosphorous with $2P^2$-orbitals of oxygen (table 3).

Such approximate equality of P_E-parameters and geometric similarity of shapes of orbitals of atoms-components shows that actual degree of their interaction ρ=100 %, thus providing the energy of formation of macroenergy bond P-O. Then bond energy of phosphorous and oxygen atoms from two different molecules of phosphoric acid necessary for structural formation during phosphorylation can be considered phosphorylation energy.

To calculate bond energies or energies of molecule reduction during photosynthesis (E) the technique previously tested [6] for 68 binary and more complicated compounds following the equation was applied:

$$\frac{1}{E} = \frac{1}{P_c} = \frac{1}{\left(P_E \frac{N}{K}\right)_1} + \frac{1}{\left(P_E \frac{N}{K}\right)_2} \qquad (10)$$

where N – bond average repetition factor, K – hybridization coefficient that usually equals the number of atom valence electrons registered.

The half of internuclear distance (for binary bond) of similar atoms or atomic, covalence or ionic radii (depending upon bond type) can be used as a dimensional characteristic of atoms.

The calculations involving anionic distances of atomic orbitals for P and O atoms were made: $3P^1$ (phosphorous)-$2P^1$ (oxygen) and for $3P^3$ (phosphorous)-$2P^2$ (oxygen). The values of E obtained appeared to be slightly greater than experimental and reference data (table 4). But actual power physiological processes during photosynthesis have the efficiency below the theoretical, being in some cases about 83% [7].

It is probable that electrostatic component of resulting interactions on anion-anion distances is registered in such a way. In fact, the calculated value 0.83E practically corresponds to the experimental bond energy values during phosphorylation (first line in table 4) and free energy for ATP in chloroplasts (second line in table 4).

The calculations of bond energy based on the same technique but on covalence distances of atoms for free molecule P...O (sesquialteral bond) and for molecule P=O in P_4O_{10} (double bond) are given in table 4 for comparison.

Sesquialteral bond was evaluated introducing the coefficient N=1.5 using the average value of oxygen P_E-parameter for single and double bonds.

It is interesting to point out that calculations of E based on covalence distances correspond to experimental data without introducing the coefficient 0.83.

ASSIMILATION OF CO_2

Binding of CO_2 takes place in aqueous medium by the carboxylation reaction of ribulose-diphosphate (RuDP) with the formation of 3-phospho-glycerine acid (PGA) – table 5. Water molecule and radical C=O at the distances of molecular interaction have quite similar values of P_E-parameters for forming the general structural grouping of dimeric composite type. Total P_E-parameter of water molecule and radical C=O nearly equals P_E-parameter of CO_2 and therefore the molecules of CO_2 and H_2O join RuBP with the formation of two radicals COOH в PGA (table 5). In ferment RuDP- carboxylase, Mg atoms and O^- ions (5.4867 eV and 4.755 eV) play an active role, their P_E-parameters similar to P_E-parameter of radical COOH.

A great difference in the number of atoms of interacting structures proves that carboxylase can play only a fermentative role, "tuned" to obtain this final product (COOH).

The further complicated way of CO_2 assimilation to form CH_2O flows through series of intermediate compounds and reactions (Calvin cycle). Let us show some results of calculations of total spatial-energy assimilation processes of CO_2. When CO_2 is reduced to the level of its structural formation in CH_2O, the chemical bonds are reconstructed on all stages of the cycle. Therefore, the additional activation energy from ATP and NADPN is required.

It is also obvious that power consumption should be rationally calculated taking into account the reconstruction processes of chemical bonds, i.e. via the values of bond energy – for binary structures, and reduction energy – for more complex molecules and radicals (E).

Thus we calculated the value E based on equation (10) for several compounds and radicals during photosynthesis – tables 4 and 5. For radical – C =O the calculations were made in two possible variants of activity of valence orbitals of carbon atoms. The compliance of calculated E values with reference data [12,13] was in the range of 5% for all bonds of covalence type without introducing the coefficient 0.83.

The main part of light energy is stored by a plant on the reduction stage to PGA. At the same time, 4.56 eV (per molecule) are spent – [12.13]. Our calculations give the reduction energy of radical COH equal to 4.487 eV. Free energy for the formation of one mole of CH_2O based on reference data [7,12, etc] is 4.96-5.07 eV. The calculations following the method of P-parameter evaluate this energy as 5.025 eV.

Table 4. Bond and reduction energies of molecules during photosynthesis (eV)

Atoms, structures, orbitals	1 component		2 component		3 component		Calculation		E by [7.8.14]	Notes
	P_E (eV)	N/K	P_E (eV)	N/K	P_E (eV)	N/K	E	0.83E		
1	2	3	4	5	6	7	8	9	10	11
P-O $3P^1$-$2P^1$	4.1862 4.1862	1/5 1/5	4.6188 4.755	1/6 1/6	— —	— —	0.400 0.405	0.33 0.34	0.34- 0.35	Phosphorylation
P-O $3P^2$-$2P^2$	8.9215	1/5	8.470	1/6	—	—	0.77	0.64	0.67- 0.59	ΔG of ATP
P...O $3S^23P^3$-$2S^22P^4$	32.403	1.5/5	70.854 63.339	1.5/6 1.5/6	— —	— —	6.277 6.024 <6.15>	—	6.14	Free PO molecule
P=O $3P^3$-$2P^2$	15.085	2/3	20.042	2/2	—	—	6.697	—	6.504	In P_4O_{10} molecule
C-H $2P^2$-$1S^1$	13.066	1/2	9.0624	1/1	—	—	3.797	—	3.772	
H_2O $1S^1$-$2S^2$	2•9.0624	1/1	17.967	1/6	—	—	2.570	—	2.476	Decompositio n of one molecule
-O-O- $2P^1$-$2P^1$	9.7979	1/1	9.7979	1/1	—	—	4.90	—	5.11	
O=O $2P^2$-$2P^2$	20.048	2/4	20.048	2/4	—	—	5.012	—	5.11	

Table 4. (Continued)

1	2	3	4	5	6	7	8	9	10	11
CO_2 $2P^2$-$2P^2$	14.581	2/4	2•20.048	2/6	—	—	4.717	—	4.56	Reduction
=C=O $2S^22P^2$-$2P^2$	31.929	2/4	20.048	2/2	—	—	8.8874	—		
C-O $2P^2$-$2P^2$	13.066	1/2	17.967	1/2	—	—	3.782	—	3.688	
(C=O)-H ($2S^22P^2$-$2P^2$)-$1S^1$	31.929	2/4	20.048	2/2	9.0624	1/1	4.487	—	4.553	
-O-H $2P^2$-$1S^1$	17.967 17.967	1/2 1/2	17.137 9?0624	1/1 1/1	— —	— —	5.894 4.511	— —	— 4.390	
CO-OH ($2S^22P^2$-$2P^2$)- ($2P^2$-$1S^1$)	8.8874	1/1	5.894	1/1	—	—	3.544	—	—	
CH_2O $2S^22P^2$-$1S^1$-$2P^2$	31.929	1.33/4	2.90624	1/1	20.048	2/2	5.025	—	4.96- 5.07	Free energy of the formation of one mole

In molecule

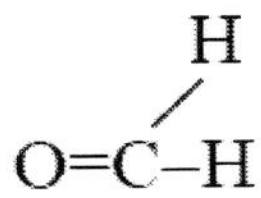

the average repetition factor for carbon atom bond was taken as equal to (2+1+1)/3=1.33.

Applying the approved approach to calculate the resulting bond energy (or the reduction energy) of structural subsystems for each stage, the values of these energies were calculated (table 5) – E_C. It is known [7] that the cycle moving energy to PGA can be 1.06 eV due to three ATP molecules (per one CO_2 molecule), one ATP molecule is consumed in the cycle to PGA.

Following our data, the cycle moving energy (ΔE_C) equals the difference of E_C values for the corresponding stages:

1) stage CO_2 – FGAK: ΔE_C=1.770-1.401=0.369 eV
Phosphorylation energy of one ATP molecule = 0.34-0.35 eV
2) stage CO_2 – FGA: ΔE_C=2.367-1.401=0.966 eV
Phosphorylation energy of three ATP molecules: 0.34x3=1.02 eV

Thus P-parameter gives the satisfactory characteristics of energetics of the CO_2 assimilation cycle main stages. Photorespiration reaction is as if "competitive" to the CO_2 assimilation reaction. Also here it is possible to reveal similar values of P_E-parameters of interacting radicals C=O and HCOH with P_E-parameters of oxygen atoms. As in assimilation reaction the ferment RuDP- carboxylase "is tuned" for the formation of final product COOH. Other ferments can participate in photosynthesis and photorespiration, for example, the substitution of Mg atoms for Fe atoms results in the formation of cytochromes, in which P_E-parameter of two-valence iron (P_E=10.093 eV) is an active spatial-energy component of photosynthesis structural interactions. Therefore, iron-sulfur proteins – ferrdoxins executing various transport functions connected with ATP synthesis are initial and secondary acceptors of electrons in the system PSI.

CONCLUSION

In this approach we give quantitative and semi-quantitative evaluation of spatial-energy interactions at main stages of complicated biophysical process of photosynthesis based on the utilization of initial atomic characteristics. The analysis of results after the application of P-parameter methodology shows that they correspond to reference data both in the direction and energetics of these processes.

Table 5. Spatial-energy characteristics of CO_2 assimilation (eV)

Reaction blocks	RuDF — C = O + H_2O + CO_2 →(Mg, RuDF-carboxylase)	PGA 2COOH →(NADPN, ATP, O^-)	PGA 2COH + O_2 →(ATP) … →(Mg)	$CH_2O + O_2 + \dots$
P_E	8.4416 9.0226; → 17,464; 17.774;	2x5.1226;	2x4.3705; 17.967x2; (8.741)	9.060 17.967x2 (7.333)
E	8.8874 2.570 4.717	3,544 3,544	4.487 5.012	5.025 5.012
E_C	E_{C1} = 1.401	E_{C2} =1.770	E_{C3} =2.367	E_{C4} =2.509

Calculation: $E_{C2}-E_{C1}$ =0.37; $E_{C3}-E_{C2} \approx 0.60$ $E_{C3}-E_{C1} \approx 0.97$

By [7]: 1 ATP molecule; 2 ATP molecules; 3 ATP molecules (1.06 eV)

Notes:

P_E – initial values of P_E-parameters, for Mg (5.4867), Mg^{2+} (8.6066), for O^- (4.755; 4.6188)

E – bond or reduction energy

E_C – resulting bond or reduction energy for groups of radicals or fragments: $1/E_C=1/E_1+1/E_2+\dots$

DESIGNATIONS

m_1 and m_2 – masses of material points (kg);
ΔU_1 and ΔU_2 – potential energies of material points (J);
ΔU – their resulting (mutual) potential energy of interaction (J);
Z* – nucleus effective charge (Cl);
n* – effective main quantum number;
W_i – bond energy of electrons on *i*-orbital (eV);
r_i – orbital radius of *i*-orbital (Å);
n_i – number of electrons on this orbital;
SEI – spatial-energy interactions;
P_0 – spatial-energy parameter (eVÅ);
P_E – effective P-parameter (eV);
R – dimensional characteristic of atom or chemical bond (Å);
$h\nu$ – light quantum energy (J, eV);
N_1, N_2, … – number of homogeneous atoms;
N – average repetition factor in the formula (10);
P_C – structural P-parameter of complex structure (eV);
α – coefficient of structural interactions, isomorphism (%);
ρ – degree of structural interaction (%);
PS – photosystem (PSI and PSII);
ATP – adenosine triphosphate;
RuDF – ribulose-diphosphate;
PGA – 3 phospho-glycerin acid;
NADP – nicotine-amide-adenine-dinucleotide-phosphate;
PGA – phosphor-glycerin aldehyde;
E – energy of bond or molecule reduction (eV);
E_C – resulting energy of bond or reduction for radical groups (eV).

REFERENCES

[1] Big medical encyclopedia. M.T.26.1985.560 p.
[2] Korablev G.A. *Spatial-Energy Principles of Complex Structures Formation.* Netherlands. Brill Academic Publishers and VSP. 2005, 426p. (Monograph).
[3] Fischer C.F. Average-Energy of Configuration Hartree-Fock Results for the Atoms Helium to Radon.//*Atomic Data*.-1972. -№ 4. -p. 301-399.
[4] Waber J.T.. Cromer D.T. Orbital Radii of Atoms and Ions//*J. Chem. Phys*. -1965. -V 42. -№12. -p. 4116-4123.
[5] Clementi E.. Raimondi D.L. Atomic Screening constants from S.C.F. Functions. 1.//*J.Chem. Phys*.-1963. -v.38. -№11. -p. 2686-2689.
[6] Korablev G.A.. Zaikov G.E. Energy of chemical bond and spatial-energy principles of hybridization of atom orbitalls.//*J. of Applied Polymer Science. USA.* 2006. V.101.n3.P.2101-2107.
[7] Photosynthesis/Edited by Govingi. M.:*Mir,* V.1-1987, 728p; V.2-1987, 460p.

[8] P. Clayton. Photosynthesis. Physical mechanisms and chemical models. M.:*Mir*-1984, 350p.

[9] S.A. Schukarev. *Inorganic chemistry*. V.2-1974, 382p.

[10] D. Hall, K. Rao. Photosynthesis. M.:*Mir*, 1983.

[11] J. Edwards, D. Walker. Photosynthesis of C3 and C4-plants: *Mechanisms and regulation*. M.: 1986.

[12] Encyclopedia in physics. M.: 1966, V.5, 576p.

[13] Kamen M.D. Primary processes in photosynthesis. L. 1963.

[14] Break-off energy of chemical bonds. Potentials of ionization and affinity to electron / Edited by V.I. Kondratjev. M.:*Nauka*, 1974, 351p.

In: Progress in Chemistry and Biochemistry
Editors: E. M. Pearce and G. E. Zaikov

ISBN: 978-1-60692-344-3

Chapter 24

NMR-INVESTIGATION OF RATS AFTER SEVERE EPILEPTIC CONVULSIONS

***A.A Mirina**[*1], **E.M. Sulejmanova**[2] **and D.A. Chepurnova**[2]*
[1] Physical Department and [2] Biological Department of Lomonosov Moscow State University. Moscow, Russia

ABSTRACT

Status Epilepticus (SE) – is a state, when epileptic seizures are so frequent and long, that stable and qualitatively another state of brain forms: acute derangements of consciousness, hemodynamics, respiration and homeostasis remain for more than 30 min. It leads to serious damage of brain structures and can have a possible fatal outcome[1].

In this work the forming of status epilepticus of the rat brain in on the lithium-pilocarpine model and ways of its inhibition was investigated. MR research of structural changes of the rat brain after repeatedly transmitted status epilepticus was made. The research was done using T_1- and T_2-weighed sequences. The wave flow of motor convulsions during status epilepticus on the lithium-pilocarpine model was displayed. The local (point) damages were found in the neocortex, thalamus and hippocampal formation, also symmetry disturbance of ventricles of the brain was shown.

Preliminary research of the influence of ginsenosides of Korean Red ginseng and ginseng cell culture Dan25 on postepileptic structural changes of the rat brain was made. Chronical injection of ginsenosides (4-6 weeks) relieves the severity of motor convulsions during SE and also decreases pathological changes in the brain. The preliminary data indicates facilities of application of ginsenosides in neurology.

Keywords: epilepsy, MRI, Status Epilepticus, ginseng, rat brain visualization.

* Address correspondence to Alexandra Mirina. Tel: +7 (926) 369-76-17; E-mail: alexandra.mirina@gmail.com

INTRODUCTION

Status Epilepticus

Epilepsy is one of the most common neuropsychic diseases. According to different data from 1 to 2 percent of the population suffers from various forms of epilepsy and epileptical syndromes. More than 50% of patients have the first attack before 20 years of age, usually at 10-15 years. Epilepsy is the result of highly complex structural, biochemical, electrophysiological changes that occur at various levels not only within the brain but also in the entire body.

SE develops during different brain lesions and diseases that can make the forecast essentially worse. SE develops either suddenly or gradually by quicker attacks. Usually after one of the convulsive attacks paroxysms begin to repeat with short intervals. Sopor and then comatose state develop after 2-3d attack.[2]

Nowadays lethality of SE that in 60th years was about 20-25% is three times less.[3] Nevertheless the problem of SE taking place during epilepsy still is of critical importance.

There are two main form of Status Epilepticus possible beginning:[3] proper SE and symptomatic SE (during current brain diseases and acute, chronic intoxications).

In this work we used the Lithium-Pilocarpine model[2] (model of temporal lobe epilepsy of brain) which rather exactly repeats clinical and neuropathological characteristics of this type of human epilepsy. Pilocarpine is muscarinic alkaloid, acts as non-selective muscarinic receptor agonist in the parasympathetic nervous system.[2] In adult rats piloparpine usually initiates SE in 50-60 min after injection, lasting for 8-12 h and is followed by recurrent seizure episodes for up to 24-48 h. This acute seizure phase is followed by a "silent" seizure-free phase of a mean duration of 14-25 days. Thereafter all animals exhibit spontaneous recurrent seizures (SRSs). This epilepsy lasts for the entire life of the animal.[4]

Ginsenosides of Korean Red Ginseng

Ginsenosides are substances (glycosides) excreted from the ginseng root. The new generation of antiepileptic drugs (tiagabine, vigabatrin, sabril etc.) don't offer much prospective for treating and preventing SE. They have such side effects such as sedative, oppression of cardiovascular and respiratory action.

Nowadays anticonvulsant activity of ginsenosides is considered proven. Antioxidant activity, cytoprotection, neuroprotection, inhibition of calcium channels, inhibition of neuromediators recapturing, modulation of specific binding of $GABA_A$ and $GABA_B$ receptors – these physiological effects of the extract were shown.[5]

Moreover there was shown the stabilizing effect of the medication on the cerebral blood flow in conditions of increased blood flow during convulsive activity or by influence of chemical impact. There is a supposition of possibly using ginsenosides in combination with other antiepileptic drugs for preventing their side effects.[5]

NMR Imaging

Magnet-resonance imaging (MRI) is based on applying methods of nuclear magnet resonance.

Unlike magnet resonance spectroscopy, where the main purpose is spectral analysis, the aim of MR-tomography is providing information about spatial distribution of spin density and relaxation parameters.

For solution of this problem there are special consecutions of impulses of radio and heterogeneous magnetic field impulses – gradient impulses which provide spatial coding of MR signals.[6,7]

MRI research was made on Bruker BioSpin 70/30 tomograph (Germany). Field B_0 = 7.05 Tl. We got images in two projections: coronal and horizontal – with slice thickness 1.5 mm.

MATERIALS AND METHODS

Lithium-Pilocarpine Model of Status Epilepticus

Objects of investigation: adult male rats about 6-8 months old (n=30). Three experimental groups were organized:

1 – everyday injection of ginsenosides (n=10) – GS group;
2 - no injection of ginsenosides (n=10) – Control group;
3 – saline injection instead of pilocarpine (n=10) – Control Saline group.

According to figure scheme (figure 1) before one day of experiment all rats got LiCl. Concentration of it in blood after 10 h is 0,69±0,03 mEq/l, and after 24 h is 0,16±0,01 mEq/l.[8] The concentration of Pilocarpine Hydrochloride («Acrus») injection was 40 mg/kg in 2 ml of saline. All medications were injected intraperitoneal.

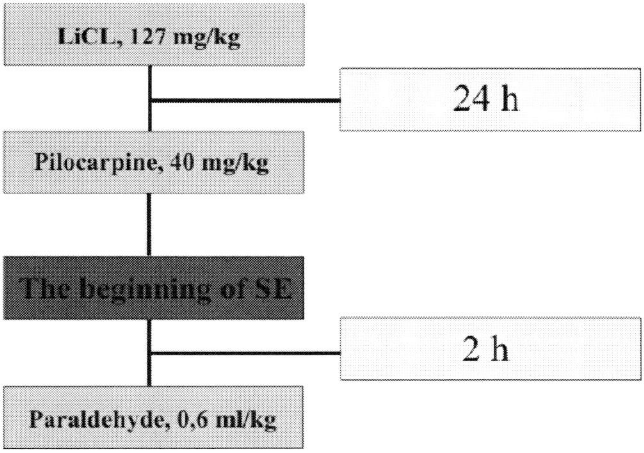

Figure 1. The Scheme of experiment with Lithium-Pilocarpine SE.

Table 1. The scores of behavior and motor seizures by P. Mares

0	Normal behavior, no change
1	Single mioclonic spasm, years jerk, head sake
2	Atypical minimal seizures, body jerk
3	Clonus head and limb muscles, jerk of body, jumping
4	"Kangaroo pose" tonic seizures body muscle
5	Very severe tonic-clonic seizures with falling

SE was observed during 2 h. For estimation of seizures severity we used 5th mark scale by P. Mares (table.1)

The behavior of animal was observed in special boxes (25x30x25 cm). Latency of motor attacks and severity of seizures were measured. After 2 h of observation convulsive activity was stopped by Paraldehyde - concentration 0.6 ml/kg (0.1 ml of Paraldehyde per 0.9 ml of saline). Mortality was determined in 24 h after experiment. After that was made MR visualization of brain structure. During this process animals were under anaesthetic. (Nembutal, 40 mg/kg).

In this work we studied the effect of ginsenosides. The medication was excrete from ginseng cell culture Dan25.[9] The butanol extract was used (BIOHIMMASH, Moscow). The content of ginsenosides (Rg1, Re, Rf, Rb1, Rc, Rb2, Rd) was measure as 7,552 mg/g dry biomass of extract. Ginsenosides were dissolved in saline (24 mg/ml), filtered and injected in concentration 18 mg/kg.

Results and Discussion

The Characteristics of Lithium-Pilocarpine Induced SE

Experiments were made on adult rats. Also we observed animal's behavior before the beginning of SE.

After injection of pilocarpine strong mastication was observed in 5-7 min, abnormal moving into the box, scratching reflex (by back pad) down to skin damage and bleeding. The salivation, defecation, urination (infrequent ejaculation) were observed in 7-15 min after injection. Sudden quick running and head jerk were not compulsory seizures precursor. The first 15 min following pilocarpine administration the animal could exhibit a brief pause in motor activity. In our experiments the aggressiveness of rats after SE was increased. If SE didn't develop the rats restored after 3 h.

After pilocarpine injection SE developed in 37 min in average. But there were different durations of latency period: from 3 min to 1.5 h.[10,11]

During 2 h of observation the severity of motor seizures was variable. It has wave characteristic. We observed severe tonic convulsions (4 stage) in the beginning of SE (30 min in average – the first wave), then there was decrease of severity – only body and head jerks (3d and 2nd stages). This period lasted for about 1 h, and then the second wave started (the last 30 min of observation on average) (figure 2).

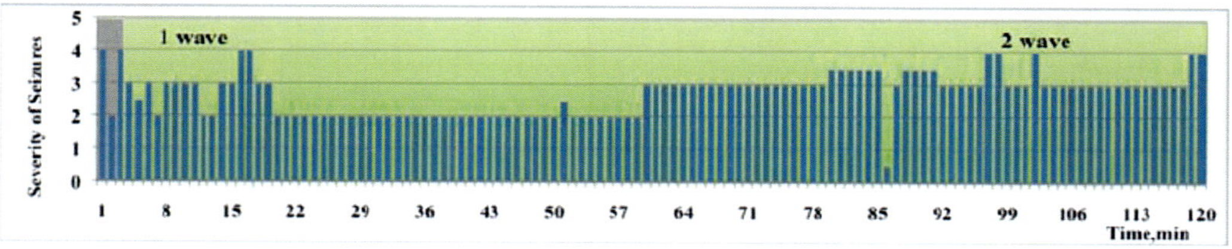

Figure 2. The two-wave character of pilocarpine SE (single animal from control group).

Effects of Chronical Ginsenosides Injection

For the experiment with chronical injection of GS we used adult male rats 6-8 months old (n=10). Rats got the medication every day during 3 weeks. Then we induced pilocarpine SE for three times (a week apart). While these 3 weeks of experiment rats also got ginsenosides. So animals received the medication for 6 weeks totally. It was shown that severity of seizures during SE was not so strong after several injections of GS (figure 3).

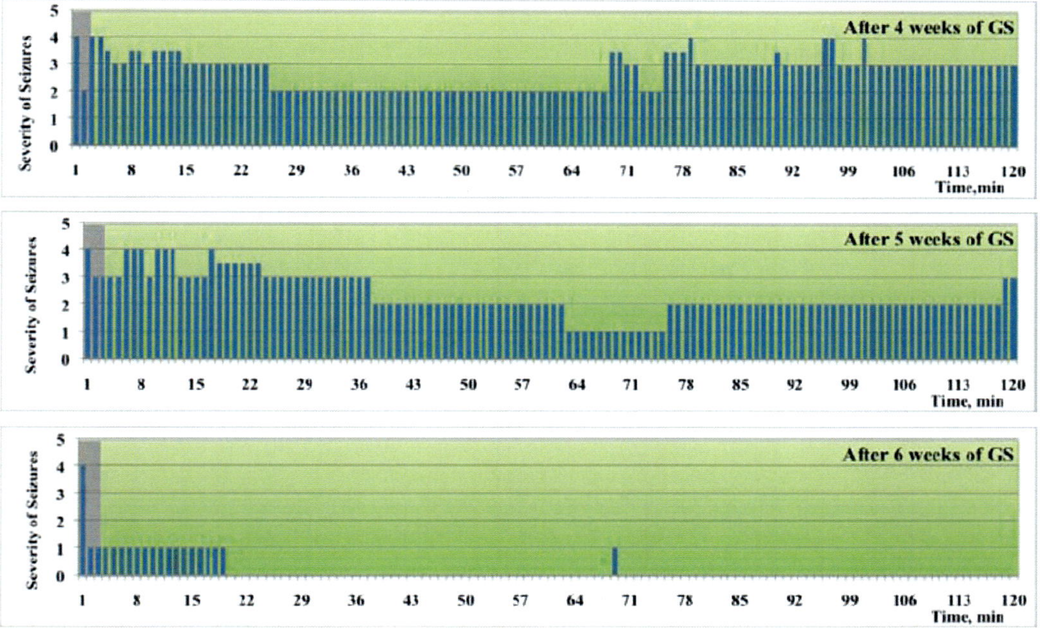

Figure 3. Comparative diagrams of SE character by the example of rat with premedication for 4, 5 and 6 weeks.

On figure 3 it's shown that the second wave of SE was decreased: the severity of seizures was diminished on 1-2 marks. During the last SE (6th week) there was only one 4th stage in the beginning of SE, with 19 s duration.

Comparison by latency period, average quantity and duration of 5th stages is shown on figures 4, 5 and 6. On the diagrams we can see that average number and duration of 5th stages in case of GS premedication is less than without it in control group. After GS injection for 3 weeks only 2 rats showed 5th stage convulsions. In control group such convulsions were observed in 5 cases. However the length of latency period in control group was more than in GS group.

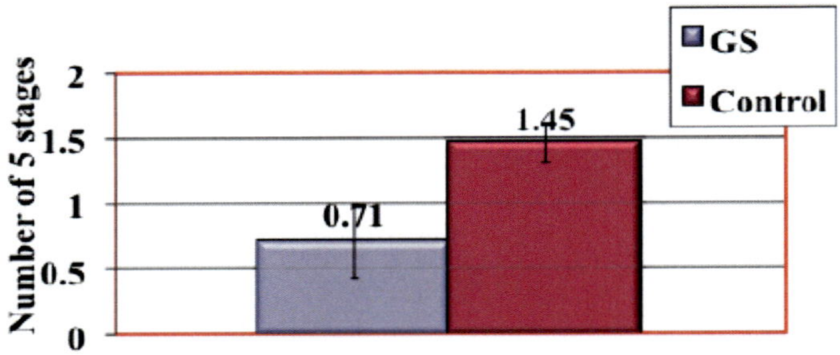

Figure 4. Average number of 5th stages during SE observation (for 2 h).

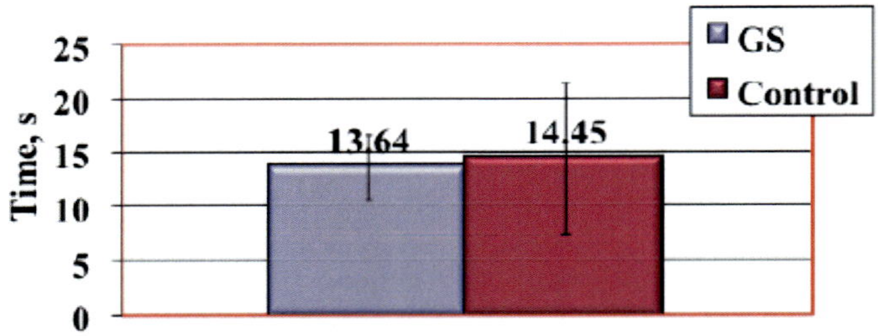

Figure 5. Average length of 5th stages during SE observation (for 2 h).

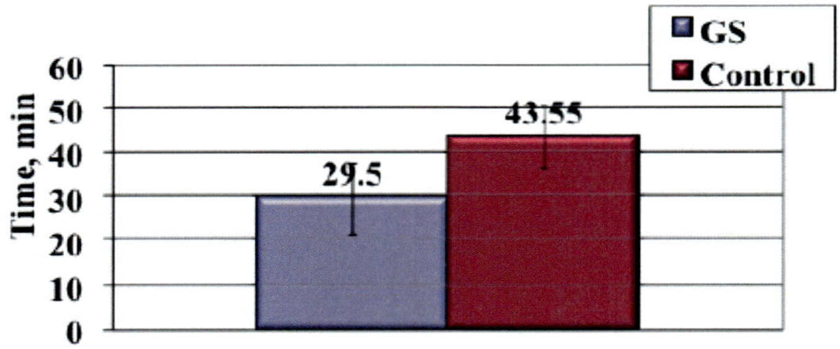

Figure 6. Average latency period of SE beginning.

Mortality was determined in 24 h after SE. The data also proves that ginsenosides have positive effect. All rats from GS group stayed alive, but 3 rats of 10 from control group died (30%).

Visualization and Diagnostics of Changes in Brain Structure after SE by MR Tomography

A number of electrophysiological and neuroimaging procedures are used for the presurgical localization of the epileptogenic focus. Together with clinical data noninvasive imaging procedures such as MRI, PET, SPECT etc., are used to provide information of the focus that can be further investigated using invasive EEG recordings.

On MRI slides were observed local damages of structure in neocortex, thalamus, hippocampus, piriform lobe cortex and different pathology of brain ventricles, that are not so severe in case of premedication by ginsenosides of ginseng cell culture Dan 25.

As an example on figure 7 and 8 there are scans of rat brain after 3 SE with GS premedication, without premedication and from control saline group.[12]

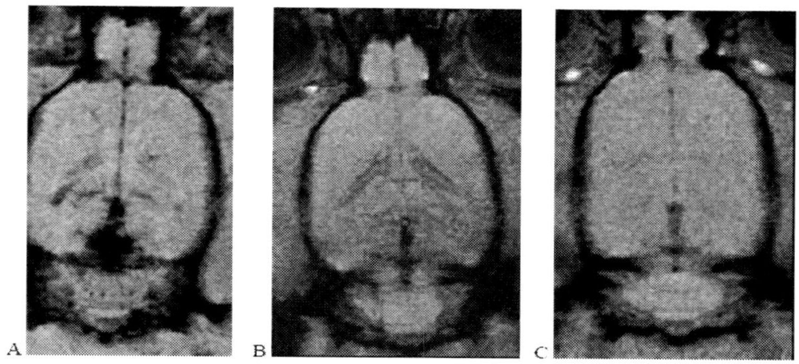

Figure 7. Horizontal projection of rat brain: A – rat experienced 3 SE; B – rat experienced 3 SE with GS premedication; C – rat from control saline group.

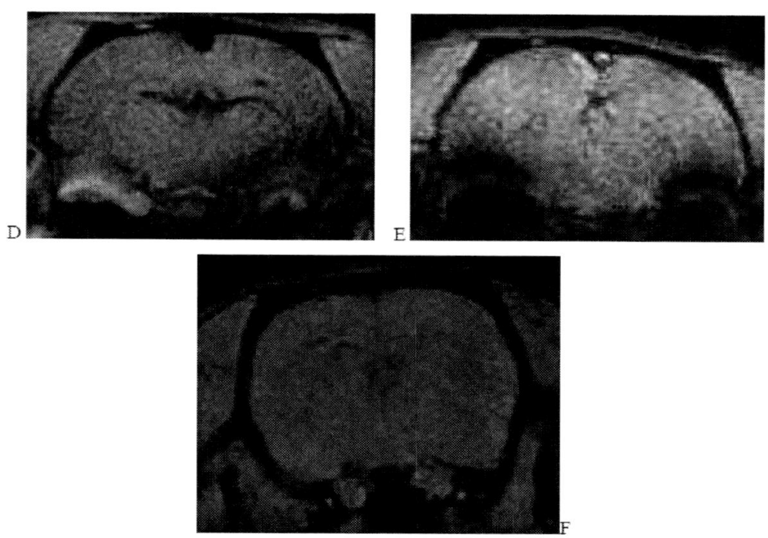

Figure 8. Coronal projection of rat brain: D – rat experienced 3 SE; E – rat experienced 3 SE with GS premedication; F – rat from control saline group.

Conclusion

These preliminary data show that application of ginsenosides in neurology is prospective. We can conclude that in the case of prolonged daily premedication, ginsenosides have on anticonvulsant effect, reduce severity of SE character. However it can't stop status epilepticus totally. So there is a possibility of using ginsenosides in combination with other antiepileptic drugs, especially because ginsenosides in addition to an anticonvulsant effect, have a restorative and tonic influence on an organism.

The mechanism of ginsenosides activity still has not been studied. GS are amphiphilic substances and they can be built into the cell membrane and change its functions and, functions of receptors on it.[13] It's possible that ginsenosides influence steroid receptors in cytoplasm and, therefore, genes expression is changing.[14] It's known that ginsenosides increase brake potential in hippocampus cells, inhibit recapture of neurotransmitters and NMDA-mediate increasing of intracellular calcium. Ginsenosides have neuroprotection effect, which can be explained by their antioxidant activity – capability to neutralize free radicals.

Since in our research we used medication that consists of 7 types of ginsenosides , there is another prospective direction of investigation: partial or total purification of the extract and excretion of each of the ginsenosides in pure form for more precise establishment of the chemical agent that provides anticonvulsant activity.

REFERENCES

[1] Karlov, V. A., *Epilepsy; Medicina:* Moscow, 1990.

[2] Chepurnov, S. A.; Chepurnova, N. E.; Redkozubova, O. M.; Saakyan, S. A., *Uspekhi Physiologicheskih Nauk* (in Russuan) 36, 68 2005.

[3] Karlov, V. A., *Convulsive epistatus*; Moscow, 2003.

[4] Roch, C.; Leroy, C.; Nehlig, A.; Namer, I. *J., Epilepsia.* 43, 325 2002.

[5] Chepurnov, S. A.; Park, J.-K.; Chepurnova, N. E.; Abbasova, K. R.; Berdiev, R. K.; Goncharov, O. B.; Tolmacheva, E. A.; Pereversev, M. O.; Strogov, S. E. Advances in ginseng research 2002. Proceedings of the 8th international symposium on ginseng., Seoul, Korea., October 28-31, 2002 2002, pp 84.

[6] Rink, P. A., *Magnetic Resonance in Medicine*; Blackwell Wissenschafts-Verlag: Berlin - Vienna, 2001.

[7] Anisimov, N. V.; http://cmts.cmm.msu.ru/site/modules.php?name=Content&pa=showpage&pid= 6: 2003.

[8] Honchar, M. P.; Olney, J. W.; Sherman, W. R., *Science.* 220, 323 1983.

[9] Konstantinova, N. F.; Makhankov, V. V.; Uvareova, N. I.; Samoshina, N. F.; Sova, V. V.; Michailova, O. M., *Biotekhnologia.* 9, 35 1995.

[10] Jones, D. M.; Esmaeil, N.; S.Maren; MacDonald, R. L., *Epilepsy Research.* 50, 301 2002.

[11] Kondratyev, A.; Gale, K., *Molecular Brain Research.* 121, 86 2004.

[12] Chepurnov S.A., C. N. E., Jin-Kyu Park, Redkozubova O.M., Pravduhina A.A., Abbasova K.R., Buzinova E.V., Mirina A.A., Chepurnova D.A., Dubina A.A., Pirogov U.A., de Curtis M., Uva L., *J. Ginseng Res.* 31, 159 2007.

[13] Lian, X. Y.; Zhang, Z. Z.; Stringer, J. L., *Epilepsia.* 46, 15 2005.

[14] Attele, A. S.; Wu, J. A.; Yuan, C. s., *Biochem. Pharmacol.* 58, 1685 1999.

INDEX

A

B

D

E

F

G

H

I

K

L

M

N

O

P

Q

R

S

T